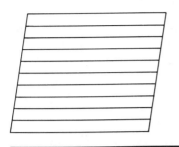

COMPUTER-
AIDED
SYSTEMS
ENGINEERING

HOWARD EISNER
The ORI Group
Atlantic Research Corporation

Prentice Hall, Englewood Cliffs, New Jersey 07632

Library of Congress Cataloging-in-Publication Data

EISNER, HOWARD.   (date)
    Computer-aided systems engineering (CASE).

    Bibliography: p.
    Includes index.
    1. Systems engineering—Data processing.
2. Computer-aided design. I. Title.
TA168.E37 1987   620'.0042   87-12424
ISBN 0-13-162918-2

Editorial/production supervision: Gretchen K. Chenenko
Cover design: Lundgren Graphics, Ltd.
Manufacturing buyer: Richard Washburn

LIMITS OF LIABILITY
AND DISCLAIMER OF WARRANTY:

Dedicated to my children

Oren David
Seth Eric
Susan Rachel

who,
while now in their twenties,
continue to stretch
and reach
for the best in themselves

 © 1988 by Prentice Hall
A Division of Simon & Schuster
Englewood Cliffs, New Jersey 07632

Printed in the United States of America

10 9 8 7 6 5 4 3 2 1

ISBN 0-13-162918-2   025

PRENTICE-HALL INTERNATIONAL (UK) LIMITED, *London*
PRENTICE-HALL OF AUSTRALIA PTY. LIMITED, *Sydney*
PRENTICE-HALL CANADA INC., *Toronto*
PRENTICE-HALL HISPANOAMERICANA, S.A., *Mexico*
PRENTICE-HALL OF INDIA PRIVATE LIMITED, *New Delhi*
PRENTICE-HALL OF JAPAN, INC., *Tokyo*
SIMON & SCHUSTER ASIA PTE. LTD., *Singapore*
EDITORA PRENTICE-HALL DO BRASIL, LTDA., *Rio de Janeiro*

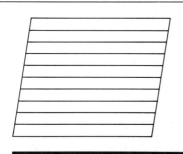

CONTENTS

Part II Tools and Techniques   48

3 BASIC COMPUTER TOOLS   48

4 DIAGRAMMING TECHNIQUES   69

Contents

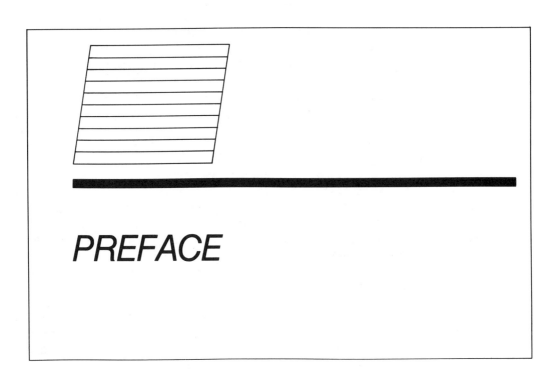

PREFACE

The primary objective of this book is to introduce the notions and elements of computer-aided systems engineering (CASE) to both the student and the practicing systems engineer. This is done by articulating the features of systems engineering and building bridges between systems engineering, various mathematical and logic-based structures, and computer systems and their associated commercially available software. The orientation with respect to the latter is the personal computer (abbreviated in this text as the p-c), since in the last several years the p-c has come upon the scene with such force that it has totally changed the way systems engineers do their jobs, and students, their assignments. Further, even more profound effects are anticipated in the future, at least by this author, especially as the systems engineer tackles larger and more complex systems and the price–performance of the p-c improves.

This text is primarily a survey book that covers a broad range of topics with limited depth of coverage in each. This approach was necessitated by the expansive scope of systems engineering as well as the numerous computer and noncomputer tools that relate to systems engineering. As a survey text for the student, the book is designed specifically to span a one-semester, 15-week course. The first two chapters can be covered in the first week and from that point on, one chapter per week may be scheduled, up through Chapter 15. The last two chapters would be covered in the last week. This would represent a reasonably intense pace, to the point where a one-year course would make sense as well, leaving twice as much time for in-depth coverage and study.

Liberal use has been made of real-world examples and references to the systems engineering literature. Hopefully, the examples will give the student a feel for practical issues and problem solving. The systems engineering literature is relatively mature, but new points of view show up on a continuing basis in the various professional journals as well as many federal-government publications. It is important for the student and the practicing engineer to maintain contact with both of these sources.

Most chapters contain a mixture of narrative, which explains and comments upon the material, as well as illustrative analyses and problems. The latter are intended to convey specific skills that are to be mastered. The notation in the text is not uniform but reflects the notation cited in the various parts of the literature. In particular, both subscripted indices and argument indices are used, the former more dominant in mathematical sources and the latter prevalent in computer materials. The student should ultimately feel comfortable working with either type of notation.

In total, the book has five parts and 17 chapters. Part I deals with introductory concepts with respect to systems engineering. Various tools and techniques that are an integral part of CASE are dealt with in the second part. In general, these tools and techniques are of two types: mathematical or logic-based, and applicable software tools. The chapter order is set up so that the early ones prepare the reader for more complex later material (e.g., probability theory is presented first so that the reader can cope with some of the later statistical computer software or information theory) or simply to provide a change of pace, especially for the student. Computer tools are considered in Chapters 3, 6, 7; diagramming techniques in Chapter 4; probability theory in Chapter 5; information and search theory in Chapter 8; and the programming languages of BASIC and Pascal in Chapter 9.

Part III returns to the elements of systems engineering in the following sequence: top-level systems engineering in Chapter 10; scheduling and costing in Chapter 11; performance modeling and simulation in Chapter 12; risk and decision analysis in Chapter 13; software development and analysis in Chapter 14; and systems engineering support functions in Chapter 15. In all of these chapters a CASE application summary is included to provide a closer linkage with the computer software tools that are available to provide assistance to the systems engineer. Next, Part IV deals with projections and directions, with Chapter 16 highlighting the areas of forecasting and artificial intelligence and Chapter 17 considering a future CASE environment.

Most chapters are followed by a list of references and problems for that particular subject area. The problems are oriented to the student solution rather than a textbook solution. That is, an attempt is made to get students thinking about how they might approach a given issue rather than giving one unique text solution.

There are four appendices that make up Part V. The first cites specific packages in the market for each of the 29 microcomputer software categories discussed in the text. Appendix B moves into the minicomputer and mainframe world for database management systems, largely since this is a particularly relevant software area. The third appendix provides a series of references to Department of Defense (DoD) standards, directives, and other important documentation that should be known to the systems engineer. Finally, Appendix D contains a set of overview guidelines for the student. These guidelines define the principal areas that should be absorbed by the student. Of course, they are also available to assist instructors who use this text.

The specific subject of computer-aided systems engineering (CASE) is considered to be new in the sense that it was first broadly introduced and defined by this author in a paper in *Signal Magazine* (see Reference 2 in Chapter 1) and is now more fully developed in this text in terms of its elements and their interrelationships. It is hoped that various readers will both comment upon and add to the structure and content of this subject in the time to come.

Over the years, particularly in relation to my long association with ORI, Inc., (now the ORI Group, consisting of ORI, Inc., ORI/Intercon, and ORI/Calculon) many colleagues and friends have been close at hand and contributed in one way or another to my understanding of the art, the science, and the practice of systems engineering. In this regard, I must cite Frank Brown, Don Burklew, Merrill Cornell, Gene Feinberg, Lee Gilford, Sam Kneale, Harvey Kushner, Art Milkes, Dave Reed, Bill Rogers, Rich Scotti, Norman Shusterman, Pat Steen, and Fred Zusman, and apologize to the many others who have helped that I may have left out. I would also like to acknowledge the guidance of Professor Solomon Kullback who, many years ago at The George Washington University, gave me whatever insight I might have into statistical information theory and related subjects. I also feel indebted to the many teachers who shared with me their wisdom, wit, and expertise during my sojourns at The City College of New York, Columbia University, Brooklyn College, and The George Washington University. I have also learned a great deal from my students at the latter three institutions. Notwithstanding the above and other people from whom I gained valuable insight and understanding, in the final analysis I must take responsibility for the structure of CASE and its various elements, as developed herein, as well as all its warts, imperfections, and omissions.

I want to thank Bernard Goodwin at Prentice Hall for his unflagging interest in and support for this book, and I also wish to acknowledge my production editor, Gretchen Chenenko, for a wonderful job.

On a more personal note, I received much support of various kinds over more than 20 years from Joan Eisner who is, among other things, a superb teacher and purveyor of excitement and enthusiasm to her students and friends. I am indebted as well to my brother Mel and my Mom for simply always being there. In more recent years, Jane Takeuchi has encouraged this particular endeavor by her interest, support, and far-ranging intellect. Finally, my friend Peter Roemer has helped me keep my eye on the ball both on the tennis court and on the various pathways along which we all seem to travel.

Howard Eisner
Potomac, MD

Part I   Introductory Concepts

The greatest ocean of truth lay all undiscovered before me

Sir Isaac Newton

SYSTEMS ENGINEERING AND THE COMPUTER

1.1 INTRODUCTION

We live now in the "information age," a time of computer terminals; bits and bytes; high-speed printouts; micros, minis, and mainframes; local area networks, pac-man, spreadsheets, workstations, decision support systems, data communications, database management systems, and much more. We live indeed within what Toffler [1] has called "The Third Wave," a period in our history that has moved us far beyond the agricultural society and through the industrial revolution. We also live embedded in a variety of large-scale systems—communication systems, transportation systems, education systems, housing systems, family systems, health systems, financial systems, environmental systems, defense systems, and a myriad of others. As such, we need tools to help us deal as effectively as possible with these large and complex systems in the new age of information that is here upon us.

The specific subject of this book, in a broad sense, is the relationship or bridge between our large-scale systems and the information age in which we live. In particular, we focus on the matter of developing large-scale systems and how computers and software can be and are being applied to the tasks and activities embodied within systems engineering. In one sense, applying computers to assist in the systems engineering process is nothing new. Both systems engineering and the digital computer, as well as their union, go back to the 1950s. With the launching of Sputnik in 1957, considerable new impetus was given to computer applications and systems engineering in the United States. In another sense, however, new ingredients have been added in the late 1970s and into the 1980s that have dramatically changed the manner in which computers can be and are being utilized within a systems engineering context. These ingredients are

- larger and more complex systems
- the computer "chip"
- commercially available software

With the steady improvements in technology as well as population increases and consequent demand for services and products, we have conceived of, built, and implemented larger and more complex systems (and their subsystems) such as the national aviation system, the national telecommunications system, the national highway system, and the national defense system. These are all very large-scale systems whose subsystems are themselves massive systems in their own right. The resultant size and complexity have required more hardware, more software, more people resources and, of course, higher costs.

Alongside this growth in the sheer size and number of large-scale systems we have the substantial growth in the computer industry. One of the key reasons for this growth has been the revolutionary digital computer "chip." The chip, simply put, has introduced more computing power and storage capacity in smaller packages at a lower cost. And with this capability has come the personal computer, profoundly a part of the process of changing how we develop systems. Indeed, the computer "at home" is changing the way we teach, the way we learn, and the way we approach our tasks at work and at home. It has also boosted the acceleration of commercially available software products at all computer size levels, moving up from the microcomputer where such products are a necessity.

Thus we seem to be influenced at this time by both a "systems pull" and a "technology push" [2]. The systems pull requires better systems engineering formalisms and tools because of the growth in size in these extremely large and complex systems. The technology push leads with the ubiquitous digital computer chip and its attendant software, firmware, hardware, and interconnecting systems. The evolution of our systems engineering capability, therefore, is proceeding within the context of revolutionary changes in computer hardware technology and the evolution of other related technology areas such as communications. Within the computer technology revolution the personal computer stands out as especially important both as a standalone device as well as a smart terminal component in larger computer systems.

The personal computer (abbreviated here as *p-c* without reference to a particular manufacturer) has become, for good reason, a basic necessity for students as well as practicing engineers. The p-c has replaced the simple multifunction calculator and slide rule to help them in their day-to-day activities. As a result, we emphasize the p-c here and its role as an aid in the systems engineering process. Thus we have on one hand the tasks and activities of systems engineering that are to be performed and, on the other hand, the p-c as an increasingly important tool to assist with these tasks and activities.

In Table 1.1 we list a general set of tasks and activities that are normally performed within the overall context of large-scale systems engineering. In order to carry out these tasks and activities, we can employ a variety of mathematical and logic-based techniques, as illustrated in Table 1.2. Many of these techniques are embedded in commercially available software packages, selected categories of which are listed in Table 1.3. We look then to the computer, in this case primarily (but by no means exclusively) the p-c, as the essential tool that facilitates the process of

---

TABLE 1.1 TYPICAL SYSTEMS ENGINEERING TASKS AND ACTIVITIES

• Definition of needs/goals/objectives	• Risk analysis
• Requirements analysis	• Integrated logistics support
• Requirements allocation	• Pre-planned product improvement
• Functional analysis	• Reliability, maintainability, and availability
• Functional allocation	• Integration
• Specification development	• Test and evaluation
• System and subsystem design	• Configuration management
• System and subsystem analysis	• Quality assurance
• Trade-off/alternatives evaluation	• Training
• Software development and analysis	• Documentation
• Interface definition	• Production
• Schedule development	• Installation
• Life cycle costing	• Operation
• Technical performance measurement	• Operations evaluation/modification
• Program and decision analysis	

TABLE 1.2 ILLUSTRATIVE MATHEMATICAL AND LOGIC-BASED TECHNIQUES

• Communication theory	• Linear systems theory
• Computer programming	• Modeling and simulation
• Control systems theory	• Network analysis
• Cost effectiveness/benefit analysis	• Nonlinear programming
• Decision analysis	• Nonlinear systems theory
• Decomposition theory	• Optimization theory
• Diagramming	• Probability theory
• Dynamic programming	• Queuing theory
• Forecasting	• Regression analysis
• Game theory	• Risk analysis
• Graph theory	• Search theory
• Heuristics	• Statistics
• Industrial dynamics	• Structured analysis
• Information theory	• Urban dynamics
• Input–output (econometric) analysis	• Utility theory
• Linear programming	• Value theory

TABLE 1.3 CATEGORIES OF MICROCOMPUTER-BASED SOFTWARE PACKAGES

- Spreadsheets
- Database management systems (DBMSs)
- Word processors
- Graphics
- Integrated packages
- Multitask applications managers
- Project management aids
- Alternatives and preference evaluators
- Decision support systems
- Toolchests/desk managers
- Idea processors
- Management evaluators/aids
- Statistical
- Mathematical
- Engineering
- Reliability
- Mathematical programming
- Simulation
- Curves and plotting
- CAD/CAM/graphics
- Workbenches
- Structured-analysis tools
- Languages
- General support tools
- Utilities
- Micro-to-mainframe links
- Forecasting
- Expert systems
- Artificial intelligence-oriented

bringing all of the above elements together. We call this process *computer-aided systems engineering* (CASE) [2].

The scope and content of CASE differ from computer-aided design (CAD), computer-aided manufacturing (CAM), and computer-aided engineering (CAE), although there are some interrelationships. CAD tends to deal with lower-level design issues such as circuit board layouts, three-dimensional graphics, and the like. In general CAM supports the manufacturing process only, including real-time control and operation of machines. CAE, perhaps a second cousin to CASE, also deals at a lower level of engineering detail and therefore is often particularized to a subbranch of engineering analysis such as the performance of a digital circuit or the power budget of a satellite communications link. Various CAE packages might become subroutines or modules in a larger CASE performance modeling or simulation application. All of the above utilize the computer extensively, but CASE is focused upon a "systems" orientation and set of related problems and issues.

A principal goal of this book, then, is to provide the user, whether student or practicing engineer, with an introduction to the means whereby the tasks and activi-

TABLE 1.4 ILLUSTRATIVE P-C APPLICATION AREAS

Systems Engineering Task or Activity	Illustrative P-C Application Area
• Requirements analysis	• Database management system tabulations
• Performance analysis	• Computer language or simulation program
• Design alternatives evaluation	• Spreadsheet overlay or decision support system
• Life-cycle costing	• Spreadsheet modeling
• System documentation	• Word processor usage

ties of systems engineering (Table 1.1) can be performed using appropriate mathematical and logical techniques (Table 1.2) with the assistance and facility afforded mainly by the p-c and its attendant software (Table 1.3). An illustrative list of p-c application areas, to be expanded upon in detail in later chapters, is provided in Table 1.4.

1.2 THE ROLE OF THE GOVERNMENT

The role and practices of the government are featured in this text for a variety of reasons. First, the government is the acquisition agent and indeed the user of many of our large-scale systems. As an example of the former, the Department of Transportation, through its Federal Aviation Administration, procures and operates our air-traffic-control system. This system is a national resource and supports a larger air-transportation system. In the latter case, an example can be drawn from our defense programs for which the government, through the Department of Defense (DoD), is both the system acquirer and user. As a consequence of this deep involvement in large and complex systems, the federal government has been the purveyor of standards that temper and guide the manner in which systems engineering is performed. These standards (see Appendix C) have had an enormous impact on the industry's point of view regarding systems engineering. Indeed, it is not unreasonable to say that, in large measure, we do systems engineering in this country in a manner strongly influenced by the various standards, directives, and handbooks promulgated by agencies of the federal government. Finally, a third reason for featuring federal involvement is the fact that it sponsors a considerable amount of research and development (R&D), a significant portion of which deals directly with computer systems and related software. Aside from a number of specific programs, as for example the DoDs STARS (software technology for adaptable, reliable systems) program [3], the government researches, compiles, and disseminates enormous amounts of information regarding computers and software. An example of the latter is shown in Table 1.5, which contains a selected list of publications sponsored by various federal agencies and is available to the general public through the Government Printing Office (GPO). As we shall see in later chapters, these types of publications are directly relevant to the body of knowledge represented by systems engineering. Thus the federal government's role and influence is far reaching, and today's systems engineer should be familiar with its key thrusts and directions.

1.3 DESIGN AND THE SYSTEMS ENGINEER

Notwithstanding the substantial number of typical systems engineering tasks and activities (Table 1.1), it is important to recognize the central role that design plays in systems engineering. In design, we are fundamentally attempting to develop a new configuration of hardware, software, and people resources to do a particular set of jobs, i.e., satisfy a set of defined requirements. This design process hopefully brings together or synthesizes pieces into a larger whole with all the desired system at-

TABLE 1.5 ILLUSTRATIVE GOVERNMENT PRINTING OFFICE PUBLICATIONS

- *Computer-Based National Information Systems: Technology and Public Policy Issues*, S/N 052-003-00852-1, 1981.
- *Computer Programs for Urban Transportation Planning*, PLANPAC/BACKPAC, *General Information Manual*, S/N 050-001-00125-0, 1977.
- *Analytic Study of a Shared Device Among Independent Computing Systems*, S/N 003-003-02271-3, 1980.
- *Architecture for Data Base Management Standards*, S/N 003-003-02383-3, 1982.
- *Computer Model Documentation: A Review and an Approach*, S/N 003-003-02020-6, 1979.
- *Computer Model Documentation Guide*, S/N 003-003-02282-9, 1981.
- *Computer Software Management: A Primer for Project Management and Quality Control*, S/N 003-003-01795-7, 1977.
- *Features of Software Development Tools*, S/N 003-003-02295-1, 1981.
- *Final Report: A Survey of Software Tools Usage*, S/N 003-003-02378-7, 1981.
- *Future Information Processing Technology*, S/N 003-003-02504-6, 1983.
- *Guide to Computer Program Directories*, S/N 003-003-01867-8, 1977.
- *Introduction of Software Tools*, S/N 003-003-02414-7, 1982.
- *Microcomputers: A Review of Federal Agency Experiences*, S/N 003-003-02492-9, 1983.
- *Microcomputers: Introduction to Features and Uses*, S/N 003-003-02560-7, 1984.
- *Software Development Tools*, S/N 003-003-02389-2, 1982.
- *Software Validation, Verification, and Testing Technique and Tool Reference Guide*, S/N 003-003-02422-8, 1982.
- *COSMIC: A Catalog of Selected Computer Programs*, S/N 033-000-00700-1, 1977.
- *Introduction to STARPAC: The Standards Time Series and Regression Package*, S/N 003-003-02525-9, 1983.
- *Reference Manual for the Ada Programming Language*, S/N 008-000-00394-7, 1983.
- *Technology Assessment and Forecast*, S/N 003-004-00542-4, 1977; S/N 003-004-00580-7, 1981.

tributes. It is a process of invention, with many inherent difficulties that are a consequence of the large number of alternatives that are or can be formulated. For example, if we have an overall system composed of just ten functional subsystems, each of which has two possible and plausible methods of implementation, then in principle there are 2^{10} or 1,024 "possible systems." The talented systems engineer can sort through these possibilities to arrive at just a handful of systems that are the most cost effective. With the further help of a computer, an engineer can develop a near-optimal system design. Thus we are attempting to aid systems engineers in what is primarily a process of synthesis or invention by assisting their logical and intuitive reasoning and sorting powers with the rigorous and rapid sorting, computational, logical, and iteration strengths of the digital computer that today is readily available in a microbased environment.

It should be noted here as well that while the computer can assist a talented systems engineer, it is unlikely to transform a pedestrian engineer into a superlative engineer. Simon Ramo, a founder of the successful systems-engineering-oriented company known as TRW, has commented [4] that there is a lack of really good systems engineers. Robert Frosch, previously the administrator of the National Aeronautics and Space Administration (NASA), similarly lamented the lack of good systems engineering [5] when he was assistant secretary of the Navy. He apparently felt that we needed a good deal more systems engineering appropriately applied to systems engineering itself. By focusing on the individual, he claimed that we must remember that someone ultimately should be in control and that such a person must exercise knowledge, understanding, and judgment, as well as management ability, in order to create a system properly. He also chided us to keep in mind that change is an integral part of the systems engineering process, implying that we must both recognize this factor and take it into account in our design, firmly reiterating that good systems engineering is an art, not a technique.

A. D. Hall [6], an investigator of the nature of systems engineering, identifies some characteristics that appear to be important to (good) systems engineers, namely

- a demonstrated affinity to the systems point of view
- the faculty of objective judgment and sound appraisal
- imagination and creativity
- a facility in human relations
- effectiveness as a broker of information

To this list we make explicit the additional and crucial ingredient of having a talent for synthesizing and inventing by sorting through large numbers of interrelated concepts, facts, and figures. We also concur with Frosch that it is important to bring a sense of art and excitement into the engineering process. Thus it appears that systems engineers are hard to find and that systems engineering is not easy to do. We pause briefly to examine the points made by some other authors who apparently agree and who have documented their perceptions of several fundamental "truths" in these areas.

1.4 SOME IRREVERANT VIEWS

In later chapters we explore in some detail the nature of the systems engineering process and the computer tools that may assist in such a process. Here we take note of the fact that it is a matter of no small difficulty to design, build, and operate large-scale systems that work well in the real world. These difficulties have been examined in a tongue-in-cheek manner by a few observers of the scene who have paused to comment upon their observations.

In 1982, N. Augustine produced an anecdotal collection of comments in a book entitled *Augustine's Laws* [7]. A participant in the military world of developing large-scale systems, Augustine coins, among others, the following "laws"

- Ninety percent of the time things will turn out worse than you expect. The other ten percent of the time you had no right to expect so much (Law Number III).

- The last ten percent of the performance sought generates one-third of the cost and two-thirds of the problems (Law Number VII).
- If a sufficient number of management layers are superimposed on top of each other, it can be assured that disaster is not left to chance (Law Number XIII).
- Software is like entropy. It is difficult to grasp, weighs nothing, and obeys the Second Law of Thermodynamics; i.e., it always increases (Law Number XVI).
- One-tenth of the people involved in a given endeavor produce at least one-third of the output, and increasing the number of participants merely serves to reduce the average performance (Law Number XIX).
- Truly simple systems are not feasible because they require infinite testing (Law Number XXV).
- Hardware works best when it matters the least (Law Number XXXII).
- It is true that complex systems may be expensive, but it must be remembered that they don't contribute much (Law Number XXXV).

In a similar vein, a book entitled *Systemantics* [8] contains strong statements regarding how systems work and especially how they fail. J. Gall, its author, cites yet another set of immutable systems-oriented "laws"

- Systems in general work poorly or not at all (Primal Scenario).
- New systems generate new problems (The Fundamental Theorem).
- Big systems either work on their own or they don't. If they don't, you can't make them (Basic Axiom of Systems Function).
- A simple system, designed from scratch, sometimes works.
- A complex system that works is invariably found to have evolved from a simple system that works.
- A complex system designed from scratch never works and cannot be made to work. You have to start over, beginning with a working simple system.
- Complex systems are beyond human capacity to evaluate.

These laws are only samples of what the systems engineering community has accepted as humorous prodding but must also consider as containing more than just a dash of wisdom. Systems engineering clearly provides us with no small measure of challenge as well as extensive opportunities to do better in the future.

1.5 SUMMARY

Recognizing the aforementioned aspects of and commentary on systems engineering and the real difficulties associated with the development of systems, we have come to a time in our history during which we can make substantial productivity gains in the systems engineering process through the proper use of the computer and its associated software. Referring to Figure 1.1, we note that through such available soft-

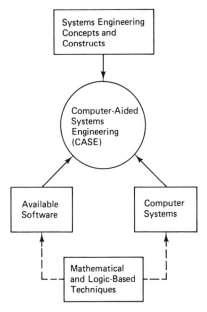

Figure 1.1 Computer-aided systems engineering relationships.

ware these computer systems now embody a wide variety of mathematical and logic-based techniques. Together, these elements form the underpinnings of CASE, computer-aided systems engineering, which represents a formal application to the field of systems engineering and its subordinate concepts and constructs.

The computer systems and related software of Figure 1.1, as briefly discussed before, are a definite part of a very fast moving technology and systems industry. Both hardware and software have been changing dramatically in very short periods of time. Many companies, flourishing just a few years ago, find themselves struggling to sustain a small market share. Table 1.6 provides an overview of some aspects of the computer industry in the form of a series of hypothetical "headlines." These illustrate the breadth of activity in the industry as well as the tendency toward change and pressure for improved performance. This is the industry in which the computer and software side of CASE finds itself firmly embedded.

The personal computer, a specialized subset of the general industry, has moved continuously toward better price-performance. This has been coupled with an ever-increasing flow of commercial software. A particularly instructive and interesting set of papers on the p-c appeared several years ago [10], dealing with the following subjects:

- an overview of future personal computers
- technology changes in regard to personal computers
- the users of personal computers
- the future impacts of personal computers
- the government role in personal computing

The interested reader, with the benefit of the passage of time and 20/20 hindsight,

TABLE 1.6 HYPOTHETICAL HEADLINES FOR THE COMPUTER-RELATED INDUSTRY

- General Motors Commits to MAP (Manufacturing Automation Protocol)
- 10 Million Computer Home Workstations Into the 90's
- Interactive Video Technology Heating Up
- Grid-Connected Minicomputers on a Chip Yield a Supercomputer
- Database Machines Enter the Market
- T. Roszak [9] Warns Of "Megahype"
- Computer Sales Hit Major Slump; Red Ink Flowing
- Corporation for Open Systems Encouraging Open Communications and Compatibility
- Personal Computer Clones Selling Like Hotcakes And Outperforming Originals
- 3 ½-in. Floppies Battling 5 ¼ in. Floppies
- Expert System Flying As Integrated Avionics Monitor
- U.S. Watching Japan; Japan Watching U.S.
- Videophone's Time Has Come!?
- Laptops Feature Disk Drives and Liquid-Crystal Displays
- 32-Bit Workstation Technology Growing
- Government Annual Budget for Computer Hardware Exceeds 2.5 Billion
- Expert Systems Help Repair Automobiles
- Software Companies Think Small, Find Niches
- Computer Spreadsheet Gives Birth to P-C Software Market
- 4 Megabit Super-RAM Chip Being Improved
- CAD/Graphics System Revolutionizing Systems Engineering Documentation
- Videotex Systems' Future Uncertain
- Stock Market Analyzing Computers; Computers Analyzing Stock Market
- Fail-Safe Computers Gaining in Sales
- Analysts Trying to Sort Out Mini-Supercomputers from Super-Minicomputers
- Cost Per Unit of Computing Power Dropping 20–30 Percent Per Year
- RISC Technology Moving Forward
- Researchers Put Expert System with Fuzzy Logic On A Chip
- P-C Sales Hit 20 Billion
- Submicron Chips the Leading Edge Of ULSI (Ultra-Large Scale Integration)
- Growth Rate for P-C Software Drops from 80 to 20 Percent Over Eight Years
- Narrow Focused P-C Software Expected to Grow at 50 Percent Annual Rate
- CAE/CAD Systems Reducing Time to Design and Test Computer Chips
- P-C Packages Understand English
- Large Size Flat Panel Screen Displays Promising
- Government's MIMIC Program Pushes Gallium Arsenide Superfast Chips
- Defense Contractors Moving More Positively Toward Ada
- Low-Cost Fast Printer Uses Copy Machine Technology
- High MIPS Chips in RISC
- CD/ROMs Moving into Market
- Networks Still Seeking Standards
- LANs Going Open
- Superlattices Promise Higher Speeds
- Quarter Micron 16M Chip Needed for "Star Wars"
- Multiprocessors May Make Von Neumann Approach Obsolete

can compare the status and projections at that time to the realities of today. In any case, the personal computer is here to stay and its role, power, and areas of application will continue to evolve with time. A thesis of this text is that an increasing domain for the current p-c and its future versions will be that of systems engineering.

We also have emphasized here the significant role of design or synthesis as part of the systems engineering process. This does not by any means negate analysis but serves to identify priorities and ways of thinking, hopefully more constructively, about how to approach systems engineering issues. In the final analysis, it would seem that systems engineering involves the sensible blending of analysis and synthesis, of subject matter expertise, of tools and techniques, of computer-based packages, of functional support areas, and a considerable amount of mature professional judgment. In Chapter 2 we continue to examine the vast subject of systems engineering, but from a more highly structured and systematic vantage point.

REFERENCES

1. Toffler, A., *The Third Wave*. New York: Bantam Books, 1981.
2. Eisner, H., "CASE: Computer-Aided Systems Engineering for C³I Systems," *SIGNAL Magazine,* Journal of the Armed Forces Communications and Electronics Association, July 1984.
3. "The DoD STARS Program," *Computer Magazine,* 16, no. 11 (November 1983).
4. Ramo, S., "The Systems Approach: Automated Common Sense," *Nation's Cities,* vol. 6, March 1968.
5. Frosch, R., "A New Look At Systems Engineering," *IEEE Spectrum,* 6, No. 9 (September 1969).
6. Hall, A. D., *A Methodology for Systems Engineering*. Princeton: Van Nostrand Reinhold, 1962.
7. Augustine, N. R., *Augustine's Laws*. © Norman R. Augustine, all rights reserved, New York: American Institute of Aeronautics and Astronautics, Inc., 1982.
8. Gall, J., *Systemantics: How Systems Work and Especially How They Fail*. © John Gall, all rights reserved, New York: Times Books, a Division of Random House, Inc., 1975.
9. Roszak, T., *The Cult of Information*. New York: Pantheon, 1986.
10. Nilles, J. M., J. P. Hayes, M. G. Holmen, P. Gray, and F. R. Carlson, Jr., and M. J. White, "Correspondence Section on Personal Computers," *IEEE Transactions on Systems, Man, and Cybernetics,* SMC-10, no. 8, August 1980.

Let the street be as wide as the height of the houses

DaVinci

FORMAL SYSTEMS ENGINEERING STRUCTURES

Chapter 1 introduced the notion of computer-aided systems engineering (CASE) as an endeavor that utilizes computer systems and software to carry out the tasks and activities of systems engineering. In addition, some perspectives were presented with respect to design and the attributes of a good systems engineer. In this second introductory chapter we describe a variety of formal systems engineering structures. First, some 13 steps of a systems approach are explored. This is followed by an examination of some of the definitional frameworks of systems engineering, to include

- a problem solving dimension
- a time dimension
- a knowledge dimension
- an evolution and usage dimension
- the phases of system design
- the tools of system design

The design or synthesis aspects of systems engineering are examined in some detail from a morphological point of view, followed by a presentation of heuristic matrix evaluations. Both systems engineering management and the systems acquisition process are also explored. A summary and road map to the main chapters of this text are provided at the end of the chapter.

2.1 A SYSTEMS APPROACH

By way of introduction, we develop at the outset a broad overview of a systems approach. This approach is not unique but serves here to define the major steps required to develop a system and some of the key relationships between these steps. In later subsections we explore several formal systems engineering constructs. Figure 2.1 shows a conventional block diagram of the approach, generally proceeding from left to right.

2.1.1 Needs Statement

The process starts with block 1 calling for a statement of the need for the system. This is normally a broad statement explaining why the system is necessary and what specific needs are to be satisfied. It may be couched in the framework of missions to be performed as well as users and beneficiaries of the system when it becomes operational. If it is a replacement for an existing system, needs may be articulated in terms of crucial deficiencies present or projected. If it is a defense system it may demonstrate that the need derives from a significant change in threat.

2.1.2 Goals and Objectives

As shown in block 2, goals and objectives are derived from the needs statement. Goals are broad and objectives may be considered more specific statements categorized under each goal, although there is no recognized and accepted convention in this area. As a general guideline, the goals and objectives should cover the accepted journalistic questions of who, what, when, where, why and how, at a broad and high level. Solid statements of goals and objectives help to provide focus to the next steps in the process.

2.1.3 System Requirements

Block 3 of Figure 2.1 calls for a definition of system requirements. These usually involve a functional breakdown of the system and a statement of the desired performance of the system in each functional area. It is in the requirements' documentation that a drive toward quantification begins in earnest. To the maximum extent possible, it is necessary here to use numbers to describe the system and subsystem requirements. It should also be as clear as possible that a definitive linkage exists between the requirements, goals, objectives, and needs. In other words, if the requirements are in fact satisfied it should then be evident that the goals, objectives, and needs will likewise be satisfied.

2.1.4 Specifications

These embody the final level of detail for the system and likewise flow from the requirements. There is usually a hierarchy of specifications covering functions and subfunctions as well as subsystems and their component elements. System-perfor-

14

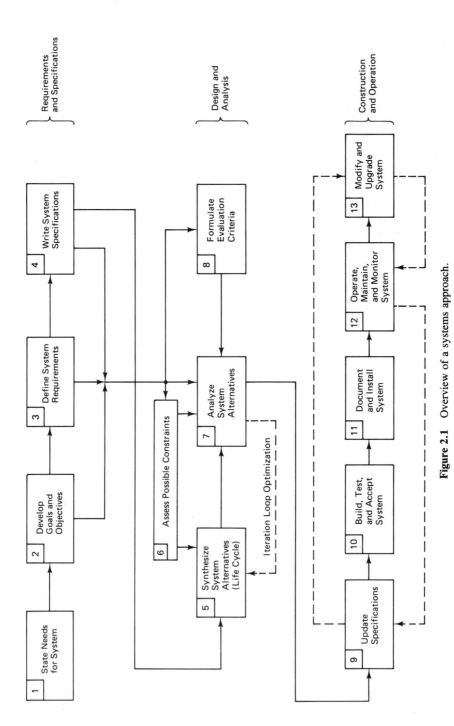

Figure 2.1 Overview of a systems approach.

mance evaluation and testing should always be referred back to the specifications to judge whether or not the system is satisfactory and will ultimately be accepted by the procurement agent or customer for the system.

2.1.5 Synthesis of Alternatives

In this crucial step (box 5 in Figure 2.1) a series of systems alternatives is defined, all of which are selected so as to nominally satisfy the requirements and specifications. A full definition must consider the entire life cycle of the system and should describe performance, costs, and schedules. As suggested in Chapter 1, this step is considered to be the most creative and difficult for the systems engineer.

2.1.6 Assess Possible System Constraints

We note in box 6 the existence of possible constraints that might serve to limit the range of design alternatives that are developed. Constraints actually could be a subset of the specifications but they are called out here so as to focus special attention on them. Examples of constraints are: continued operability of the system with changing and uncertain external factors such as user shifts, economic perturbations, and changes in threat; existing or modified union rules; maximum use of off-the-shelf components; and physical limitations having to do with available space.

2.1.7 Analysis of Alternatives

Analysis of the postulated alternatives proceeds by attempting to measure or predict the level of performance of the alternatives, together with an assessment of costs and schedules. In all cases, the standards for this analysis or evaluation process are contained in the system specifications and requirements and are also formally stated at a top level in a set of evaluation criteria (box 8). In principle, each alternative postulated has different performance levels as well as potentially different costs and schedules. Thus the evaluation procedure continues in an iterative manner in an attempt to optimize (which is rarely achieved in any real system).

2.1.8 Formulation of Evaluation Criteria

The evaluation criteria, alluded to in the previous subsection, include such items as: satisfaction of requirements, cost, schedule, and performance measures of merit. The latter may deal with reliability, interoperability, throughput, maintainability, environmental compatibility, and others.

2.1.9 Update of Specifications

Flowing out of the above iterations between synthesis and analysis is a *base-line system* that hopefully represents a best solution, considering all the evaluation criteria, specifications, requirements, and constraints. More often than not, arriving at a base-line system will result in some modifications in the formal specifications. This will necessitate an update in the specifications so that, at this crucial point in the process, the base-line system is believed to satisfy all of the system specifications.

2.1.10 Building, Testing, and Acceptance of System

With a firm system design and related set of specifications, a long process of building, testing, and acceptance of the system is undertaken. This may take several years and contains internal processes of iteration that usually involve a return through the earlier steps. Acceptance of the system is done in pieces, from elements to components to subsystems to system functional areas.

2.1.11 Documentation and Installation

The accepted system must be fully documented so that operators and maintainers of the system can keep the system operational. Installation proceeds according to agreed-upon schedules.

2.1.12 Operation of System

In addition to operation and maintenance of the system, performance is continually monitored to assure continued satisfaction of specifications since real-world operation over time is the final test of the system's ability to perform as specified.

2.1.13 Modification and Upgrade of System

As the monitoring process continues, it may be desirable to modify and upgrade the system as new and initially unforeseen circumstances affect the system or as the environment in which the system is operating changes. Even with all the previous iterations and updates, changes are made in virtually all systems in order to fine-tune their capabilities in the real world.

We note in Figure 2.1 the many iterations that are part of the systems approach. As suggested, these are part of the development of virtually all systems. The professional systems engineer understands the concept of continual change (see Reference 6, Chapter 1) and attempts to develop a "robust" system, one that can be adjusted to accommodate a changing environment without having to scrap the system or institute major design changes that would be extremely costly or time consuming. Robust design is another one of the features that is considerably easier to talk about than actually achieve in real-world systems engineering.

2.2 ELEMENTS OF SYSTEMS ENGINEERING

We now explore in some detail the elements of systems engineering. We also examine some of the ways in which systems engineering concepts have been developed in an institutional sense, showing how systems engineering is practiced primarily through our federal government procurement of large-scale systems. Indeed, a large percentage of such systems have been developed and funded by the government (e.g., the air-traffic-control system) in its attempt to satisfy our national needs. Through such efforts many of the developers and implementors of systems engineering constructs have been supported, directly or indirectly, by federal funds.

We may broadly define systems engineering as *an iterative process of top-down synthesis, development, and operation of a real-world system that satisfies, in a near-optimal manner, the full range of requirements for the system.* In this definition, note first that systems engineering is viewed as an *iterative process.* As a process it occurs over a period of time, namely, over the full life cycle of the system. The notion of iteration implies a back-and-forth interaction between synthesis and analysis, first postulating or inventing alternative designs and then analyzing them to see how well they will fit and perform. The analysis step, although a crucial part of the process, is not placed in the definition in order to emphasize by exclusion the primary character of synthesis or design in systems engineering. The words *top-down synthesis* further reinforce the role of design or invention and also suggest decomposition of a system from its large pieces (the so-called system level) to its subordinate parts (its subsystems and their components). Development and operation are included in the definition to cover the real-world introduction and use of the system. The phrase *near-optimal manner* is important to a system in that it indicates the need to attempt to converge to a system that has the best possible attributes. This, of course, leads us to develop tools and techniques of optimization that can be applied to the systems-engineering process. And finally, *the full range of requirements* suggests an articulated need for the system that is to be satisfied. Often it is extremely difficult to develop such a set of requirements but it is nevertheless a crucial part of the process of systems engineering.

In a collection of papers on the subject of methodology and applications with respect to systems engineering [1], A. P. Sage sets forth a seven-step framework of systems engineering, as outlined here:

1. problem definition
2. value-system design
3. system synthesis
4. systems analysis and modeling
5. optimization or alternative ranking
6. decision making
7. planning for action

These seven steps are drawn from the work of A. D. Hall [2,3,4] who appears to have formulated this framework, which he also calls the steps or problem-solving dimension of systems engineering.

Commenting upon these steps, first we have the matter of problem definition. Although perhaps this is a vague set of words, they certainly imply a good understanding of the needs for the system, the environment in which the system is to operate, and a first-order statement of requirements for the system, drawn from the definition. The issue of value-system design requires the definition of the objectives of the system and the set of criteria by which the ultimate "goodness" of the system is to be judged. In many real-world systems, various users of the system have different viewpoints as to what constitutes a good system, and in many cases these criteria are irreconcilable. For example, a rush-hour commuter trying to get to work every day by subway may have an entirely different view than a casual tourist or a handicapped person of what is important.

The third step in the framework, systems synthesis, has already been emphasized in terms of its central role in systems engineering and will be examined in greater detail later in this chapter. Systems analysis is the evaluation of how good a particular design alternative is in respect to the system requirements and the criteria by which the system is to be judged. Indeed, a considerable number of our more formal mathematical and logic-based tools reside within the category of systems analysis and form the basis for later full chapters of this book. The same is true for step 5 dealing with optimization of the suggested alternative designs.

In step 6, decisions are made as to the preferred system design, applying evaluations 1 through 5 including the application of whatever value system may have been articulated. Finally, planning for action covers the implementation of the next phase of the system if such is to be needed.

The above seven-step framework for systems engineering develops a problem-solving or logic dimension to the construct of the subject. As such it is perfectly sound as a top-level definition of a complete systems engineering procedure. However, in the practicing world of systems engineering, a subordinate set of tasks and activities has evolved that is recognized as constituting the specific way in which systems engineering is to be carried out. Again, this evolution has largely flowed from the process of federal development of large-scale systems and is reintroduced here (from Chapter 1, Table 1.1) in relation to the types of questions that each task or activity addresses. For purposes of this discussion, these are called elements of systems engineering and they represent, explicitly or otherwise, all of the seven steps shown as the framework of systems engineering.

Another dimension in the matter of attempting to understand the purview of systems engineering is the so-called *time* dimension that describes the various life-cycle phases of a system. This dimension [3] can be characterized by

1. program planning
2. project planning
3. system development
4. production
5. distribution
6. operation
7. retirement

Certainly these seven descriptors suggest the time sequence in the life cycle (birth to death process) of the system. The seven-step framework or elements of systems engineering, as noted earlier, have more or less emphasis or purpose during each of the life-cycle steps. Today's practice of large-scale systems engineering takes into account these phases. In recent years more emphasis has been placed on the need to address more explicitly the operational performance and readiness of our systems, including our ability to maintain and support these systems over their full life cycles.

A third dimension relates to the body of knowledge that is needed in order to adequately consider the other two dimensions. Hall [3] has identified engineering, social science, law, medicine, architecture, business, and so forth as fields that represent such bodies of knowledge. In this book, we recognize the relevance of these subject areas but focus on the engineering aspects. Clearly, in a broader context, systems are developed with other areas of subject matter expertise but most of the tools and techniques we describe here are applicable whatever the particular knowledge dimension happens to be.

Yet another way of describing systems engineering [4] is in relation to what might be described as an *evolution and usage dimension,* as follows:

1. the way in which it has evolved (history)
2. its internal procedures (generalized from case histories)
3. its objectives (goals achieved by the above procedures)
4. kinds of work previously performed (from case histories)
5. organizations performing it
6. the kinds of people doing it
7. its relationship to other fields (such as operations research, and industrial engineering)
8. its most usable tools and techniques

Although Hall does not appear to use the latter description as strongly as he does the other three discussed, we have this additional perspective in his overview of the subject. In particular, the last element in the latter description seems to be especially relevant since we are interested, within the context of computer-aided systems engineering (CASE), in ways in which the computer can be used as a tool and the kinds of techniques represented by the software that is available to carry out the tasks of systems engineering.

Two of the earliest investigators of systems engineering, Goode and Machol [5], examined both the process itself and documented in a tutorial fashion what they considered at the time (1957) to be the major tools of systems engineering. With respect to the process, they focused upon what they called an integrated approach to design, implicitly emphasizing the element of synthesis. They broke this element into six phases

1. initiation
2. organization
3. preliminary design
4. principal design

5. prototype construction

6. test, training, and evaluation

They also defined 13 tools of system design, namely

1. computing

2. cybernetics

3. game theory

4. group dynamics

5. human engineering

6. information theory

7. linear programming

8. mathematical statistics

9. probability

10. queueing theory

11. servomechanism theory

12. simulation

13. system logic

This early work set the stage for the evolution of both the process and the tools of systems engineering.

Chapter 1 contained a brief listing (Table 1.2) of some of the mathematical and logical techniques often used in systems engineering. As indicated, many of these techniques are now or will be available to the student and practicing engineer using the p-c as a resident tool. Table 1.1 is a listing of the elements of systems engineering that defines the tasks and activities that are to be performed. For purposes of CASE and this particular exposition, we adopt as a baseline structure the two dimensions listed here

1. the elements (tasks and activities of systems engineering)—Table 1.1

2. the tools (computer) and techniques (mathematical and logic-based procedures of systems engineering)—Table 1.2

It is left as a challenge to the reader to map the above two dimensions and their subordinate elements into the various descriptions and dimensions provided earlier (see the problems at the end of this chapter) as well as those discussed in Section 2.5, Systems Engineering Management.

2.3 THE MORPHOLOGICAL APPROACH TO SYNTHESIS

If we take Hall's [3] three dimensions of systems engineering—the seven-step, problem-solving dimension, the seven-step, life-cycle phase dimension, and the body of subject-area-knowledge dimension—we can form three axes and map each dimension on a separate axis. If we assume eight knowledge elements, we have a total of $7 \times 7 \times 8$ or 392 activities that might represent individual cells within the broad context of systems engineering. Hall has taken such a view, adopting a *morphological* or *taxonomic* approach to defining all the dimensions and elements of systems engineering. Such an approach is useful for such a purpose and is based upon the work of Zwicky [6] who has been important in exploring the process of synthesis and invention. It is interesting to observe that Hall has used this morphological approach to design or synthesize his view of the elements of systems engineering, which itself contains the subelement of systems synthesis, as the third step in the seven-step, problem-solving procedure or dimension.

We continue to emphasize synthesis or design because of its importance in systems engineering and due to the fact that it is difficult to do and perhaps even more difficult to understand how to do it well. Talented systems engineers appear to have a storeload of facts and system elements or components that they are able to sort through very quickly, using experience, intuition, and logic, in order to formulate new and often near-optimal design configurations. Since the number of possible alternatives in a design or synthesis process is usually extremely large, we can imagine that the computer should become increasingly important in assisting in this process if we can understand how to instruct the computer to do so.

In Zwicky's morphological approach, he suggests that the key dimensions or parameters of a design problem need first to be understood and defined and then these dimensions must be mapped against one another. By so doing, a "space" is created such that each cell in the space requires more detailed investigation as to its viability and characteristics as a design alternative. As we have seen in Hall's adaptation of the morphological procedure in regard to systems constructs, he has formulated three dimensions leading to a total of the order of 400 cells. If the problem at hand has more than three dimensions, the number of cells increases in this multiplicative fashion. In the general case of n dimensions, we are exploring the cells in n-dimensional space and hoping to be able to analyze the attributes of each of the cells. If n is ten, for example, and each dimension has, for example, ten elements, we have ten billion possibilities! Such is the potential difficulty of this problem area.

A key element in Zwicky's approach, as alluded to in Hall's application to systems engineering, is the construction of the morphological box. If developed properly, this box contains all solutions and considerations relative to the given problem. According to Zwicky, it also embodies an unbiased evaluation of the various solutions that may be found. The following steps are necessary to establish Zwicky's morphological box and its associate analyses:

1. concisely formulate the problem
2. define and analyze all of the important parameters
3. construct the morphological box (multidimensional matrix)
4. evaluate all cells in the box in relation to the purposes of the system
5. select and apply optimally suitable solutions

This relatively concise description, Zwicky notes, may require what he calls a supplemental morphological study. However, the concept of developing a systematic way of constructing a space that should contain all possible solutions is indeed a very important one. In step 4, Zwicky suggests that each cell be analyzed, in effect, against the preestablished evaluation criteria for the problem or system at hand. We note here that this does not circumvent the combinatorial aspect of the essential issue; as the number of dimensions and the elements within each dimension increase, the number of cells to be examined increases exponentially. Thus we are ultimately in need of efficient search algorithms that will assist us in both cutting down on and moving through this vast array of alternatives. In addition, as we shall explicitly examine later, the evaluation criteria generally will not be commensurable, which will complicate the evaluation in step 4 as well as the selection of the so-called optimally suitable solutions called for under step 5.

By way of example, Zwicky explores types of energy and energy transformations. He constructs a two-way morphological box of energy transformations by mapping the following ten types of energy against one another

1. $E(1)$ = kinetic energy
2. $E(2)$ = elastic energy
3. $E(3)$ = gravitational energy
4. $E(4)$ = heat
5. $E(5)$ = electrical energy
6. $E(6)$ = magnetic energy
7. $E(7)$ = chemical energy
8. $E(8)$ = electromagnetic energy
9. $E(9)$ = nuclear (atomic) energy
10. $E(10)$ = rest energy of matter

Given this morphological box, practical applications of each type of energy transformation (in the one hundred cells) can systematically be sought with respect to medicine, communications, agriculture, science, and various other fields. Although Zwicky did not mention transportation, we invite the reader (in the problems at the end of the chapter) to explore this field from a morphological point of view in regard to different trip lengths, motive powers, and modes of transportation.

2.4 HEURISTIC MATRIX EVALUATIONS

The morphological approach to synthesis was introduced in the previous section as a potentially powerful way of exploring all the various aspects of a design problem. It relies on the identification of the problem dimensions and the subsequent mapping of these dimensions against one another to form a set of "cells" at all intersections, each of which requires detailed analysis and evaluation. In the energy example cited, a two-way table or matrix is created since we have only two dimensions that evolve from the pairwise consideration of energy types.

The two-way table or matrix, limited as it is to two dimensions, is a very effective and under-valued tool of the systems engineer. We note the extraordinary success of computer-based, two-way tables known as spreadsheets, which we will discuss and utilize in some detail in later chapters. However, the two-way table or matrix can also be used to examine problems and relationships in more than two dimensions by considering various combinations of the rows and columns, mapped against one another. This can be particularly helpful in the analysis and evaluation process in distinction to the synthesis problem as represented by the morphological box. We have established the term heuristic matrix (HEMAT) evaluations to refer to this exploration of two-at-a-time relationships. The heuristic designation goes back to the original meaning of the word and suggests the stimulation of further investigation. Thus we wish to use the matrix format, a well-developed, computer-aided tool, to serve as a nonrigorous means of sorting through alternatives, possibilities, and relationships, helping us to formulate and converge to fewer alternatives and more formal investigations that may be feasible once the field of examination has been narrowed.

As we have seen, the systems engineering approach and process begin with statements of needs and requirements that lead to a formal set of specifications for the system. A decomposition of the system into its functions is an implicit part of the process, evolving to the definition of the system and subsystem alternatives. Through an evaluation procedure in relation to a variety of criteria, a preferred system alternative is selected. Examining this limited portion of the process, we have defined, for example, five intermediate products

- requirements
- specifications
- functional capabilities
- system and subsystem alternatives
- the preferred system and its subsystems

If we wish to formally explore the interrelationships between these products, we can construct a series of matrices by considering all the ways in which they can be mapped against one another. For five items, we have $nCr = {}_5C_2 = (5)(4)/2$ or ten combinations of five items taken two at a time. These resulting matrices are illustrated in Figure 2.2 [7]. This approach recognizes that all combinations are considered for the sake of completeness (as in the case of the morphological box) but that ultimately some of these matrices may be collapsed if large numbers of cells for any given matrix turn out to be of marginal or no interest. This format allows us eventually to employ the computer power resident in the two-way tabulation of the spreadsheet and database management system. At the most elemental level, the cells in a HEMAT contain checks or crosses to indicate that a relationship exists between a particular row and column. For example, if we are looking at the mapping between requirements and specifications we can show cases for which several specifications (columns) have been developed to relate to a single requirement (rows). By examining a given specification we can see if it applies to more than one requirement. If we analyze the preferred system versus system alternative matrix, we may identify the systems and components thereof that have been selected to be part of the preferred

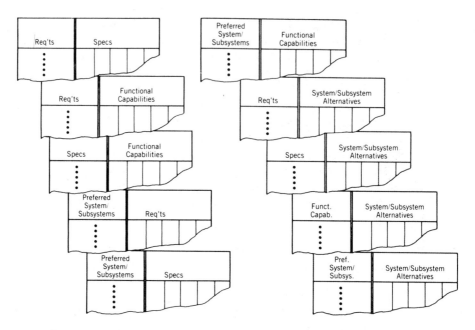

Figure 2.2 Illustrative heuristic matrix evaluation examples. (From H. Eisner, "CASE: Computer Aided Systems Engineering for C³I Systems," *Signal,* July 1984. Reprinted with the Permission of the Armed Forces Communications and Electronics Assoc., Burke, VA, © 1984.)

system. At another level, we can attempt to quantify the appropriate cells by listing relevant performance parameters or cost-related information. Illustrations of questions that might be addressed by this type of quantification are

- What are the parameter values in the specifications that will (a) satisfy a given requirement, (b) satisfy a given functional capability?
- What is the performance capability of the system/subsystem alternatives versus the preferred system?
- What is the cost of implementing each of the functional capabilities for (a) the system alternatives, (b) the preferred system?

The morphological approach as applied to all combinations of parameters, as alluded to in the aforementioned list, may lead to sparse matrices, which may then be collapsed initially or by command at a computer terminal. As an example, in mapping requirements against specifications either or both may be entered under the subcategory of the system function to which it applies. Thus we may wish to categorize specifications for a command, control, and communications system by the functional area of communications or search, but also be mindful of the possibility that a given specification may apply to more than one system function. This may be the case, for example, when a system contains both a search mode and a tracking mode such that common sets of equipment are used in both modes of operation as the system switches from the search to the tracking function.

In summary, the point of the two-dimensional HEMAT structure of all combinations is (a) to assure completeness at the outset, recognizing that collapsing of data may be appropriate later, (b) to encourage systematic and heuristic exploration of meaningful relationships, and (c) to place the information in a computer-compatible format so that further manipulation and examination can be exercised automatically.

2.5 SYSTEMS ENGINEERING MANAGEMENT

Systems engineering management is generally carried out for large-scale systems in a relatively structured manner. Often, various standards, instructions, and specifications guide this management process. Documentation of the methods to be employed as part of such a process can be contained within what is called a systems engineering management plan (SEMP). For example, the DoD, in Military Standard 499A [8], details the structure of systems engineering management and what the contents of a SEMP are to be. This approach will be examined in some detail in this section.

A structure for systems engineering management is shown in Figure 2.3, representing an adaptation of Military Standard 499A. We note that in this structure, systems engineering management is composed of the following three types of activities:

1. technical program planning and control
2. the systems engineering process
3. engineering specialty areas

Each of these activities in turn, has subordinate areas, which in total define the management process and what may be expected to be included in a systems engineering management plan. Note the relationship between the areas in Figure 2.3, the set of systems engineering tasks and activities defined in Table 1.1, and the elements of systems engineering described in Section 2.2. These formulations have many aspects in common and their differences reflect the points of view taken by the various authors and investigators as well as some differences in the intended areas of application. They also demonstrate that there is currently not one accepted, universal definition of the ingredients of systems engineering and its management.

The areas shown in Figure 2.3 have become well known to the developers of large-scale systems for the federal government. The CASE approach, in this context, is to provide the computer resources to assist in carrying out the necessary tasks implied by the defined areas. Through an understanding of the contents of these areas on one hand, and the availability of computer-software tools on the other, a CASE capability can be created that will improve the effectiveness and productivity of systems engineering as a whole.

2.5.1 Technical Program Planning and Control

This area has been defined [8] as *the management of those design, development, test, and evaluation tasks required to progress from an operational need to the de-*

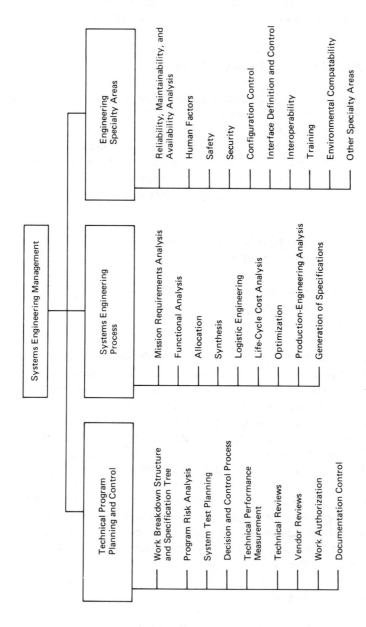

Figure 2.3 A systems engineering management structure.

26

velopment and operation of the system by the user. We briefly comment here upon each of the subordinate aspects of technical program planning and control as delineated in Figure 2.3.

Work breakdown structure and specification tree. A *work breakdown structure* (WBS), as the name implies, is usually a hierarchical layout of the work to be performed by the organization responsible for the systems engineering activities. An illustrative WBS is shown in abbreviated form in Figure 2.4; other examples of a WBS are provided in the literature [9]. Since the WBS is hierarchical it will map easily into many database management systems that are available. The elements of a WBS will also have to be assigned to various parts of the organization and will have to be correlated with specific items to be delivered to the customer.

At the highest level the specification tree is also a hierarchical structure, but it also contains the various types and levels of specifications for the system [10]. Within each level of specification there is a substructure that is also hierarchical in nature. For example, in a system-level specification there may be a section on requirements that has a set of elements such as

- system definition
- characteristics
- design and construction
- documentation
- logistics
- personnel and training
- functional-area characteristics

Reference to the standards developed for the area of specifications will provide further details on these structures [10].

The coupling of the WBS and the specification tree more or less suggests the need to correlate one with the other. The WBS can be constructed in such a way as to facilitate the mapping of work elements into items that appear in the specification, but this need not be the case. However, a correlation between WBS work elements and specification elements can be set forth in a simple database management system.

Program risk analysis. This area calls for a continuing assessment of risks normally associated with cost, schedule, and technical performance. Risk analysis tools have been developing and for that reason, as well as its role in systems engineering, risk analysis is considered separately in Chapter 13. The primary emphasis in relation to program development is normally given to cost, schedule, and performance-related issues in the context of satisfying the system requirements and specifications. If a program manager sees risks in these areas, they need to be revealed to the sponsor, and the program manager should attempt to put them in some priority order or quantify them in some other fashion. Unfortunately, it is sometimes the case that longer range societal risks are not seriously considered except where the potential issues are overwhelming. Such situations have included the building of

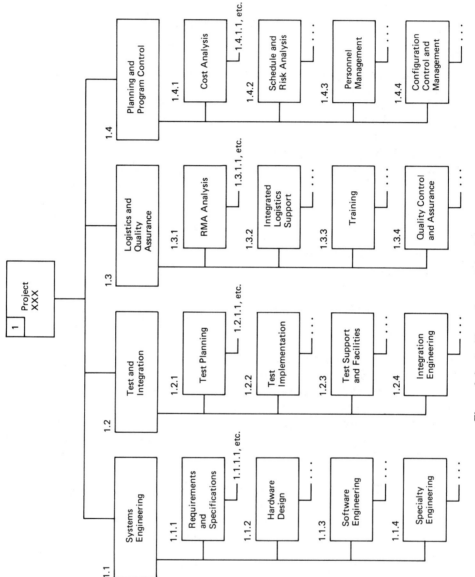

Figure 2.4 An illustrative work breakdown structure.

nuclear power plants, the handling of carcinogenic materials, and the potential effects of ozone-layer depletion resulting from the operation of supersonic transports. In these cases we are dealing with risk from the point of view of low probability events coupled with disastrous results should they occur. Program risk, on the other hand, usually considers much more likely situations with less ominous implications if they happen. In either case it is important for the systems engineer to have some tools available that will assist in performing the necessary risk analyses.

System test planning. The most reliable way to determine if a system has satisfied its requirements and specifications is through test. In the early stages of development, system performance is a prediction; as subsystems are implemented in hardware and software, they may be tested to verify performance. Thus there is always a strong linkage of testing parameters to the system specifications and, in general, an increase in the cost of a test as more and more subsystems are integrated. Test planning is thus a crucial area, particularly in regard to establishing efficient test procedures, solid acceptance criteria, and ties back to the appropriate specifications. Such linkages can be maintained in a database management system and related as well to requirements tracking and traceability. Testing is a critical part of the early life-cycle stages of research, development, test, and evaluation (RDT&E).

Decision and control process. Here the emphasis is on the process of making decisions that control the systems engineering effort and its risks. Clearly, a risk analysis by itself is not sufficient. It is necessary to find alternative solutions should the risk be unacceptable or have to be reduced. Decisions need to be made to implement such solutions where necessary. Thus, this area is a very active part of systems engineering management where alternatives to the current course of action are continually under evaluation and new courses charted where appropriate. Some of the formal means by which decision alternatives can be considered (e.g., decision trees) resemble the risk analysis procedures. For this reason they are both considered in greater detail in Chapter 13.

Technical performance measurements. Technical performance measurements (TPMs) take their place alongside cost and schedule as a critical aspect of systems engineering. Considerable attention is therefore given to TPMs as part of systems engineering management. The first problem is the selection of the technical performance parameters that represent the key indicators of system success. In turn, these parameters are usually interrelated so the next step is to determine and display the nature of these interrelationships. Various graphical techniques such as parametric dependency diagramming, as discussed and explained in Chapter 4, can be employed for this purpose.

A further issue with respect to TPMs is the need to continuously track the achievement of key TPM values as compared with the required values. In principle, as we move through a program, we increase our assurance that the required technical performance values of the system will be achieved. A time profile for each parameter can thus be envisioned, as illustrated in Figure 2.5, together with other notions related to the requirement for technical performance measurement, that is, the partic-

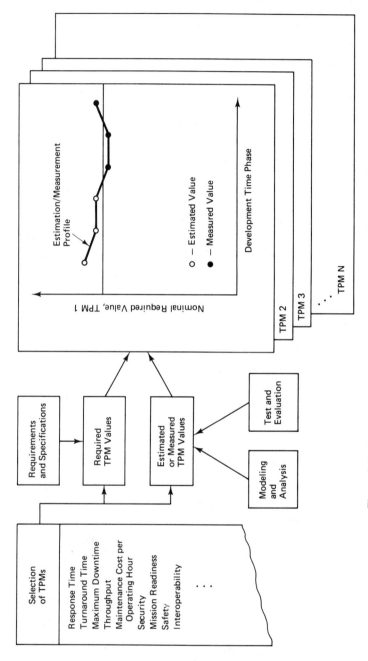

Figure 2.5 Technical performance measurement relationships and tracking.

ular ways in which a technical performance measurement is made in the real world. In practice, such time profiles are not easily developed since the measurement process is often not well formulated or maintained. In any case, numerous computer aids are available to assist in the implementation of a TPM program, which is revisited in some detail in Chapter 12.

Technical reviews. Specific types of technical review are required, corresponding to clearly identifiable milestones during the system development process. These review points are listed here and discussed in connection with Section 2.6 dealing with the systems acquisition process

- system requirements review(s) (SRR)
- system design review (SDR)
- preliminary design review (PDR)
- critical design review (CDR)

Vendor reviews. This area simply addresses the need to assess the cost, schedule and technical performance status of system components or subsystems that are being provided by outside vendors.

Work authorization. Within this area it is necessary that all aspects of the designated WBS have been carefully planned and staffed before authorization to start on any work package is given.

Documentation control. The matter of control over documentation for large-scale systems' development is not a trivial one. Indeed, separate military standards (Mil. Std. 480–483) deal with configuration management and control, all of which must be properly documented. The principal notion here is that the system is undergoing change and as it does so its documentation must also change and be updated. Computer-assisted tools are a very powerful mechanism for maintaining up-to-date documents and assuring that changes can only be made through authorized change procedures.

2.5.2 Systems Engineering Process

The second major category of systems engineering management, as shown in Figure 2.3, is the systems engineering process itself. We can see that the areas that are part of this process correlate more closely with those elements, tasks, and activities discussed as being ingredients of systems engineering. The process is defined [8] as *a logical sequence of activities and decisions transforming an operational need into a description of systems performance parameters and a preferred system configuration.*

Mission requirements analysis. Requirements analysis, in this context, refers to a concept of continuous assessment to assure validity, consistency, desirability, and attainability. The mission requirements are viewed as potentially changeable thereby calling for either verification of the current set or the development of new requirements that are more appropriate. As part of this continuous evaluation process, one is urged to consider impacts of the stated system operational characteristics, the mission objectives, possible changes in threat, environmental factors, minimum acceptable system functional requirements, technical performance, and system figures of merit. Whereas this interpretation appears to imply a world of ever-changing requirements, it should be kept in mind that at any point in time the requirements are fixed and should be considered inviolate from the point of view of the system-design engineer.

Functional analysis. Generically, the functions to be carried out by a system include all those activities necessary to satisfy the overall system mission or set of needs postulated for the system. As an example, some of the functions to be carried out by a command, control, and communication system might include the following:

- command
- communication
- control
- damage assessment
- data communications
- data processing
- detection
- guidance
- identification
- networking
- programming
- reconnaissance
- search
- surveillance
- tracking

Under the area of functional analysis, it is necessary to identify all primary functions of the system first and then progressively break these down into subfunctions and their subordinate functions, if appropriate. This constitutes a *functional decomposition* of the system and is discussed further in Chapter 4. Some of the functions and subfunctions also represent alternative operational modes for the system as, for example, with a radar system moving from a search mode to a tracking mode.

Allocation. This is a critical and often misunderstood area. It basically deals with the allocation of requirements to both the system functions and subfunctions as well as to the hardware, software, and personnel or organizational units that are to implement these functions and subfunctions. If the mission requirements are known, then in this step they are broken down so that they can be levied upon the items (often referred to as configuration items, CIs, or computer program configuration items, CPCIs, or computer software configuration items, CSCIs) to be developed by the design engineers. Without such a step the subsystem design team has no specific requirements upon which they must focus.

However, the process of allocation remains a very difficult and often murky one. The reason is that good algorithms for suballocation are conceptually complex and nonunique. They are often done, therefore, by combining state-of-the-art evaluations (i.e., what can current equipment actually deliver by way of performance?) with seat-of-the-pants judgments about allocation (what equipments get how much requirement and how do the requirement values "add" up?). This is perhaps best illustrated by the specific case of error analysis for a large system. If the total error for a given function has a specified required value, (e.g., tracking error), then the errors to be allocated to the subordinate configuration items is a complex function, in general, of the design of the equipments and how they are hooked together. Often a root-sum-square approach is taken in the absence of a better formulation for such an assignment. This acknowledges the possible statistical character of the problem but not the specific functional dependencies or nonlinearities. This area will be discussed with some rigor and illustrations in Chapter 10.

Synthesis.   Here we return again to the central design element of systems engineering. This is described in the standard [8] as *the performance, configuration, and arrangement of a chosen system and its elements and the technique for their test, support, and operation shall be portrayed in a suitable form. . . .* Examples of portrayals considered important include

- schematic diagrams
- physical and mathematical models
- computer simulations
- layouts and detailed drawings
- engineering graphics
- intra- and intersystem interface definition
- system, CI, and critical item specifications
- consolidated facility requirements
- procedural handbooks and instructions
- personnel task loading
- operational computer programs
- specification trees
- dependent WBS elements

Logistic engineering.   This area addresses the design of a logistics process and system to support the deployment and operational phases of the system. It includes such subareas as maintenance engineering, repair-level analysis, logistics-support modeling, spares provisioning, transportation, and personnel selection and assignments. This represents an extremely important area, sometimes known as integrated logistics support (ILS). Even a well-designed system will be unable to maintain its operational readiness unless it is properly supported logistically.

Life-cycle cost analysis. Total life-cycle costs for a system normally include the costs in each of the three major phases of the system, namely

1. research, development, test and evaluation (RDT&E)
2. procurement or acquisition
3. operations and maintenance (O&M)

Generally, these costs vary for alternative design configurations so we should envision a parametric set of life-cycle cost estimates. Techniques and tools for system-cost analysis are provided in some detail in Chapter 11.

Optimization. This one word embodies all of the analysis and iteration implied by Figure 2.1, the systems-approach diagram. Primary factors to be considered include technical performance, risk, schedule, and life-cycle costs. Subordinate activities to be carried out in an attempt to develop so-called optimum configurations include

- trade-off studies
- cost-effectiveness evaluations
- effectiveness-analysis modeling

These areas are likewise explored in later chapters, particularly 10 and 12.

Production-engineering analysis. The key issue here is producibility, its assurance, and what it will require in terms of facilities, equipment, personnel resources, and cost. Particular attention is focused on special production needs that may lead to increased program risk, measured in respect to cost, schedule, or technical performance.

Generation of specifications. As alluded to earlier, specifications play a key role in systems engineering. They flow from the set of requirements, and their updates evolve (see Figure 2.1) from the definition of the preferred system configuration. Thus there are "levels" of specifications, starting with what is sometimes referred to as an *A* spec and moving successively into a *B* and *C* spec. A military standard [10] has been adopted that provides guidelines as to how to generate appropriate sets of specifications. There is normally a close coupling between the requirements and specifications documentation for a system, as well as to the WBS.

2.5.3 Engineering Specialty Areas

These areas, as subject matter specialties, tend to be treated as support to the mainline systems engineering process as well as technical program planning and control areas. Specialty areas are selected based upon the needs of each particular program, and emphasis is placed upon the appropriate intermeshing of specialty areas with the systems engineering process areas so as to have the maximum positive influence on the system design. The reference document [8] alludes to some specialty areas but does not develop a standard list of such areas. In this discussion we refer to the items in Figure 2.3.

Reliability, maintainability, and availability analysis. The reliability, maintainability, and availability (RMA) areas are almost always an important and integral part of a systems engineering program. They are sometimes considered as separate engineering specialties and at times included under performance or effectiveness analysis. In complex high-performance systems, RMA considerations have become crucial as we have discovered more conditions of operational unavailability and lack of readiness. Chapter 15 contains a more detailed and quantitative examination of these areas.

Human factors. This subject involves consideration of all human interaction within the system including the human as operator, maintainer, and user. Systems in which the human's life may be in jeopardy, such as the space shuttle and station, have received special human-factors attention over the years. Examinations of human factors have been categorized also under the title *man–machine interfaces* (*MMI*). In addition, Military Standard 1472 deals with the area of human-engineering design criteria.

Safety. Safety is usually a rather complex area for most systems. Those systems that involve the safety of relatively large numbers of people normally have extensive safety requirements, rules, and regulations, but it has been difficult to quantify and demonstrate the actual levels of safety achieved. Notable examples have included commercial aircraft service and our air-traffic-control system and the issue of potential catastrophe associated with mishaps at nuclear power plants. There are a variety of formal safety analysis tools such as fault-tree analysis and relative-accident-probability (RAP) analysis and the reader is referred to Chapter 13 and its references for additional information regarding these techniques.

Security. The security aspects of large-scale systems have many dimensions including such items as physical and computer security. The former was introduced into all air travelers' lives when the aircraft hijacker problem came up. The latter case will receive more and more attention as we move through the information age and maintain accounts and transactions of our possessions through computer systems. Essentially all of our military systems now carry a requirement for significant planning and implementation of computer security measures as a result of the infusion of both support and embedded computer systems.

Configuration control. The issue of maintaining control and records regarding the current and past configurations for the system is a very important one. Here again, the military has developed several standards related to this issue. Military Standards 480 and 481 deal with configuration control in regard to engineering changes, deviations, and waivers, the latter being the short form. Standard 483, dealing with configuration management practices for systems, equipment, munitions, and computer programs [11], is used a great deal in identifying all the management steps needed to assure configuration control. Computer-aided packages have been developed and are being used by the industry to implement the required configuration-management functions.

Interface definition and control. A requirements document or specification will normally contain an explicit definition of both intra- and intersystem interfaces, at least at an overview level. The task here is to provide the next levels of detail in such definitions, assure their completeness, and maintain control over them so that they are not changed without proper authorization. As the system and its subsystems become more completely defined, we arrive at the configuration item level and it is necessary then to have full definition and control over the interfaces at these levels. It is natural to consider keeping track of the interfaces with a database management system.

Interoperability. As we have experienced the proliferation of independently developed systems, we have observed that there have been considerable problems in having them interoperate efficiently with one another. A good example is our national aviation system, which may be thought of as consisting of the following major subsystems:

- airports
- air traffic controls
- air vehicles
- supporting access systems

Clearly there is a need for these subsystems to interoperate with one another although, in general, they are produced and run by a large number of relatively independent parties. We have all experienced or heard about problems in these areas such as difficulties in getting to and from airports and waiting for long periods on runways before take off from a particular airport. These can all be viewed as interoperability issues and greater attention is being paid to them as we have noticed their importance in the overall efficiency of a system.

Training. As we define and allocate requirements, we soon come down to the personnel or human element in the system. Operators of any system clearly need to be trained in order to have the system run properly. In addition, the maintenance people require training to keep the system in operating condition. Sometimes neglected is the arena of the user who is not formally trained in many situations but nevertheless needs instruction on how to use the system. An example is a bus or subway system for which it is necessary to provide effective signs and brochures to allow users to find their way on the system. At least initially, training materials can be prepared on word processors—the computer can be a valuable aid even in this context.

Environmental compatibility. Many systems today still require the preparation of a formal *environmental impact statement* before approval is given to the development of the system itself. Environmental hazards can take numerous forms such as unwanted radiation; pollution of our air and water; detrimental effects on our animal life, including human beings. Some effects are simply aesthetic but have a major impact on the tendencies of users to want to make use of the system. Often, environmental-compatibility analysis requires the preparation of rather complex models of our environment in order to be able to make such assessments. Here again, computers can be invaluable in carrying out such modeling efforts.

Other specialty areas. As indicated, there are a large number of engineering specialty areas that can be selected that will be relevant to the particular needs of a given program. Some of the more common ones have been listed and briefly discussed. Additional areas include quality assurance, electromagnetic compatibility, standardization, and value engineering. Some of these may be construed as being subsumed within the other areas that are part of program planning and control or the systems engineering process.

In broad terms, all of these areas are part of what is termed systems engineering management, in other words, they represent the full set of activities that need to be performed and managed in order to develop, install, and operate a system. With this as a base line, we look to the power and capability of the computer to help us with these activities. To the extent that we can create and use this power, we are also creating a computer-aided systems engineering (CASE) capability.

2.6 THE SYSTEMS ACQUISITION PROCESS

This process tracks the dimension of the system that has been referred to earlier as the *time* or *life-cycle* phase. Much of this, as we have seen, has been formalized so that workers in the field recognize and use certain terminology in this context. Figure 2.6 shows the systems acquisition process in diagrammatic form from the formulation of the need for the system to its operational phase. Thus, the distinguishable phases of the development cycle may be construed as

- need development
- concept definition
- concept validation
- engineering development
- production
- operations

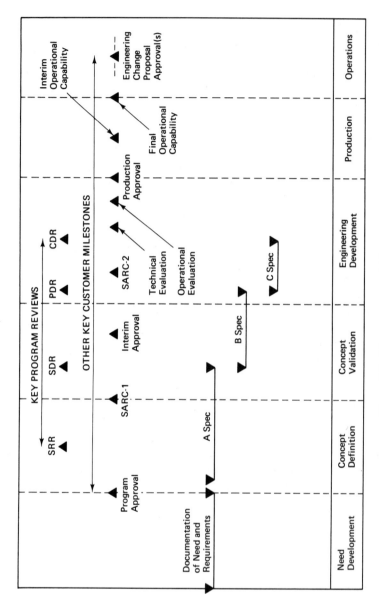

Figure 2.6 The systems acquisition process.

We see that the four key reviews of the program—SRR, SDR, PDR, and CDR—span the three key phases, that is, concept definition, concept validation, and engineering development. We also note that the customer has several major milestones

- program approval
- system acquisition review council (SARC-1)
- interim approval
- SARC-2
- technical evaluation
- operational evaluation
- production approval
- interim operational capability (IOC)
- final operational capability (FOC)
- engineering change proposal (ECP) approval

The initial program approval confirms that there is a bonafide need for the system. The SARC-1 milestone shows a generic acquisition council review that assures that the system requirements and functional definition have been properly executed, and that the program may then proceed into the concept validation phase. The interim approval occurs during the concept-validation phase and after the system has gone through SDR at which time the A specification has been completed and is therefore available for review and approval. Going into the engineering-development phase, the B specification is completed and the PDR is held. After that, SARC-2 is held and a further go-ahead is confirmed. The CDR is hopefully carried out with the benefit of the C specification. The system then goes into technical and operational evaluation after which approval is given to leave the engineering-development phase and enter the production phase. Production proceeds to an IOC and then to a final operational capability FOC. During operations, engineering change proposals are considered and implemented to improve the capabilities of the system. This process of improvement usually continues throughout the operational phase. The systems-engineering management process and set of activities receive emphasis during the concept definition and validation phases as well as through the engineering-development phase.

2.6.1 Air-Traffic-Control Example

In 1981, the Federal Aviation Administration (FAA) realized they needed to replace the mainframe computers that support the air-traffic-control function. They produced an order [12] that served as a requirements-definition statement for this replacement program. This requirements definition by the customer, in effect, established the end of the need-development phase and gave approval to proceed into the concept-definition phase. To illustrate the scope of such a needs- or overview-requirements statement, in Table 2.1 we show the subject areas that were included in that statement. The statement is designed to firmly establish the need, briefly describing the expected benefits, costs, and major milestones for the system.

TABLE 2.1 SCOPE OF AIR-TRAFFIC-CONTROL-REQUIREMENTS STATEMENT

I. Definition of the Problem and Opportunity
II. Mission Need and Operational Requirements
   A. General System Requirements
      1. System Accuracy
      2. System Reliability
      3. System Size
      4. System Performance and Maintainability
      5. Central Support Facility
      6. Human-Factors Engineering
   B. Functional Requirements
      1. Initial Operations Changeover Capability
         a. Data Processing
         b. Data Display
         c. Simulation of Operational Functions
      2. Future Requirements
         a. Expansion of Direct Interfaces
         b. Automated Decision Making
         c. Accommodation of Unique Aircraft Characteristics
         d. High-Speed Computer-to-Computer Interface
         e. Flow of Traffic Organization
         f. Search for Aircraft Future Path
         g. Conflict-Resolution Alternatives
         h. Increased Controller Productivity
         i. Digital Two-Way Communications
         j. New Feature Interfaces
         k. Actual and Forecast Weather
         l. Automated Planning and Control Functions
         m. Data Link to Pilot
         n. Consolidated Terminal Automation
         o. Expansion of Direct Routings
         p. Changing Weather and Surveillance Data Sources
         q. Expansion of Processing
         r. Lateral and Longitudinal Relationships
         s. Flow-Management Function
         t. Training Capability
         u. Simulation Training
         v. Interface with Field Reporting
         w. Aviation Safety Data
III. Potential Rulemaking
IV. Potential Benefits
   A. Direct Benefits
      1. Improved Safety
      2. Maintenance Cost Reduction
      3. User Cost Avoidance
   B. Enabling Benefits
      1. Insufficient Capacity
      2. O&M Cost Reduction
      3. Increased Safety
         a. Improved Reliability
         b. Avoidance of Sector Size Reduction

TABLE 2.1 *(cont.)*

          c. Avoidance of Conflicts
          d. Hazardous-Weather Information
        4. Delay and Fuel Consumption Reduction
        5. Flexible System Architecture
        6. Improved Controller Skills
   V. Alternative Selection
  VI. Estimated Costs
 VII. Applicable Guidelines
      A. Reliability and Maintainability
      B. Productivity and Staffing
      C. Energy and Environmental
VIII. References
  IX. Authorization

2.6.2 Space Operations Communications Example

The Air Force confirmed the essential requirements by 1983, and then entered the concept definition phase for the Consolidated Space Operations Center (CSOC) program. In particular, they issued a request for proposal (RFP) to industry [13] for the communications segment of that program, whose basic purpose is to support the operations of various satellites in space. The key systems engineering tasks called for in that requested effort are listed here in Table 2.2 [7]. These tasks describe the emphasis given to the systems engineering management activities discussed in this and later chapters.

2.7 SUMMARY AND ROAD MAP

A summary of several of the main points of the first two introductory chapters on systems engineering concepts and the way they relate to the other parts and chapters of this text is depicted in Figure 2.7. The seven boxes on the left of this figure show various tools and elements of system design, four dimensions of systems engineering, and Zwicky's morphological approach, as cited in the literature and referred to in this chapter. These form a backdrop and basis for the selection (as shown in the figure) in this book of

- systems engineering tasks and activities (Table 1.1)
- mathematical and logic-based tools and techniques (Table 1.2)
- categories of software packages (Table 1.3)
- the systems approach (Figure 2.1)
- systems engineering management (Figure 2.3)
- the systems acquisition process (Figure 2.6)
- heuristic matrix evaluations (Figure 2.2)

TABLE 2.2 KEY SYSTEMS ENGINEERING TASKS FOR CONSOLIDATED
SPACE OPERATIONS CENTER (CSOC)

I. Functional Allocation/Analysis
   A. Allocation of System Requirements to Configuration Items
   B. Specification Trees
      1. Prime and Critical Items
      2. Computer Program Configuration Items
      3. Hardware Configuration Items
   C. Improvements To Stated Functions
   D. Design Approaches For Each Function
   E. Key Risk Areas/Reductions
   F. Allocation of Functions and Subfunctions
   G. Support of Allocation Decisions
      1. Time-Line Analysis
      2. Synthesis of System Design
      3. Cost Effectiveness of Trade-Offs
      4. Traceability of Completeness
      5. Design Approach
II. Interface Definition
   A. External Interface Requirements
   B. Critical Internal Interfaces
   C. Interfaces and Specification Trees
III. Technical Performance Measures (TPMs)
   A. Twelve TPMs
   B. Describe and Justify
   C. Relate to System Performance Measures (SPMs)
   D. Tracking of TPMs
   E. Active Systems Engineering Discipline
   F. Minimization of Design, Implementation, and Cost Problems
IV. Specialty Engineering
   A. Integrated Specialty Engineering
      1. Reliability and Maintainability
      2. Computer Resources
      3. Human Engineering
      4. Manufacturing and Production
      5. Quality Assurance
      6. Parts, Materials, and Processes Control
      7. Corrosion Control
      8. Contamination Control
      9. Safety and Security
      10. Specialty Engineering and Design
      11. Specialty Engineering and Allocation Process
      12. Design to Life-Cycle Cost
V. System Test
   A. Test Methodology
   B. Incremental Testing for Allocated Configurations
   C. Testing and Correct Interfaces
   D. Support of Test Agencies

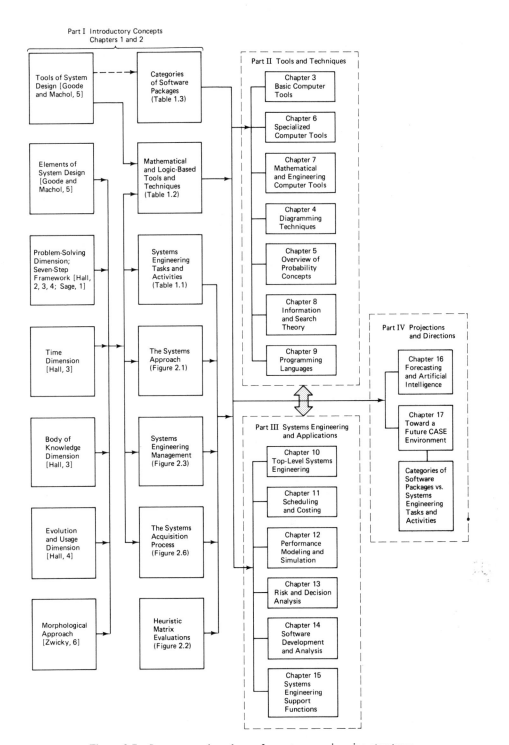

Figure 2.7 Summary and road map for systems engineering structures.

Sec. 2.7   Summary and Road Map

All of these are contained within the first two chapters that constitute Part I. Chapters 1 and 2 are considered introductory to the main body of text provided in Parts II, III, and IV, which deal with, respectively

- Tools and techniques
- Systems engineering and applications
- Projections and directions

The tools and techniques are generally influenced by the categories of software packages as well as the definition of mathematical and logic-based tools and techniques that are considered to be most important in systems engineering. As discussed in this text, the tools and techniques are covered in seven chapters, as listed here:

- Chapter 3   Basic Computer Tools
- Chapter 6   Specialized Computer Tools
- Chapter 7   Mathematical and Engineering Computer Tools
- Chapter 4   Diagramming Techniques
- Chapter 5   Overview of Probability Concepts
- Chapter 8   Information and Search Theory
- Chapter 9   Programming Languages

The first three in the list are grouped together since they all deal with computer-based tools as embodied in existing classes of software packages. They are not consecutive in the text since the notions of Chapters 4 and 5 are needed in order to fully understand the computer tools considered in Chapters 6 and 7. Chapters 4, 5, 8, and 9 are grouped together since they are largely analytical in nature, leaning heavily upon mathematical or logical formalisms. The student who masters the material in these seven chapters of Part II will be in an excellent position to apply these tools and techniques to the systems engineering aspects of Parts III and IV. The latter, in turn, contain the following chapters:

PART III:

- Chapter 10   Top-Level Systems Engineering
- Chapter 11   Scheduling and Costing
- Chapter 12   Performance Modeling and Simulation
- Chapter 13   Risk and Decision Analysis
- Chapter 14   Software Development and Analysis
- Chapter 15   Systems Engineering Support Functions

PART IV:

- Chapter 16   Forecasting and Artificial Intelligence
- Chapter 17   Toward A Future CASE Environment

Chapter 17 represents a type of summing up in the sense that a possible future environment for CASE is described. A key element in this exposition, as shown in Figure 2.7, is a matrix of 29 categories of software packages (Table 1.3) mapped against 29 systems engineering tasks and activities (Table 1.1). Through this matrix and related discussion it is possible to see in a summary fashion how the software packages can and have been used to address the issues and elements of systems engineering. This same subject is considered as well in the various chapters of Parts III and IV in that they contain CASE application discussions. The reader who is familiar with the tools and techniques of Part II can move directly into the main discussions of the individual aspects of systems engineering in Part III.

REFERENCES

1. Sage, A. P., "Introduction to Systems Engineering Methodology and Applications," from *Systems Engineering: Methodology and Applications*. New York: IEEE Press, John Wiley, 1977.

2. Hall, A. D., "Systems Engineering from an Engineering Viewpoint," from *Systems Engineering: Methodology and Applications*. New York: IEEE Press, John Wiley, 1977.

3. Hall, A. D., "Three-Dimensional Morphology of Systems Engineering," from *Systems Engineering: Methodology and Applications*. New York: IEEE Press, John Wiley, 1977.

4. Hall, A. D., *A Methodology for Systems Engineering*. Princeton: Van Nostrand Reinhold, 1962.

5. Goode, H. H., and R. E. Machol, *System Engineering*. New York: McGraw-Hill, 1957.

6. Zwicky, F., *Discovery, Invention, Research—Through the Morphological Approach*. New York: Macmillan, 1969.

7. Eisner, H., "CASE: Computer-Aided Systems Engineering for C³I Systems," *SIGNAL Magazine,* Journal of the Armed Forces Communications and Electronics Association (AFCEA), July 1984.

8. *Engineering Management*. Military Standard 499A, Department of Defense, May 1, 1974.

9. Blanchard, B. S., and W. J. Fabrycky, *Systems Engineering and Analysis*. Englewood Cliffs, N.J.: Prentice-Hall, 1981.

10. *Specification Practices*. Military Standard 490, Department of Defense, May 18, 1972.

11. *Configuration Management Practices for Systems, Equipment, Munitions and Computer Programs*. Military Standard 483, Department of Defense, December 31, 1970; second notice: March 21, 1979.

12. *System Requirements Statement for Air Traffic Control Computer Replacement*. Order 1812.4, Federal Aviation Administration, Department of Transportation, December 3, 1981.

13. *Consolidated Space Operations Center (CSOC) Communications (Phase 1)*. Request for Proposal (RFP) F04701-82-R-0017, Department of the Air Force, Headquarters Space Division, Los Angeles, CA, February 18, 1983.

*14. Meredith, D. D., K. W. Wong, R. W. Woodhead, and R. H. Wortman, *Design and Planning of Engineering Systems*. 2nd ed., Englewood Cliffs, N.J.: Prentice-Hall, 1985.

*15. Sage, A. P., *Methodology for Large Scale Systems*. New York: McGraw-Hill, 1977.

*16. Chestnut, H., *Systems Engineering Tools*. New York: John Wiley, 1965.

*17. Chestnut, H., *Systems Engineering Methods*. New York: John Wiley, 1967.

PROBLEMS

P2.1. Describe two other approaches to the seven-step or problem-solving dimension to systems engineering as represented by Hall in the chapter text and Reference 3.

P2.2. Determine common and differing elements of the tasks and activities of systems engineering as defined in Table 1.1 in reference to
   a. Sage's [1] and Hall's [3] definition of the problem solving dimension
   b. Hall's [3] time dimension
   c. the systems approach of Section 2.1
   d. Goode and Machol's [5] approach
   e. the structure of systems engineering management, as discussed in Section 2.5
   f. the systems acquisition process, as discussed in Section 2.6
   Explain the main reasons for the differences where they occur.

P2.3. Develop and explain the structure of a morphological box in the context of a transportation system for
   1. four trip lengths
   2. five modes of transportation (e.g., air, rail)
   3. three different motive powers

P2.4. Develop and explain the structure of a morphological box in the context of Tables 1.1 and 1.2. Select five cells that are of particular relevance and discuss their meaning.

P2.5. Select two of the ten diagrams in Figure 2.2 and fill them out.

P2.6. Construct a heuristic matrix for evaluation of a set of system characteristics other than those shown in Figure 2.2 and fill it out. Explain its significance.

P2.7. Explain the differences and similarities between the morphological box and the heuristic matrix.

P2.8. Discuss your understanding of the combinatorial problem associated with the synthesis aspect of systems engineering. Explain your approach to how to address a possible solution to that problem, i.e., how you would narrow down the number of possible solutions in order to find a limited number of good solutions.

P2.9. Compare the systems engineering tools defined in Reference 5 with those presented in Table 1.2. Discuss your interpretation of similarities and differences.

P2.10. Compare the systems engineering process that is defined as part of systems engineering management specifically with the tasks and activities of systems engineering shown in Table 1.1.

P2.11. Compare the specialty engineering areas that are defined as part of systems engineering management specifically with the tasks and activities of systems engineering shown in Table 1.1.

*Reference is relevant to the text but is not specifically cited.

P2.12. Develop a functional hierarchy for
    a. an air-transportation system
    b. a rail-transportation system
    Discuss their differences and similarities.

P2.13. Develop a set of ten high-level requirements for a general communications system. Define three subrequirements for each of the ten you have selected.

P2.14. Develop your own preferred set of systems engineering elements and explain your reasons for the selections you have made.

Part II Tools and Techniques

Very unlike a divine man would he be who is unable to count one, two, three

Plato

BASIC COMPUTER TOOLS

3.1 OVERVIEW

There are literally thousands of available software packages that can assist in the systems engineering process. We consider in this chapter five main functional classes of such packages, namely

- spreadsheets
- database management systems (DBMSs)
- word processors
- graphics packages
- integrated packages

All of these are abundantly available through commercial vendors and all have specific and rather powerful versions available for the p-c (see Appendix A). It is also the case that such capabilities can easily be found for larger computers as well (e.g., see Appendix B with respect to DBMSs).

The intent here is first to identify the essential format and capability represented by the five software categories and then illustrate in a simple manner how they can be related to one or more of the systems engineering elements. Further discussion of the application of these tools to systems engineering tasks is also contained in later chapters.

It is assumed that the reader, whether student or practicing systems engineer, may not have immediate access to a computer. We are interested, however, in conveying the concepts and the essential nature of these classes of computer tools without delving into particular commands used by various software packages. Thus we attempt to focus on a functional understanding of each class of package and how it is related to systems engineering. A list of specific packages that have been available for the p-c is provided in Appendix A. More specialized commercially available software packages that relate to systems engineering are examined in Chapters 6 and 7. These tools, which we will explore in this and the next several chapters, are summarized in Figure 3.1.

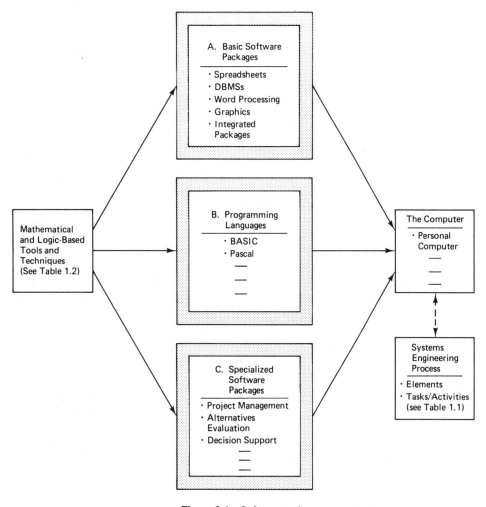

Figure 3.1 Software tools.

3.2 THE SPREADSHEET

The advent of the spreadsheet is generally associated with the introduction in 1979 of a software package named VISICALC. Since that time the specific capabilities of spreadsheets have been greatly enhanced but, with the exception of extensions into integrated packages, the essential concept remains intact.

The spreadsheet, as the name implies, is a set of rows (designated by 1,2,3, and so on) and columns (identified as A,B,C, and so on) with each intersection representing a *cell,* as illustrated in Figure 3.2. Spreadsheet capacities, for example, run from the 254 row × 63 column = 16,012 cell structure of VISICALC to the much larger 2,048 row × 256 column = 524,288 cell structure of the Lotus 1-2-3 integrated package. The essence of the spreadsheet popularity is its utter simplicity and the fact that its format is so prevalent in our day-to-day activities at work and at home.

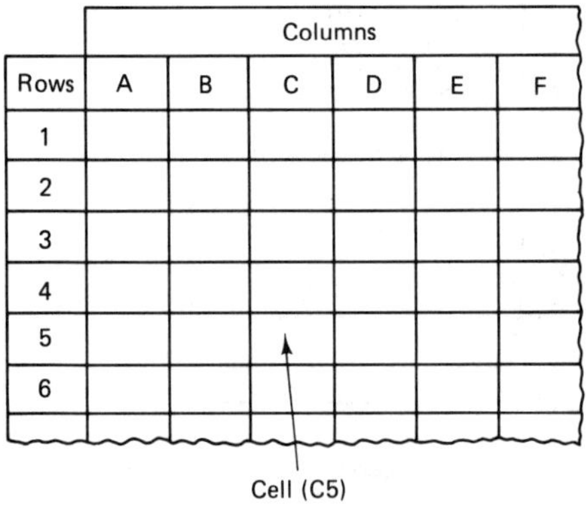

Cell (C5)

Figure 3.2   Spreadsheet format.

Figure 3.3 illustrates simply a completed spreadsheet called a *time runout* in which the columns represent years, and the rows may be drawn from a variety of application areas such as system costs or budgets. The figure shows a ten-year runout of current or projected costs for a system in three major categories

1. RDT&E
2. procurement
3. O&M

The rows are the categories and subcategories of cost and the columns represent the ten-year runout with the final column showing the total ten-year costs for each category and the total cost of the system. The column summations are the costs by year. If this spreadsheet representation is extended to the full life of the system, then the resultant tabulation is normally called the system *life-cycle cost.*

Setting up a spreadsheet in a p-c environment requires that the relationship between the cells be specifically and unambiguously stated. The spreadsheet information is entered on a cell-by-cell basis. The headings must be included in the spreadsheet for which we will reserve rows 1 and 2 as well as columns A, B, and C.

	Cost Category	1990	1991	1992	1993	1994	1995	1996	1997	1998	1999	10-Year Total
3	1.1 Research and Development	8,000										
4	1.2 Test and Evaluation	2,000										
5	1. RDT & E (Subtotal)	D5 = +D3 +D4	E5 = +E3 +E4	F5 = +F3 +F4								N5 = @SUM (D5 . . . M5)
6	2.1 Installations											
7	2.2 Equipment											
8	2.3 Stocks											
9	2.4 Initial Training											
10	2.5 Other Procurement											
11	2. Procurement (Subtotal)											
12	3.1 Equipment Replacement											
13	3.2 Maintenance											
14	3.3 Training											
15	3.4 Salaries											
16	3.5 Material											
17	3.6 Other O&M											
18	3. Operations and Maintenance											
19	TOTAL	D19 = +D5 +D11 +D18										

Figure 3.3 Spreadsheet time run-out illustration.

Therefore, the first cell in the example would be designated as D3 and would correspond to the first element of cost (i.e., research and development) in the first year. This could be entered as a specific number, as shown in Figure 3.3. We then would normally move to cell E3 or D4, depending upon the way in which costs are related to each other and how they are aggregated. In this example we next enter the cost number for cell D4, representing the test and evaluation cost for the first year. Proceeding down the first-year column, we next wish to aggregate the RDT&E costs and provide a subtotal row for each year. We instruct the spreadsheet to calculate in cell D5 the sum of the costs in cells D3 and D4. This is done by entering into cell D5 the formula

$$D5 = +D3 +D4$$

This means that whatever numbers are placed in cells D3 and D4, cell D5 will always calculate and display their sum.

Now if we look to the right along row five we note that we wish to have the same calculation made for each of the columns (i.e., the years). We can go to each of those cells and enter the formula $E5 = +E3 +E4; F5 = +F3 +F4$, and so on. In most spreadsheets we can simplify this process through the use of the *replicate* command that will tell the spreadsheet to use the same basic formula across the entire row. This is an example of a time-saving command, many of which are available in the popular spreadsheet packages. The process continues for the procurement and O&M costs until we get to the bottom of the first row where we wish to display the total cost for the first year. For this example, the entry for that cell, namely D19, would be the formula

$$D19 = +D5 +D11 +D18$$

Spreadsheets also allow for the use of functions that are built in to their structure, such as those listed in Table 3.1, which are considered to be a minimum set of avail-

---

TABLE 3.1 TYPICAL BUILT-IN
SPREADSHEET FUNCTIONS

- Sum
- Absolute value
- Exponential
- Logarithm
- Square root
- Sine, cosine, tangent
- Arcsine, arccosine, arctangent
- Counting
- Minimum and maximum numbers
- Mean, standard deviation
- Net present value
- Data management functions (several)
- Logical functions (several)
- Error (rule break) function

---

able functions. As a simple example of the use of a function, referring again to Figure 3.3, we may calculate the ten-year costs by aggregating yearly costs across the columns instead of summing from top to bottom in the fourteenth column, N. To use this same example, the ten-year RDT&E costs would be calculated in cell N5 by using the *sum function* as

$$N5 = @ \text{ SUM } (D5 \ldots M5)$$

In this situation, the @ signifies a function and *SUM* is the name of the function, showing that we wish to add all the elements across the row from D5 to M5 inclusive. The value of the sum is computed and placed in cell N5.

Suppose we wish to expand the spreadsheet by considering the number of systems that are to be acquired and the total cost per system at the end of the ten-year period. We can then add two rows to the spreadsheet, as illustrated in Figure 3.4. The number of systems acquired each year is entered as data and the total number of systems is obtained in column N by using the above SUM function. In cell N21 we can then compute the total cost per system acquired over the ten-year time frame by entering

$$N21 = N19/N20$$

Numerous other expansion possibilities are available depending upon the application and the specific data required by the user. For example, if a certain constrained budget is available on a year-by-year basis, the user may wish to enter the available budget in another row and compare the year-by-year projected costs with the available budget and calculate the *within budget* or *exceeding budget* amounts. This process is illustrated in Figure 3.4 in rows 22 and 23. The total yearly-estimated costs would then be subtracted from the available budget amounts in row 23. Changes can then be made to the programming of cost expenditures each year in an attempt to live within the budget constraints, or the budgeted amounts may be adjusted to conform to the expected cost run outs if such flexibility in budgeting for the system is available as an option.

We have only touched upon the power and flexibility available in the spreadsheet. Through the following example we explore a simplified matrix format, which might be used to evaluate the attributes of a set of alternative system configurations. A similar framework can be used in a situation in which several evaluators are attempting to judge the merits of a given single system against a set of criteria or, for example, several judges scoring a given performance at an Olympic event (see problems at the end of this chapter).

In Figure 3.5 we have set up a generic system-evaluation framework. The columns represent the systems to be evaluated, designated here as systems I, II, and III. The rows list a set of illustrative evaluation criteria. We associate with these criteria a series of weighting factors, $w(i)$, such that the sum of these weights is equal to unity. The weighting factors are indicators of the relative importance that the evaluator places upon each criterion. Two columns are reserved for each of the three systems under evaluation: the first contains a score or rating given to the system against the given criterion, and the second provides the product of the rating and the weight established for each criterion. Numerous scoring systems are available

	A	B	C	D	E	F	G	H	I	J	K	L	M	N
1		Cost Category						Year						10-Year Total
2														
18	3. Operations and Maintenance													
19	TOTAL													N19 = @SUM (D19...M19)
20	Number of Systems													N20 = @SUM (D20...M20)
21	Cost Per System													N21 = N19/N20
22	Available Budget													
23	Within (+); Exceed (−)			D23 = +D22 −D19	E23 = +E22 −E19									

Figure 3.4 Expanded time run-out example.

	A	B	C	D	E	F	G
		Systems					
		I		II		III	
Illustrative-Evaluation Criteria	Criteria Weights w(i)	Rating r(i, I)	Weight × Rating w(i) × r(i, I)	Rating r(i, II)	Weight × Rating w(i) × r(i, II)	Rating r(i, III)	Weight × Rating w(i) × r(i, III)
1 Overall Performance			C1 = A1 * B1		E1 = A1 * D1		G1 = A1 * F1
2 Environmental Compatibility							
3 Robustness							
4 Reliability							
5 Maintainability							
6 User Service							
7 Ease of Operation							
8 Safety							
9 Security							
10 Cost							
11 SUMs	1.00		C1 = @SUM (C1 . . . C10)				

Figure 3.5 Generic system-evaluation framework.

and we adopt the numeric scores in Table 3.2, which correspond to a college grading system. In this illustrative example the scores will be applied to the merits or deficiencies of each of the given systems. We may set up the weights for the ten illustrative criteria as data to be entered in the ten cells of the first column and calculate the sum in the eleventh cell (row) as a check to assure that the sum of the weights is unity, as

$$A\,11 = @\ \mathrm{SUM}\ (A\,1\ .\ .\ .\ A\,10)$$

For the first column shown under each system, we provide room for the user to insert a score number against each criterion. There is no particular significance, in this example, to the sum of the scores although such a total may not exceed forty (4×10) and the user may wish to see the sum as a check cell. In the second column under each system (columns C, E, and G) it is necessary to calculate the product of the weight and the score, designated as $w(i)r(i,\ j)$, where $r(i,\ j)$ is the score of the jth system against the ith criterion. As an example of this calculation, cell C1 would contain the entry

$$C\,1 = A\,1 * B\,1$$

In the last row the weighted scores are added so that for column C we have the entry

$$C\,11 = @\ \mathrm{SUM}\ (C\,1\ .\ .\ .\ C\,10)$$

This value represents the final weighted score that will be applied for each system for comparative evaluation purposes. If desired, the user can further multiply these weighted scores by twenty five to normalize them to a base of one hundred in cells C12, E12, and G12.

TABLE 3.2 ILLUSTRATIVE SCORING SYSTEM

Score	Numeric Value	Interpretation
A	4	Excellent (meets all requirements with margin)
B	3	Good (meets requirements, some with margin)
C	2	Average (barely meets all requirements)
D	1	Poor (meets most but not all requirements)
E	0	Unacceptable (misses all or most requirements)

The aforementioned weighting and rating scheme can be utilized within the systems engineering process as a means of numerically and subjectively assessing alternative system designs. It also has numerous other applications such as providing a framework for company personnel performance evaluation when the criteria embody the various elements of desired job performance. Several variations on this generic weighting and rating procedure are also possible using different scoring systems and multiple evaluators attempting to arrive at a consensus judgment as to the merits of a system. Some of these considerations are broached in Chapters 10 and 13.

The built-in spreadsheet functions previously cited in Table 3.1, including several data management and logical functions, provide the user with a *modeling* or *performance* analysis capability that should not be overlooked or underestimated. The lean matrix format of the spreadsheet can be deceptive and with appropriate algorithms combining available functions, complex relationships can be constructed such as power budgets for a satellite communications link, radar system performance, and demand analysis for transportation systems. If the functions required are contained within the spreadsheet repertoire, or can be approximated by this repertoire (as with a series expansion), then the appropriate calculations can be strung together to constitute a model of system performance. Note also that as new data are entered into the spreadsheet by a user, all computations related to that data are made automatically when a run is made. This requires careful planning in formatting a spreadsheet so that the maximum set of results is obtained for each run.

In a particular systems engineering context, the user must evaluate the various spreadsheets available on the market on the basis of their characteristics in relation to the specific systems engineering application or set of applications. Each spreadsheet has positives and negatives in this regard as well as features that may or may not be pleasing to the user.

The overall size of the spreadsheet is an important characteristic and is determined by the number of rows and columns whose product defines the total number of cells available. Each cell has a capacity that reflects the total number of characters that can be placed in the cell. From a hardware storage point of view, the user should investigate the minimum required memory (K = kilobytes) for the spreadsheet in relation to the random access memory (K of RAM) available for the machine that is being used. In addition, there is a concept of maximum usable memory in the sense of what is actually accessible by the software as the spreadsheet is run. In some spreadsheets, a *virtual memory* is available, allowing the user to place some of the spreadsheet on a disk. This provides more spreadsheet capacity.

Relative to the screen display, some spreadsheets use 40 columns and others use 80 columns, which can affect the ease of application by the user. In addition, there are differing user-friendly implementations, reflecting themselves as availability of a tutorial, the use of help screens, the availability of windows that allow several portions of a display to be seen simultaneously, and various methods of formatting the contents of the individual cells. Print options also vary in terms of margins, page numbering, and other related features. Other significant characteristics include speed, extent of programmability, and the number and type of built-in functions.

The extensive power of spreadsheets for systems engineering purposes is only touched upon here. The reader is referred to the problems posed at the end of this chapter as well as tutorial texts on particular spreadsheets in order to gain facility with the menus, commands, functions, and other capabilities represented by these spreadsheets. Since our intent is to deal with systems engineering applications, we focus on such areas for which spreadsheet constructions can be developed without defining the precise rules and conventions used by any one spreadsheet software package or vendor.

3.3 DATABASE MANAGEMENT SYSTEMS

Database management systems (DBMSs) allow a user to enter alphanumeric information and then process and retrieve it so that the interrelationships can be identified and explored. Underlying DBMSs are various internal data structures, models, and even languages that facilitate the processing, sorting, and retrieving capabilities of these systems [1]. These structures include relational models and hierarchical models. Whatever the underlying structure, all the available DBMS software packages provide a specific set of commands that allow for the key functions of data entry, field definition, sorting, search, and retrieval. In all cases, an important element is that of utilizing the sort capability to place in evidence the interrelationships between data elements and to allow the aggregation and slicing of data into categories and subcategories.

DBMS software is available for the p-c as stand-alone packages (such as dBASE III Plus) or as a part of an integrated package (such as the DBMS portion of Lotus 1-2-3). Information regarding such software is abundantly accessible at many microcomputer stores.

Since our interest is the application of the DBMS to systems engineering, we will explore some examples that deal with

- personnel assignments and utilization
- system requirements
- system types and costs

3.3.1 Personnel Example

Personnel assignments and utilization are of obvious importance in the context of large-scale systems engineering. Engineering companies or their divisions will normally keep track of their personnel in their human resource departments as well as within divisions and departments. To illustrate the use of the DBMS in this application, we display in Figure 3.6 an alphabetic master list of personnel showing a lim-

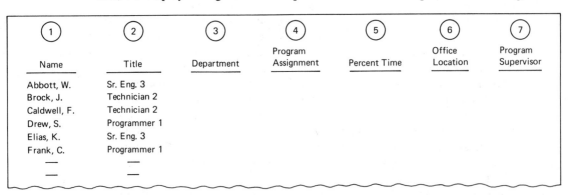

Figure 3.6 Personnel listing example.

ited amount of data related to each individual. The individual's name, title, and department assignment are listed as the first three fields. If each department carries out a series of programs, which is a common practice in systems engineering, the current program assignment of each individual is shown in the fourth column. In the fifth column the percent of time that the person is currently assigned to the given program is also displayed. Ancillary data as to the person's current office location and program supervisor are listed in the last two columns.

The essential power of the DBMS lies in its ability to sort and aggregate the illustrative data in Figure 3.6 in a variety of ways, such as

- listing all people (alphabetically) by their title
- listing all people by title who are assigned to a particular program
- listing all people by program who work at a particular office
- showing the aggregated number of people (percent time sum) assigned to a particular program

An example of listing people alphabetically by title is shown in Figure 3.7. Clearly, a variety of other sorts and subsorts can be visualized (see problems at end of chapter).

We note the ability of the DBMS to provide multilevel sorts. This is an important characteristic of DBMSs although they do not, in general, provide an open-ended capability to sort through multiple levels of files and records.

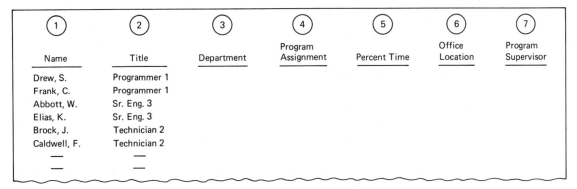

Figure 3.7   Personnel list sorted by title.

3.3.2 System Requirements Example

A critical part of the systems engineering process is the detailed analysis of requirements. Within this framework we need to look at the system requirements in relation to the specifications for the system, the application breakdown of the system into CIs, and CPCIs (or CSCIs), and the overall test and evaluation (T&E) program. These various elements can be related to one another by means of a DBMS. In Figure 3.8 we illustrate one simple way of doing this. First, we list the requirement

Requirement Paragraph Number	Name of Requirement Parameter	Required Value of Parameter	Specification Paragraph Number	Value of Parameter in Spec	Applicable CI/CPCI	Test Parameter	Test Value
4.2.7.1	Bandwidth	6 MHz	3.4.2.1	6 MHz	C7.8	BW3.7.7	6 MHz
4.2.7.4	Bit Error Rate	10^{-8}	3.4.2.5	10^{-8}	C7.8	BE3.9.2	10^{-8}
6.3.8.3	No. Analog Channels	48	5.5.3.2	48	H4.7	AC5.5.4	48
9.4.7.2	No. Operator Positions	8	8.4.6.1	8	P6.3	—	—

Figure 3.8 Requirements-correlation information.

paragraph number, the name of the requirement parameter, and the required value of the parameter if a value is given in the requirements document. The specification for the system can be correlated to the requirements document by listing, as a minimum, the specification paragraph number and the specification value for the parameter referred to in the set of requirements. In addition, we can also show the applicable configuration item that will implement the required and specified parameter. Finally, all of this can be tied to the test and evaluation process by listing in the same DBMS the test parameter that is the same as or contains the required and specified parameter together with the test or measured value when it is obtained through the test program.

By using the DBMS, it is then possible to correlate the requirements documentation and the specification documentation with the hardware and software that implements the system functions for the test and evaluation program. This is always an important part of maintaining requirement traceability through the systems engineering process and assuring that all the system requirements are ultimately being satisfied.

3.3.3 System Type and Cost Example

We can also use the DBMS to examine and sort the costs between various types of systems. In the previous example discussed in relation to Figures 3.3 and 3.4, we showed a spreadsheet cost format for a hypothetical system. In that case, the major cost categories were RDT&E, procurement, and O&M. If we consider several systems, each broken down into the same major cost categories, we have the tabulation provided here as Figure 3.9. This figure shows ten systems with the aforementioned three cost categories and three systems types, namely, command and control, communications, and navigation. If we now wish to sort these so as to determine the expenditures by system type, we can instruct the DBMS to do so. Figure 3.9(a) shows the original data listed by system number. By a simple sort command, we can obtain the sort by system type as shown in Figure 3.9(b). We can also aggregate the costs by type to show subtotals for each of the three types of this example. In addition, the process can include a summation over all types to verify that the total costs are the same for the three cost categories.

On large programs information is often presented in matrix or tabulation form, which shows relationships between two or three parameters at a time. The following list, taken from a real-world, large-scale, communications system program design

System Number	System Type	RDT & E ($)	Procurement ($)	O & M ($)	Total ($)
1	Communications	2,000	31,000	47,000	80,000
2	Command and Control	17,000	49,000	82,000	148,000
3	Communications	5,000	54,000	74,000	133,000
4	Navigation	8,000	25,000	56,000	89,000
5	Command and Control	12,000	32,000	72,000	116,000
6	Navigation	7,000	22,000	68,000	97,000
7	Command and Control	15,000	44,000	93,000	152,000
8	Navigation	6,000	21,000	64,000	91,000
9	Navigation	9,000	28,000	85,000	122,000
10	Communications	13,000	37,000	97,000	147,000
	Grand Total	94,000	343,000	738,000	1,175,000

(a)

System Number	System Type	RDT & E ($)	Procurement ($)	O & M ($)	Total ($)
2	Command and Control	17,000	49,000	82,000	148,000
5	Command and Control	12,000	32,000	72,000	116,000
7	Command and Control	15,000	44,000	93,000	152,000
	Subtotal: C & C	44,000	125,000	247,000	416,000
1	Communications	2,000	31,000	47,000	80,000
3	Communications	5,000	54,000	74,000	133,000
10	Communications	13,000	37,000	97,000	147,000
	Subtotal: Communications	20,000	122,000	218,000	360,000
4	Navigation	8,000	25,000	56,000	89,000
6	Navigation	7,000	22,000	68,000	97,000
8	Navigation	6,000	21,000	64,000	91,000
9	Navigation	9,000	28,000	85,000	122,000
	Subtotal: Navigation	30,000	96,000	273,000	399,000
	Grand Total	94,000	343,000	738,000	1,175,000

(b)

Figure 3.9 Sorting costs by system type.

review, indicates some of the parameters that have been related in two- or three-way tables

- performance requirements
- systems functions
- hardware items
- software items
- personnel operators
- interfaces
- processor types
- lines of code
- cost

- evaluation criteria
- test parameters
- milestone dates

A DBMS can help to integrate this information and produce all the needed tables with the use of simple commands. It can also provide a means for exploring various tabulations that may not be obvious when the information is kept in separated formats. Thus the DBMS is a powerful tool for synthesizing and examining new relationships between important system parameters.

These sorting examples illustrate the capability of the DBMS to perform simple sorting functions. This can be expanded to multilevel sorts and other data manipulation and correlation tasks, all of which are done automatically and save the user considerable amounts of time. With the use of these commands and a printer, the user can employ the full power of the DBMS and obtain the results in hard copy whenever it is necessary to do so.

As is the case with spreadsheets, the user must assess the various features of DBMSs in relation to the systems engineering application at hand. Although the DBMS carries out a relatively small number of generic functions (i.e., placing data into the database, editing this data, deleting data, sorting and searching the database, and printing), each software package does these differently. In almost all of the packages, the user must assure compatibility with the machine, an important element of which is the minimum RAM required as well as the RAM utilized by the program itself (in relation to what remains to be used by the data stored in the database). In short, users must be assured that the data they need to store can be handled in terms of size by the given package. Speed of operation is also an important feature of DBMSs as well as the other types of packages. In the case of the DBMS, one looks as well for user-friendly features, which depend to a large extent upon the idiosyncracies of the individual user. For example, some DBMSs are command driven in distinction to being menu driven. The command structure can itself be viewed as a development "language" peculiar to that DBMS. Once the language is learned, however, it can be extremely powerful and productive to use.

Other characteristics to be explored include the sort capability in terms of the number of levels of sort (e.g., one to eight), the maximum number of fields and records, the search capability in regard to the number of search criteria, and the database statistical functions contained in the package.

3.4 WORD PROCESSORS

Word-processor software packages allow the user to create, edit, store, retrieve, transmit, and print text in alphanumeric form, and provide for great flexibility and speed in producing reports, letters, and similar materials. Examples of p-c-based word processors, of which there are dozens on the market, are provided in Appendix A.

Word processors are generally menu driven on the screen, including "help" menus which enable the user to learn and operate the systems quickly, with minimum reference to hard copy manuals. Menu and command formats differ from package to package and the most successful ones have good eye appeal and a logical structure of menus and commands.

The applications of word processors, from a systems engineering point of view, are numerous. All the required reports and systems documentation can be done with a word processing package, including such items as the

- systems-engineering management plan (SEMP)
- program-implementation plan (PIP)
- software-development plan
- technical-performance-measurement plan
- test-and-evaluation-management plan (TEMP)
- integrated logistics support plan
- training plan
- program-risk analysis
- environmental-impact statement

It is also possible, with the appropriate hardware and software, for the systems engineer to work at home with a p-c word processor and then either transmit the report over telephone lines to an office-based word processing system or to take the p-c disk to work and have its files integrated into a larger document. Down-loading from a larger office system to a p-c is also an available configuration for word processing as well as other functions. Often the word processing capability is the first obtained by the systems engineer as a time-saving and productivity improvement step.

All word processors contain features of size, speed, editing, and printing capability as well as others that are more or less pleasing to the user. As in the case of spreadsheets and DBMSs, the user should try out a word processor, perhaps by using its tutorial if available, before committing to one since they appear to be so varied in terms of their appeal to users.

3.5 GRAPHICS

Graphics packages for the p-c can be thought of as being available in two categories—general purpose and specialized. The general-purpose graphics capability is embedded in the spreadsheets and to some extent in the database management systems. The various types of general purpose graphics include

- line
- bar
- stacked bar
- side-by-side bar
- x–y plots
- scatter diagrams

- hi-lo spread charts
- pie charts

Examples of some of these graphics are illustrated in Figure 3.10. If they are built in to the given software package they are drawn automatically through simple commands at the keyboard, using the data provided in the package by the user and the computations that are made.

Specialized graphics are normally available in packages that are written for the particular purpose of providing graphics capability beyond the examples shown in Figure 3.10. The Macintosh computer was the first to embody specialized graphics

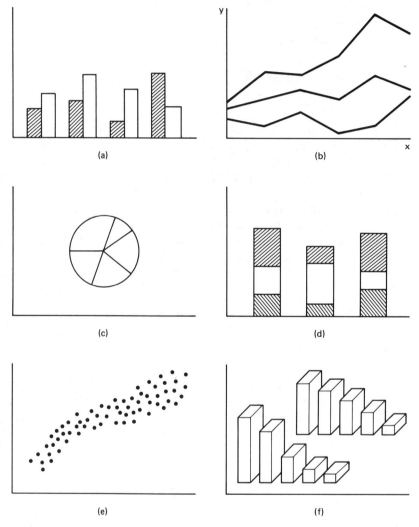

Figure 3.10 Illustrative graphics: (a) bar chart, (b) graph—X–Y, (c) pie chart, (d) stacked bar chart, (e) scatter chart, (f) 3-D bar chart.

through the use of icons on the screen as well as unusual software such as Macpaint. Examples of the kinds of graphics that specialized graphics packages provide include

- "painting" on the screen as with a paintbrush
- setting up different typesets and fonts
- using various patterns (hatches, dots, shadings, and so on)
- drawing arbitrary and regular figure shapes (e.g., polygons)
- changing sizes and shapes (e.g., zooming)
- integrating all of the above to construct almost any kind of drawing or picture (e.g., cutting and pasting)

In the systems engineering world there is a need for both general and specialized graphics. In the former category, bar and pie charts as well as x–y plots are extensively used to compare data and show functional relationships. In the latter category, graphics are used to produce such displays as

- Gantt chart schedules
- PERT-type schedules
- organization charts
- work-breakdown-structure charts
- computer flow charts
- a variety of process flow charts
- drafting and blueprint-type diagrams

These specialized graphics packages and capabilities are discussed and illustrated further in Chapter 6.

3.6 INTEGRATED PACKAGES

Integrated-software packages combine the spreadsheet, DBMS, word processing, and graphics capabilities into a single commercially available product. Some integrated software also includes other functions such as data communications that allow a user, by means of a modem (modulator–demodulator), to communicate with computer systems and their files at an off-site location.

Indeed, the first available integrated package included a spreadsheet, a file manager, word processor, graphics, telecommunications, and a forms-handling system. The very popular and successful Lotus 1-2-3 likewise contained multiple functions but also possessed superior speed, an interactive tutorial, screen splitting, and a user-friendly command structure. Symphony, an upgraded version of Lotus 1-2-3, added additional features but requires additional RAM capacity. Other specific integrated packages are cited in Appendix A.

From a systems engineering point of view, having all of these capabilities in one package can be a distinct advantage. However, compromises were made in each function so as to be able to fit the package into reasonable RAM capacities to be able to run on a variety of available p-c's. As a single investment for the individual systems engineer, therefore, it is difficult to do better than to obtain an integrated package. However, if the user has a very large database to deal with, for example, and no need for the spreadsheet and other functions of the integrated package, it might be better to go with a single-purpose package. Again, this is a choice that must be made based upon the type and variety of systems engineering tasks to be undertaken. Whether or not integrated software is chosen, the systems engineer has a vast capability on which to draw to obtain computer aids to perform systems engineering tasks.

3.7 SUMMARY

This chapter has introduced five extremely useful and popular categories of commercially available software packages, namely

- spreadsheets
- DBMSs
- word processors
- graphics packages
- integrated packages

This software is readily accessible by the p-c user and has numerous applications in systems engineering. Specific software produced and marketed by various vendors is cited in Appendix A. Some 24 additional categories of microbased software for systems engineering applications are presented and discussed in later portions of this text, starting with Chapters 6 and 7. Prior to this presentation and discussion, however, are Chapters 4 and 5, which deal with diagramming techniques and probability theory. Both help the reader gain a more complete understanding of several of the categories of software as well as important tools of systems engineering.

REFERENCES

1. Gallagher, L. J., and J. M. Draper, *Guide on Data Models in the Selection and Use of Database Management Systems*. NBS Special Publication 500-108, U.S. Dept. of Commerce, National Bureau of Standards, January 1984.
*2. Henderson, T. B., D. F. Cobb, and G. B. Cobb, *Spreadsheet Software From Visicalc to 1-2-3*. Indianapolis, IN: Que Corporation, 1983.
*3. Hayes, G., "How Spreadsheet Software Makes You More Efficient," *LIST Magazine*, May 1984.
*4. Nesbit, I. S., "Big Power In A Small Package," *Computer Decisions*, April 1984.

*Reference is relevant to the text but is not specifically cited.

*5. Simpson, A., *Understanding dBASE II*. Berkeley, CA: Sybex, Inc., 1984.

*6. Gilder, J. H., *The Integrated Software Book*. Reading, MA: Addison-Wesley, 1985.

*7. *The Software Catalog in Microcomputers*. New York: Elsevier North-Holland, Fall 1983.

*8. Stanton, J., R. P. Wells, and S. Roshowansky, eds., *Apple Computer Software 1983*. Reading, MA: Addison-Wesley, 1983.

*9. Naiman, A., "Word Processing Buyers Guide", *BYTE*/McGraw-Hill, 1983.

*10. Kelly, B. W., and D. J. Grimes, *IBM PC Compatible Computer Directory*. New York: John Wiley, 1985.

*11. McMullen, B. E., and J. F. McMullen, "The Super Spreadsheets: How Do They Compare?" *Popular Computing*, June 1983.

*12. Brand, S., ed., *Whole Earth Software Catalog*. Garden City, NY: Quantam Press/ Doubleday, 1985.

*13. Mellin, M., and N. Hays, ed., *The Book of IBM Software 1985*. Los Angeles, CA: Arrays, Inc., 1985.

*14. *IBM PC Expansion & Software Guide*. Indianapolis, IN: Que Corporation, 1985.

*15. *PC Magazine*, Ziff-Davis, New York.

PROBLEMS

For all of the spreadsheet problems, and where appropriate, show representative numbers where raw data is to be entered by the user. Where calculations are to be made in a cell, place the formula to be used in the cell with the resultant numerical value in parentheses.

P3.1. Construct a spreadsheet for five years based upon the cost categories in Figure 3.3. Postulate the acquisition of systems in various years and expand the spreadsheet to calculate the five-year cost per system. Postulate a yearly budget and calculate the "over" and "within" budget for each year as well as over the total five-year period.

P3.2. Expand the spreadsheet from P3.1 to include a net present value column as well as any other columns needed to make such a calculation. Assume a 10% discount rate.

P3.3. Design a spreadsheet that
   a. calculates the values of sine, cosine, and tangent of all angles from 0 to 360°, in increments of 30°. Convert these angles to radians
   b. uses the functions of logarithms, exponentials, and square roots

P3.4. Design a spreadsheet using the evaluation framework shown in Figure 3.5 for which three evaluators are judging the merits of a single system against a set of criteria. Select eight of your own criteria for evaluation.

P3.5. Design a spreadsheet using the evaluation framework shown in Figure 3.5 for which five judges are scoring a given performance at an Olympic event. Select five of your own criteria for evaluation. Use the concept of ease or difficulty of the event and show how the spreadsheet can be expanded to scoring several performances by an individual or team in a given event.

P3.6. Design a spreadsheet using the evaluation framework shown in Figure 3.5 to evaluate personnel in the context of a performance review in a company. Develop your own scoring system and use no less than 15 criteria.

P3.7. List ten criteria in priority order that you would use to decide upon which spreadsheet you might purchase and briefly give your reasons for the selection.

P3.8. Design a spreadsheet to evaluate the features of various audio amplifier-tuner systems that you might consider for purchase. Develop a set of ten criteria for the evaluation and illustrate the process for three systems under evaluation.

For all the database management system problems listed next, and where appropriate, define a set of six fields and as many rows as is necessary to illustrate the point of the problem.

P3.9. Develop a one-level and two-level sort based upon the example shown in Figure 3.6 but different from the listing shown in Figure 3.7. Make up your own names and associated records.

P3.10. Develop a one-level and two-level sort based upon the example shown in Figure 3.8. Make up your own records, as appropriate.

P3.11. Develop a one-level and two-level sort based upon the example shown in Figure 3.9. Make up your own records, as appropriate.

P3.12. Design a spreadsheet that you would use to evaluate the merits and deficiencies of word processor software packages, using the evaluation framework of Figure 3.5 and your own set of ten evaluation criteria. Use three word processor alternatives.

P3.13. Construct bar charts showing the results for the spreadsheet evaluation frameworks developed for the above problems.

P3.14. Research the literature and place in a database management system format the results for
  a. spreadsheets
  b. database management systems
  c. word processors
  d. integrated packages
including the name of the package, the manufacturer's name and address, the cost of the package, and one other data element that you find relevant. Cite at least eight packages for each of the above four categories.

One picture is worth more than ten thousand words

Chinese Proverb

DIAGRAMMING TECHNIQUES

Diagramming techniques have become increasingly prevalent in numerous aspects of the field of systems engineering. The most likely reason for this has been the increasing complexity of systems, leading to the need for better methods of decomposition and representation of relationships. The first level of system decomposition, both for synthesis and analysis purposes, is functional where an overall system is broken down into its major functions and subfunctions. Once this is performed, some obvious questions arise

- What is the interrelationship, including the data flow, between the resultant functions (and subfunctions)?
- What functions have common elements, namely, common subfunctions and/or common data flow and/or common program implementations?
- How is the systems engineer to grasp and deal with very large numbers of functions (and subfunctions) and their various interrelationships?
- How is the systems manager to deal with the ever-increasing complexity and cost of software development?

The answers have been supported, at least in part, by diagramming techniques. Even at the functional level (in distinction to the detailed design level), the strengths of diagramming techniques have been in their ability to depict interrelationships between decomposed elements and to do it so that a user has a visual representation that facilitates both synthesis and analysis. An illustration can allow for a faster and more complete understanding in much the same way that a graph of a function can be more easily grasped than can a column of numbers.

Thus there is great power in the use of diagrams, if properly applied. However, there is at least one major difficulty with them that requires careful examination. That is, they are hard to process, especially in the context of using modern computer systems. The issue is one largely of converting the diagram into a form that allows entry into the computer system and subsequent manipulation with data that may reside in a spreadsheet, a DBMS, or other type of accepted software packages. This problem area, although completely understood, does not have thoroughly satisfactory solutions even at the mainframe level, much less for the microcomputer. Notwithstanding this impediment, which is constantly being worked on, diagramming procedures of various kinds continue to emerge and continue to be used within the systems engineering process. In this chapter we cite and briefly explore some of the more popular and useful of these procedures.

A particularly prolific writer in computer-related subjects, James Martin, has addressed the matter of diagramming techniques with particular reference to the development of software systems [1]. In his survey of the field (together with C. Mc-Clure), the following types of diagrams are cited:

- decomposition diagrams
- dependency diagrams
- data flow diagrams
- structure diagrams
- Hierarchical input-process-output (HIPO) diagrams
- Warnier–Orr diagrams
- Michael Jackson diagrams
- flowcharts
- structured English and pseudocode
- Nassi–Shneiderman charts
- action diagrams
- decision trees and tables
- state-transition diagrams
- HOS (Higher-order-software) charts
- data structure diagrams
- entity-relationship diagrams
- data navigation diagrams

Since we are concerned with the broader issue of systems engineering and also have limited text allocated to the subject of diagramming techniques, we will address a fewer number of techniques. Those addressed here are judged to be of special interest to the systems engineer. There is no implication that techniques not referred to here cannot be used to considerable advantage by either the systems or the software engineer.

4.1 SIMPLE PROCESS FLOW CHARTS

This type of chart is an unsophisticated representation of a sequence of steps that define some process, as illustrated by the system-approach flow chart in Chapter 2, Figure 2.1. The steps are shown as rectangular boxes that are connected by means of arrow lines depicting the overall direction of flow of the process. Feedback and iterations are likewise displayed using the arrow lines. Such a chart is generic which means that it can apply to any process ranging from a high functional level to a detailed equipment level. Boxes are numbered sequentially, from left to right. For parallel processing of steps, subnumbering schemes can be devised, such as breaking a box numbered 4 into two parallel boxes numbered 4.1 and 4.2. Complex serial and parallel processing, in general, can be handled although numbering schemes can become cumbersome when the process "topology" becomes complex.

The process flow chart in Figure 2.1 makes it clear that a representation of the systems approach involves the following sequential steps:

1. state needs for system
2. develop goals and objectives
3. define system requirements
4. write system specifications
5. synthesize system alternatives
6. assess possible constraints
7. analyze system alternatives
8. develop evaluation criteria
9. update specifications
10. build, test, and accept system
11. document and install system
12. operate, maintain, and monitor system
13. modify and upgrade system

Iterations and interrelationships show directly on the chart and are not easily depicted on this linear list. Indeed, the linearity of text is what is circumvented by using the two or more dimensions of a diagram.

4.2 SIGNAL-FLOW BLOCK DIAGRAMS

A signal-flow block diagram is a more rigorous version of the simple process flow chart. An example is shown in Figure 4.1, which represents the flow of a signal from input to transmission via some type of antenna. This is viewed as a very con-

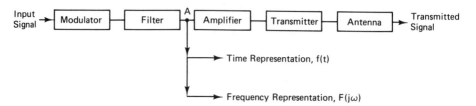

Figure 4.1   A signal-flow block diagram.

ventional diagram that is used extensively in electrical engineering, with analogs in other engineering (e.g., mechanical, civil) contexts. At each point along the signal flow (as for point *A*) the signal is traced both in the time domain as well as the (transformed) frequency domain. If the processes are linear, the entire body of literature of linear systems theory may be invoked to analyze the signals. If the processes are nonlinear, linear approximations can be utilized, if appropriate, or nonlinear analysis (such as phase–plane plot) techniques can be used. In any case, the blocks transform the signals from input to output and one desires to represent the signals at all points in both time and frequency characterizations. The *transfer function* of each block is of particular interest. For a linear system this is the response of the block to a so-called *delta* or *impulse* input, a pulse of infinite height and zero time but with finite energy. The utility of such a representation lies in the fact that one can multiply the transfer function by the transformed (frequency domain) input to obtain the transformed output. Operation in the time domain requires the process of convolution, which is more complex.

    The signal flow diagram is extremely common in basic engineering practice. So far it has not been integrated with most of the other diagramming techniques discussed in this chapter. However, it is very fundamental and likely to be used in all basic engineering analyses for the foreseeable future.

4.3 HIERARCHICAL DECOMPOSITION DIAGRAMS

The hierarchical decomposition diagrams (HDDs) are among the simplest and also among the most widely used in systems engineering. Such a diagram depicts a hierarchical breakdown of the system into subordinate parts, usually without interrelationships between these parts. A simple example is a company's organization chart. Another example is the WBS chart (shown in Figure 2.4), which is extremely common in systems engineering. In fact, the vagaries and requirements of such a structure are described in a military standard (Mil. Std. 881A) in order to provide guidance and uniformity to the various companies working on large-scale systems.

    An HDD implicitly represents a Dewey-decimal type of breakdown such as the structure shown in Figure 2.4. It can also be applied to documentation such as that represented in Tables 2.1 and 2.2. An HDD therefore can be used to display a tree-like taxonomic or morphologic structure. Without further modification and coding of elements, however, they are limited in their ability to show multiple sets of relationships other than the primary top-down decomposition. For example, if engineering and test are two separate blocks on a WBS chart, one cannot easily show how the products of engineering are tested without resorting to another type of chart. To get around such a difficulty, some practitioners have "coded" the various blocks of an HDD to show further relationships such as functional flow, product flow, process flow, data flow, logic flow (e.g., and, or, etc., operators), time sequencing, and other interrelationships. Another concept is to map all elements of an HDD against themselves in a matrix so that the cells in the matrix can depict such interrelationships. Thus both the rows and columns embody the hierarchical relationship through the numbering scheme (3.1, 3.1.1, and so on), and the cell intersections may be reserved for other relationships.

Another variation on the general theme of HDDs is one in which some of the elements at the bottom (or other) level of the hierarchy are common to two or more elements above them. This can be thought of, for example, in the use of HDDs to represent computer programs. As the programs are decomposed it may be that several programs use the same subroutine or subordinate procedure. The same situation may be true for a WBS for which certain work elements (such as a simulation activity) support several higher level work elements. This more complex HDD that departs from a pure tree-like arrangement is sometimes called a structure chart.

4.4 DATA FLOW DIAGRAMS

The data flow diagram (DFD) is considered to be an integral part of what has become known as structured analysis (SA) or structured design (SD), or a generic class of structured techniques (see also Chapter 14). Although in many cases practitioners have moved beyond the original DFD in the sense of added features and improvements, it continues to be extensively used for both systems and software engineering.

The basic element of a DFD is the block or box, shown in Figure 4.2, which represents some type of operating process. The data flow is shown as input data from the left into the box and output data emanating from the box. The output data may flow into another process box and become its input data, or it may flow into a data store, shown by two parallel lines in the figure. The process representation of the box itself should have a set of words that explain what happens to the input data in order to produce the output data.

Another simple variation or addition to the DFD, part of the structured analysis and design technique (SADT, a trademark of Softech, Inc. [2]) is shown in Figure 4.3. Here the input and output remain the same but the top and bottom arrows have a different meaning. For the former, the arrow represents a control on the process.

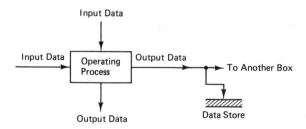

Figure 4.2 A data flow diagram box.

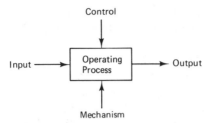

Control

Input → Operating Process → Output

Mechanism

Figure 4.3 A modified data-flow diagram.

This may be a coefficient in an equation or a modulating influence on the flow through the box. The bottom arrow, shown as pointed into the box, represents some type of mechanism, which is the "how" of the process. An example of a mechanism would be the actual formula for the box that shows how the input is transformed into an output.

Whether one is dealing with the DFDs of Figures 4.2 or 4.3, when various blocks are put together a full DFD is constructed as depicted in Figure 4.4. Experience with these diagrams has shown that the number of blocks in a single diagram should optimally be no more than 5 to 10. Larger diagrams become too complex and begin to lose their utility.

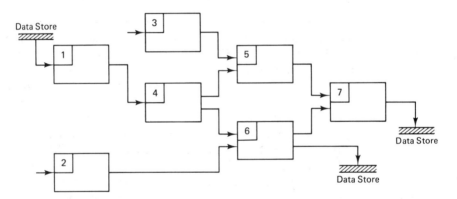

Figure 4.4 A sample data-flow diagram.

The DFD is also used to show hierarchical structures, although it does so in a manner quite different from the HDD. If one envisions a DFD at a top level for a system, then successive levels or tiers are depicted by successive breakdowns of data flow diagrams, one box at a time. That is, each box in the top level DFD is taken and broken down into its own separate and subordinate DFD, as shown in Figure 4.5. The subordinate DFD shows more detail within the box in terms of internal data flows. Interrelationships of data between DFDs are generally maintained at the same level rather than between levels, except where a DFD is a "child" or "parent" of another DFD. As an example of size, if we have a six-box, top-level DFD, a second-level representation would contain six separate DFDs. If each of these had six subordinate DFDs for each of six boxes, we would have

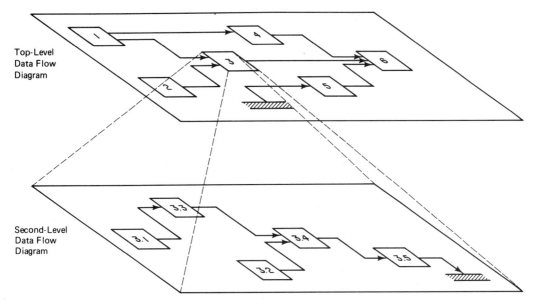

Figure 4.5 Multiple levels or tiers for data flow diagrams.

- one DFD at the top (first) level
- six DFDs at the next (second) level
- 36 DFDs at the next (third) level

Further decomposition is certainly possible and practical for large-scale systems but must be exercised judiciously.

The aforementioned hierarchy of DFDs is supported by two descriptive elements: the process specification and the data dictionary. The former describes the specific sequence of steps in the lowest level box of a DFD hierarchy, usually in a type of pseudocode. The data dictionary is normally an alphabetic reference book that precisely defines all data terms including relationships between such terms, for example as sums or products of data.

Additional information regarding DFD structures and conventions, which differ in detail such as further coding of boxes, can be found in references by some of the originators and practitioners. This includes Yourdon [3], DeMarco [4], and Gane and Sarson [5]. Computer tools are available that will produce DFDs (see Chapter 7, Workbenches). Although such tools are quite limited at present, they are essential if further productivity gains in systems and software engineering are to be achieved.

4.5 FUNCTIONAL FLOW DIAGRAMS AND DESCRIPTIONS

Functional flow diagrams and descriptions (F^2D^2) are process-oriented, as described in Section 4.4, and have a considerable amount of formality built into them [6]. The orientation of such charts is functional, as shown in the top-level (tier 0) F^2D^2 chart of Figure 4.6. The main thrust is a sequence of numbered functional blocks con-

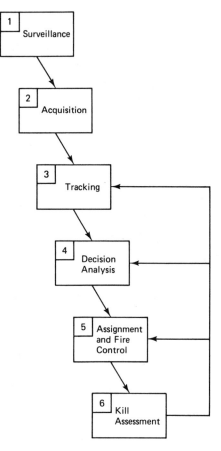

Figure 4.6 Top-level (tier 0) functional flow diagram.

nected by arrow lines. Each block represents a top-level function for a system with a general flow from left to right. Each functional block principally depicts data flow so that the input data enters from the left or top and exits to the right or bottom, as in a DFD.

The top-level (tier 0) F^2D^2 is thus a single chart that shows all major functional elements of a system, which typically might include 5 to 15 functions. The particular functions shown in Figure 4.6 apply broadly to the DoDs strategic defense initiative (SDI or "Star Wars"). The surveillance function is maintained on a constant basis (block 1) and targets following a particular type of trajectory are acquired by the system, shown as the second functional block. Potential threat targets are then tracked (block 3) and an extensive decision-analysis function is engaged to make a final determination as to whether or not a bonafide threat exists. If so, a defensive-weapon assignment and fire-control function is brought into play. Kill assessments (block 6) are made that are used to modify the three prior functions, shown as iteration or feedback loops in the chart. This process continues until the battle is over.

It is noted that particular means of implementing these functions, in general, are not depicted in the top-level tier 0 F^2D^2 chart. The top-level chart simply shows the decomposition of the overall system into its major functions. The next level (tier

1) breaks each function into subfunctions. At the requirements level, the orientation of this breakdown would be in terms of requirements statements. At the next level (tier 2), we are dealing with system and subsystem functions that are broken down into major equipments (hardware), computer programs (software), and personnel requirements for an operational system. Finally, the third tier further partitions these into the next level of detail. In general terms, the top level (tier 0) may be thought of as corresponding to the highest level functional flow diagram of the system, as referred to in the Mil. St. 490 in its interpretation of a Type A system specification.

Some further formal rules that may be applied to the construction of an F^2D^2 diagram are

- function definition is oriented to process and data flow
- the box numbering scheme is the Dewey-decimal type system or some variation thereof
- different tiers are not shown on the same diagram; separate pages are used at the various tier levels
- symbols are shown on the blocks to code them for such elements as hardware, software, personnel, decisions, actions, as well as classes of data such as stored data, support data, and criteria.

As implied by the title, the F^2D^2 methodology is supported by a functional description chart. That is, for each block or box on the diagram there is a description chart that provides

- an abstract of the function
- a listing of incoming and outgoing data
- a definition of criteria, which is interpreted as the process by which the inputs are converted into outputs
- an allocation that correlates system documentation (requirements document, specification, design document, etc.) with the particular functional block

4.6 HIERARCHICAL INPUT-PROCESS-OUTPUT DIAGRAMS

The Hierarchical Input-Process-Output (HIPO) diagram [7] in some sense is a synthesis of the generic HDD and the DFD. In general terms, if the HDD represents a hierarchy of system functions, elements, and so on, and the DFD represents the flow of data and processes inherent at any one level of such a hierarchy, then the HIPO diagram can be construed as containing both.

The HIPO diagramming technique can be thought of as consisting of two diagrams: an H-diagram and an IPO-diagram. The H-diagram shows a hierarchical structure such as the HDD. It is specifically oriented to the system functions and it is numbered hierarchically. The IPO diagram is drawn for each functional box of the H-diagram and contains input (I), process (P), and output (O) boxes, as reflected in Figure 4.7. Here the overall process box is number 5 (see Figure 4.6) and refers to the process of "assignment and fire control." Within that function is a series of func-

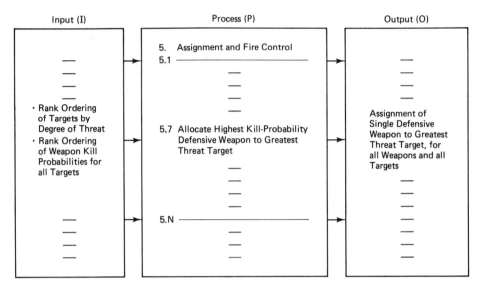

| Input (I) | Process (P) | Output (O) |

—	5. Assignment and Fire Control	—
—	5.1 ————————————	—
—	—	—
—	—	—
· Rank Ordering of Targets by Degree of Threat	—	Assignment of Single Defensive
· Rank Ordering of Weapon Kill Probabilities for all Targets	5.7 Allocate Highest Kill-Probability Defensive Weapon to Greatest Threat Target	Weapon to Greatest Threat Target, for all Weapons and all Targets

Figure 4.7 Portion of HIPO diagramming trechnique showing input-process-output format.

tions that would normally be numbered 5.1 through 5.N. Shown in Figure 4.7 is an intermediate process (5.7) that allocates the highest kill-probability defensive weapon to the target that is determined to represent the greatest threat. The inputs to that process are shown on the left in the overall input chart; the outputs are presented on the right in the output chart. As with the other diagramming techniques, codings of the inputs, outputs, or processes can be added where desirable.

The HIPO diagram thus implicitly recognizes the necessity to represent both the hierarchy and the data flow of a set of system functions. It does so in a straightforward manner, first constructing a hierarchy diagram and then capturing the flow between functions in the separate input-process-output diagram.

4.7 WARNIER–ORR DIAGRAMS

The Warnier–Orr diagrams (WODs), named after their developers Jean-Dominique Warnier [8] and Ken Orr [9], show a hierarchical structure of a program, a system, a function, a data file, a series of parameters, and so on. They were originally oriented to the software development process, but their basic structure is more broadly applicable. The Warnier–Orr diagram can be put into a one-to-one correspondence with an HDD diagram, as shown in Figure 4.8. In that sense its basic structure is that of containing and displaying hierarchies. However, the Warnier–Orr diagram has been designed to contain additional information with emphasis on relationships and data flow between elements of the diagram.

Figure 4.9 shows a Warnier–Orr diagram that contains an overall hierarchy with its bracket structure but also displays relationships between the various data elements in the diagram. The bracket precisely defines a set relationship, with all

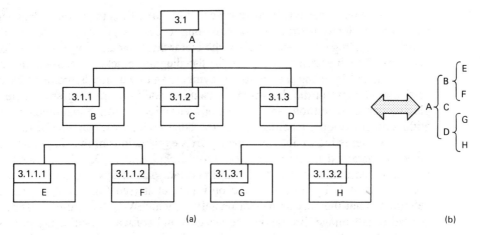

Figure 4.8 Correspondence between (a) hierarchical decomposition diagram and (b) Warnier–Orr diagram.

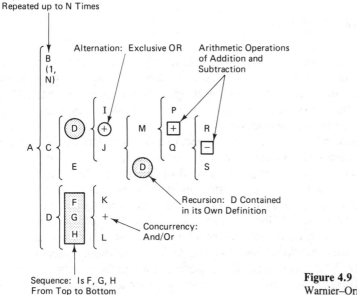

Sequence: Is F, G, H
From Top to Bottom

Figure 4.9 Additional features of Warnier–Orr diagrams.

bracketed elements being the set (of data, for example) that makes up the parameter to which the bracket is pointing. Within a given set, the top to bottom listing is the sequence in which the element is to be used (e. g., processed or printed). Such a listing implicitly represents a logical *and* relationship. The *exclusive or* is depicted by using the ⊕ symbol between elements and the concurrent *or*, meaning and/or, is represented by the + symbol. If an element is repeated, a parenthesis is placed around the element such as (1,12), which means that the element can be repeated from one to twelve times. This might apply to a record of monthly results in which the number of months in the record might vary from 1 to 12 months.

The focus of the WOD, as represented by Orr [9], is to provide a language that has sufficient internal structure to allow one to develop a model of real-world processes. In this regard, he emphasizes the hierarchy, concurrency, and recursion relationships. The hierarchy is implicit in the diagram-bracket construct, the concurrency is added by using the and/or + symbol, and recursion is provided by using a particular element in the definition of itself. In addition, the diagram is able to display sequence, repetition, and alternation. Variations on these main themes, such as arithmetic operations, labeling, keys, and other conditions, can also be displayed.

It is interesting to note that one of Orr's explanations of this diagramming technique is contained in his book, *Structured Requirements Definition*. Requirements definition and analysis is certainly an important part of systems and software engineering, but it is by no means the only part of it (see Chapter 2). Requirements definition is not the only arena, fortunately, in which WODs are useful. They can be applied to advantage throughout the systems and software engineering processes in any situation where one wishes to display hierarchical and other relationships between key elements of the system.

4.8 MICHAEL JACKSON DIAGRAMS

Michael Jackson diagrams (MIJACs) [10,11] have largely been thought of in terms of software development but like most of the other techniques, they can more broadly be applied to other systems engineering tasks and activities. MIJACs are tree-like hierarchical charts that have particularly been applied to program and data structure aspects of software development.

The major feature of a MIJAC with respect to a data structure can be illustrated by reference to Figure 4.10. An employee file might refer to all employees in a company while a program manager might only need to determine who is working on a communications system that is their responsibility. The managers, as well as the financial department, are also interested in the direct charges incurred by each person during a particular period, for example, two weeks. The data structure is first sorted into those working on the communications system and those that are not. A small circle, ○, in these blocks designates an *exclusive or* situation, similar to the, ⊕ symbol in a WOD. Next a sort is made for those that are in the communication and navigation (ComNav) division and those that are not, with the same exclusive or annotation. Finally, for those employees that are working on the particular communications system and are also in the ComNav division, the manager wishes to know the

- employee name
- employee identification number
- hourly rate
- hours charged during the two-week period

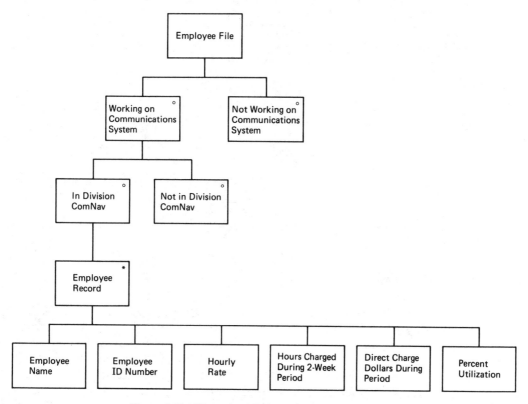

Figure 4.10 Illustrative Michael Jackson diagram of a data structure.

- direct charge dollars during the period
- employee's percent utilization

The asterisk (*) in the employee record block denotes repetition or iteration as does the (1, *N*) notation in the WOD.

Thus the MIJAC clearly shows sequence, repetition, and selection. The sequence for the desired employee record is read from left to right in Figure 4.10 in the sequence listed previously. The repetition, designated by the asterisk, means that the data are desired for all employees satisfying the stated conditions. Selection is designated by the circle ○. All of these features are likewise represented in the WOD. Other symbols can be used for special-purpose codings.

Michael Jackson has also devised a *system-network diagram* that connects data flow and structure to program structure. Given the tree-like hierarchy diagram for data and for the program that is to process this data, the two may be interrelated by means of a network diagram, as illustrated in Figure 4.11. This chart shows generic data flows (shown by a circle) and subprograms of the overall program (shown by rectangular blocks). This is an interesting aspect of the MIJAC presumably not found in the techniques discussed previously.

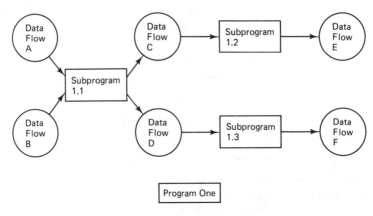

Figure 4.11 System-network diagram showing data flows and computer subprograms.

4.9 ACTION DIAGRAMS

The apparent inventors of action diagrams (ADs) are J. Martin and C. McClure [1, 12] who argue that such a technique has been designed to handle both overview high-level program structures as well as detailed low-level program logic. Further, the high-level logic should flow into the low-level logic as an integrated and well-defined procedure. As with the WODs and the MIJACs, the orientation is toward the overall process of software development. Thus, through the use of ADs one can begin with a top-level overview of a program structure and using the same AD technique applied at increasing levels of detail, wind up with a set of diagrams written in a fourth generation language (4GL), such as IDEAL, RAMIS, FOCUS, or NATURAL or a pseudocode convertible to a more conventional language such as Ada, BASIC, Pascal, C, or Fortran.

The overall structure of an AD shows hierarchy as do many of the previous diagramming techniques. A generic hierarchy for an AD is depicted in Figure 4.12, using the same hierarchy from Figure 4.8. In the AD, A contains B, C, and D; B contains E and F; D contains G and H. In the AD, we note that if an element is pointed at from the left, as A, B, and D are, they contain subelements, otherwise, they do not.

We will examine four major features of the AD, namely: sequencing, repetition/looping, selection/conditioning, and termination. These features are basically all that is necessary to write any program (see also Chapter 9). Other features are also available and are cited later without detailed explanations.

A block of steps is shown from top to bottom within a bracket, as in the WOD. The top of Figure 4.13 shows such a sequence as steps A.1, A.2, . . . , B.1, B.2, and so on. Repetition/looping is developed as the REPEAT and WHILE . . . DO statements. The interior steps for these statements are repeated until a particular condition is satisfied. The repetition/looping shows a double line at the top of the

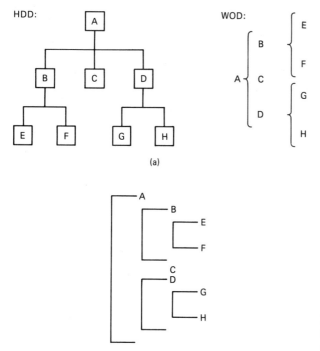

Figure 4.12 An action-diagram representation of a basic hierarchy: (a) two other representations of the same basic hierarchy, (b) an action diagram.

bracket and a dark line constituting the side of the bracket. The selection/conditioning feature is represented by the IF . . . ELSE as well as the WHEN statements. This is the conventional *branching process* built into programming languages including the 4GLs. Finally, an IF . . . statement in module or subprogram D will terminate the program if the conditional statement is true. Thus these four features are part of an AD structure and allow the user to move from a high-level analysis to the detailed program-coding structure.

The AD also contains several other features that will not be explored in detail here. These include

- rounded-corner box elements to show procedures and subprocedures
- rectangle expansions to depict input (top right of rectangle) and output (bottom right of rectangle) data
- additional coding options such as colors and other forms of labeling for different features and functions

It is expected that due to its ability to integrate top-level and detailed-level program analysis and design, the AD will become more prevalent in terms of the overall software development process.

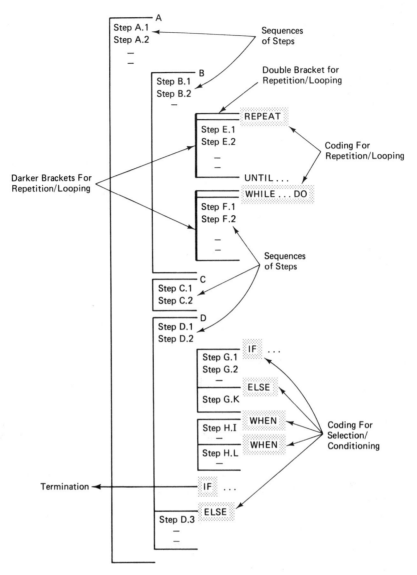

Figure 4.13 An illustrative action diagram showing sequencing, repetition/looping, selection/conditioning, and termination.

4.10 SEQUENCE AND TIMING DIAGRAMS

The sequence and timing diagram (STD), as the name implies, focuses on series and parallel activities and their associated time lines. This is of particular importance when attempting to design and balance a large number of activities that occur in sequence or in parallel such as assembly-line balancing, and real-time image processing. The issue, from a systems engineering point of view, is to achieve a time-line balance so that there is maximum throughput and no waiting or queuing. At the same time, costs must be accounted for in the trade-off between overall processing efficiency and life-cycle cost.

Although there are several variations on the STD, one implementation is constructed as an adjunct to the F^2D^2 diagram discussed previously [6]. Figure 4.14 shows a simple example based upon the top-level (tier 0) functional flow diagram in Figure 4.6. The latter roughly applies to the so-called Star Wars scenario. We will limit ourselves to the boost phase of this scenario, which has an overall time line of the order of 5 min. (300 sec.). Time is portrayed in the STD along the x-axis and the

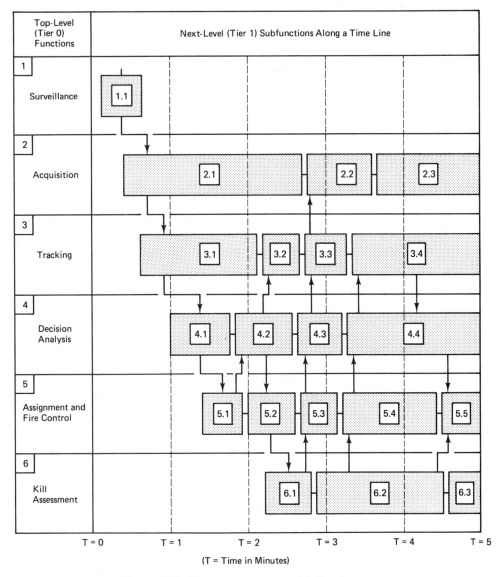

Figure 4.14 Illustrative sequence and timing diagram.

Sec. 4.10   Sequence and Timing Diagrams

six major functions (see Figure 4.6) are constructed along the y-axis, from the first at the top to the last at the bottom.

The subfunctions are developed so that the length of the rectangular box lays out along the time line over which it is expected or programmed to occur. Each discrete subfunction is numbered as such (e.g., 2.1, 3.2, 4.3) and represents a separate or different mode of operation under the overall tier 0 function. Key data that flow between subfunctions are represented by an arrow at the times they occur. Leaving a subfunction at one time and entering a different subfunction at another denotes start and finish times of the data flow. As before, a box represents a process that operates upon the data flow. Horizontal connections between subfunctions are shown and are normally quite usual unless the modes of operation within a function are distinct and unrelated.

The STD shows an approximation as to either or both expected and required time lines. If an STD representation is inadequate in terms of performing the functions within an overall time-line requirement, other analytic tools (modeling, simulation, etc.) may have to be brought to bear to satisfy the requirements. If these tools can be used to optimize the time lines, the results can be used to update the STDs. In any case the STDs give the designer and analyst a good pictorial representation of the serial and parallel activities at the functional level under consideration.

4.11 PARAMETER DEPENDENCY DIAGRAMS

The parameter dependency diagram (PDD), attributable to this author [13], shows interrelationships between key parameters and attributes of a system. The basic block in the PDD, shown in Figure 4.15, accepts parameters on the left, processes them in some fashion, and produces another set of parameters as an output. Thus the block represents some functional transformation or transfer-function procedure.

A principal purpose of the PDD is to model the performance capabilities of a system so as to ultimately determine if the system meets its sets of requirements and specifications. The process of constructing the PDD in this context is to start by defining the highest-level end products or outputs of the model. Guidance as to what these outputs should be comes from a review of the system-level requirements and specifications. The modeling process may be performed by system function such as search, acquisition, tracking. In such a case, a separate PDD is constructed for each function and the diagrams are tied together by showing parameter inputs and outputs for each function as a block.

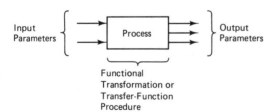

Input Parameters — Process — Output Parameters

Functional Transformation or Transfer-Function Procedure

Figure 4.15 Basic block of the parameter dependency diagram.

The process may be illustrated by a simple example. We assume the subfunction of interest to be the search mode (function) of a fire-control system and that the two key parameters are the probability of detection of a target (P_d) and the false-alarm probability (P_{fa}). These, then, are to be the principal outputs of the modeling process. For each of these key output parameters we then ask the question—what does this parameter depend upon? This successive questioning process leads ultimately to the PDD in Figure 4.16. The P_d and P_{fa} parameters are shown as outputs to the right (boxes 8 and 9), and the other boxes and parameters visually display the interrelationship between parameters from a computation-and-analysis point of view. Supporting the PDD are two *analysis dictionaries*—one that lists each parameter in the PDD (see Figure 4.17) and one that defines as precisely as possible the various functional relationships represented by each box in the PDD (see Figure 4.18). Referring to the latter we observe that boxes 8 and 9 are conventional radar calculations for single-pulse threshold detection (see Chapter 5) and that boxes 1 and 2 are the familiar radar-range equation and noise-spectrum computations, respectively. Thus we have a complete road map of the key performance parameters of the system operating in the search mode, the manner in which they are interrelated, and the functional equations that will allow us to compute the key outputs as well as other intermediate outputs as a function of parameters upon which they depend. Since we thus have a model of the system, we can program these equations and use the computer to assist in performing a variety of trade-off analyses. As an example, we can explore the manner in which the detection and false-alarm probabilities vary as we change the range values for other parameters such as

- power transmitted
- range to target
- i-f (intermediate frequency) bandwidth
- cutoff (3 db) frequencies of post-discriminator filter

All of this can be done in a computer-aided mode if the equations can be developed properly and drivers are written to cycle the parameters through appropriate ranges of values.

Experience has shown that it is often desirable to develop a signal-flow block diagram (see Section 4.2) to support the PDD construction and subsequent analysis. As an example, the signal-flow block diagram in Figure 4.19 applies directly to the PDD in Figure 4.17 and its associated analysis dictionaries. Also, the PDD technique implicitly relates to the hierarchy of system functions by modeling each function separately and then interconnecting functional PDDs through the specific parameters that are common to more than one function.

Thus the PDD is principally a performance-modeling and simulation aid (see also Chapter 12) and provides a graphical overview of the key system parameters and the manner in which they interrelate. In this role the PDD has numerous systems engineering application areas, including

- technical performance measures
- verification of performance against requirements and specifications

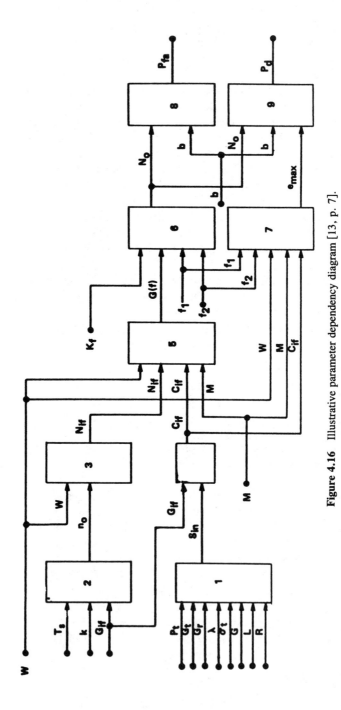

Figure 4.16 Illustrative parameter dependency diagram [13, p. 7].

b	detection threshold
C_{if}	i-f output carrier power
e_{max}	maximum output voltage
f_1	post discriminator filter low frequency cutoff
f_2	post discriminator filter high frequency cutoff
G	receiver processing power gain
G_{if}	i-f amplifier gain
G_r	receiving antenna gain
G_t	transmitting antenna gain
$G(f)$	post discriminator noise power spectrum
k	Boltzmann's constant
K_f	post discriminator filter voltage gain
L	receiver processing power losses
M	discriminator gain
n_o	i-f output noise power spectrum
N_{if}	i-f output noise power
N_o	output noise power
P_d	detection probability
P_{fa}	false alarm probability
P_t	transmitted power
R	range to target
S_{in}	i-f output signal power
T_s	system noise temperature
W	i-f bandwidth
λ	wavelength
σ_t	target cross section

Figure 4.17 Parameter-definitions analysis dictionary [13, p. 6].

- trade-off analysis
- test and evaluation
- engineering change proposal (ECP) evaluation

As such, the PDD is viewed as an important diagramming tool with respect to the many tasks and activities inherent to systems engineering.

$$(1) \quad S_{in} = \frac{P_t G_t G_r \lambda^2 G\sigma_t}{(4\pi)^3 R^4 L}$$

$$(2) \quad n_o = kT_s G_{if}$$

$$(3) \quad N_{if} = n_o W$$

$$\cdot$$
$$\cdot$$
$$\cdot$$
$$\cdot$$

$$(8) \quad P_{fa} = \frac{1}{2}\left[1 - erf\left(\frac{b}{\sqrt{2N_o}}\right)\right]$$

$$(9) \quad P_d = \frac{1}{2}\left[1 - erf\left(\frac{b - e_{max}}{\sqrt{2N_o}}\right)\right]$$

Figure 4.18 Analysis dictionary; functional relationships related to parameter dependency diagram [13, p. 9].

4.12 LOGIC FLOW CHARTS

These diagramming techniques are rather old and well known and to some extent are losing popularity since they are nonstandardized, sometimes difficult to follow, and generally do not promote structured design. They are included here for the sake of completeness, but for the reasons cited they will not be examined in any detail.

Flow charts can be used to display any logical process. Although they have in the past been used extensively as a step prior to the actual writing of code for computer programs, they can and have been applied to descriptions of methods of

- program management
- configuration management
- formulating system acquisition strategies
- preplanned product improvement (P³I)
- integrated logistics support
- maintenance planning
- quality control and assurance

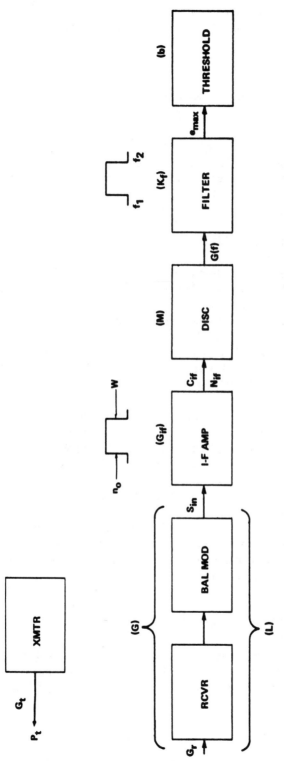

Figure 4.19 Signal-flow block diagram for illustrative parameter dependency diagram [13, p. 16].

In broad terms there are three fundamental constructs represented by logic flow charts, which have been alluded to and incorporated in diagramming techniques described earlier. These are

1. sequence
2. repetition, looping, or iteration
3. selection, conditioning, or branching

Since logic flow charts were originally popularized as a support to computer programming, examples of each of these can be seen by reference to Figure 9.1. A technique of growing usage that incorporates logic flowchart information in a compact manner is discussed in the following text.

4.13 NASSI–SHNEIDERMAN CHARTS

The Nassi–Shneiderman chart (NSC) [14] and similar diagrams appear to be replacing the logic flow chart since the former are more highly structured, lead to better standardization, and facilitate detailed program design and documentation.

In Figure 4.20 we show a simple logic flow chart and an equivalent NSC, broadly related to the Michael Jackson diagram in Figure 4.10. Although these diagrams are not rigorously drawn, they will illustrate the correspondence between the techniques and the three key features of the NSC of sequence, selection, and repetition. The first two steps in the figure portray a simple sequence in both charts. The diamond symbols and questions in the flow chart are represented by inverted triangles and corresponding questions in the NSC. This is a selection or conditioning statement. The feedback from the end-of-file question in the flow chart depicts a repetition or iteration process. The same process is achieved in the NSC through the use of the symbol shown and its associated DOWHILE statement. The particular language used in both the flow chart and the NSC is not significant and indeed is a matter of choice among the many alternatives and levels of detail. The uniqueness of the chart lies in the symbology and format used rather than the selection of words.

It is thus seen that the NSC is capable of presenting the information with considerably greater uniformity of structure than is normally contained in the flow chart. Variations in the aforementioned themes for more complex logic can be observed by consulting the references [1, 14].

4.14 DECISION-NETWORK DIAGRAMS

The decision-network diagram (DND) under consideration here represents series and parallel sequences of interconnected events. The interconnections are normally activities and therefore are associated with the expenditure of resources such as time

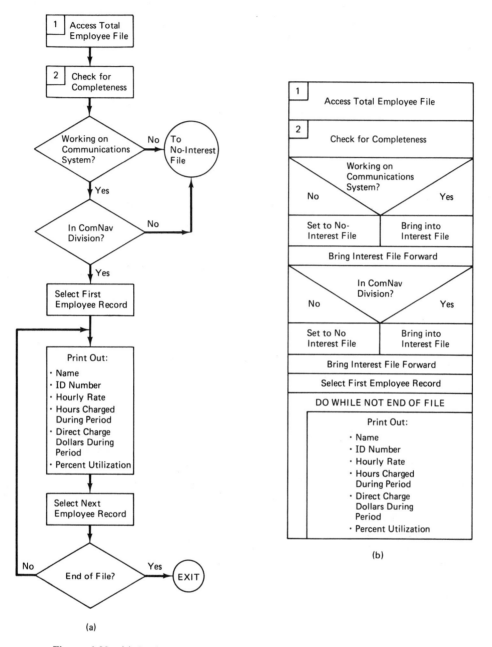

Figure 4.20 (a) Logic flow chart, and (b) corresponding Nassi–Shneiderman chart.

and person loading. Thus one type of network is a program evaluation and review technique (PERT) chart (see Chapter 11); another can be construed as a generalized simulation of almost any kind of process. Once the decision element is added, as illustrated in Figure 4.21, the network becomes a decision network, sometimes also called a stochastic network.

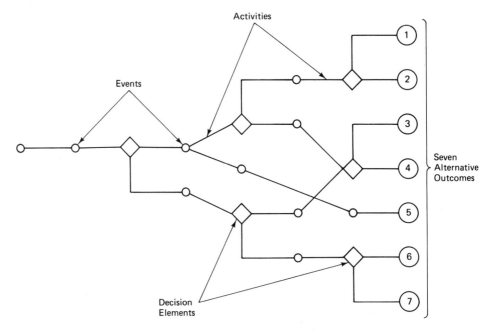

Figure 4.21   A decision-network diagram.

The notion of activity–event types of decision networks was first introduced by this author in 1962 [15]. S. Elmaghraby later developed an algebra for such networks [16] and various extensions and refinements have been made by various investigators over the years, leading to the Q-GERT procedure [17]. A dedicated practitioner with respect to the latter is A. Pritsker [18] who has brought this procedure to broad simulation-application areas (see Chapter 12).

The decision element in the network fundamentally leads to the explicit consideration of alternatives and the attributes of these alternatives, including such parameters as

- time/schedule
- cost
- technical performance measures
- measures of merit and effectiveness

In the sense that systems engineering synthesis and analysis is the definition and evaluation of alternatives (e.g., design alternatives, program alternatives, schedule alternatives), the (DND) and associated methodologies (e.g., Q-GERT) are broadly applicable and potentially very powerful. The DND, in particular, facilitates the construction and visualization of these alternatives; the methodologies are explicit means by which analyses of various types can be carrried out. The decision tree, a particular kind of decision representation normally without such elements as merge points and path dependencies [15], is viewed as a subset of the broader class of decision networks. Thus decision-tree analysis methodologies can be subsumed within the analytic techniques associated with the decision network. References 15 to 18 as well as Chapters 11, 12, and 13 provide additional information with respect to decision-network diagramming and analysis techniques.

4.15 SUMMARY

The diagramming techniques cited here are viewed as significant systems engineering and software engineering tools, with Table 4.1 highlighting some of the potential application areas. To save space many techniques were omitted such as several suggested by Martin [1] (see chapter beginning) and others such as N^2 diagrams [19] that assist in the definition of system functional interfaces, and Petri Nets [20] that can represent computing systems and their processes as well as various types of decision models. Diagrams are important visualization tools as well as heuristic tools, the latter in the original sense of the word, that is, encouraging further investigation and analysis. However, it is crucially important that we devise better methods of capturing the "information content" of diagrams into our computer-based systems in such a manner as to facilitate merging and manipulation of such information with the data now normally contained in available software packages (i.e., spreadsheets, DBMSs). Progress in this area will more fully integrate diagramming into the systems engineering process and promote increases in systems engineering productivity.

TABLE 4.1 SELECTED SYSTEMS ENGINEERING APPLICATION AREAS FOR DIAGRAMMING TECHNIQUES

Simple Process Flow Charts
- The systems engineering process to be employed on a particular program
- The software development process to be employed on a particular program

Signal-Flow Block Diagrams
- The flow of signals through a telecommunications system
- The flow of signals through an automatic-control (servomechanism) system

Hierarchical Decomposition Diagrams
- The organizational structure of a project team
- The decomposition of a program's functions into its subfunctions, and so on

Data Flow Diagrams
- The flow of data through a systems engineering project team
- The data flow requirements for a computerized information system (e.g., airline reservations, inventory control, financial accounting)

Functional Flow Diagrams and Descriptions
- The hierarchy of functions and their descriptions for
    a command, control, and communications system
    an artificial intelligence application area, (e.g., maintenance)

TABLE 4.1 (*cont.*)

Hierarchical Input-Process-Output Diagrams
- The hierarchy of functions as well as input-output flow and processing for
  a battlefield war game
  a rapid transit system

Warnier–Orr Diagrams
- The hierarchy and data flow for a set of computer programs
- The hierarchy and data flow for a set of reports to be received by a systems engineering manager

Michael Jackson Diagrams
- The tree-like hierarchy and data structure for
  a high-level set of computer programs
  the reports of a project manager

Action Diagrams
- The high-level structure of a set of computer programs
- the low-level structure and code (pseudocode, 4GL) for the above set of computer programs

Sequence and Timing Diagrams
- The decomposition of functions and their time lines for
  1. a satellite ground-based telemetry, tracking, and command (TT&C) system
  2. an air-defense system

Parameter Dependency Diagrams
- The interdependency of key parameters and attributes of our national aviation system
- The simulation of the performance of a search and tracking radar

Logic Flow Charts
- The structure of a set of computer programs
- The sequence of activities associated with a quality control and assurance function

Nassi–Shneiderman Charts
- The structure of a set of computer programs
- The sequence of activities associated with a configuration-management function

Decision-Network Diagrams
- The definition of project schedule, resource alternatives, and overall potential outcomes
- The definition of system-design alternatives and associated hardware and software configurations

REFERENCES

1. Martin, J., and C. McClure, *Diagramming Techniques for Analysts and Programmers.* Englewood Cliffs, NJ: Prentice-Hall, 1985.

2. Ross, D. T., "Applications and Extensions of SADT." *IEEE Computer Magazine*, April 1985.

3. Yourdon, E., and L. Constantine, *Structured Design.* New York: Yourdon Press, 1978.

4. DeMarco, T., *Structured Analysis and System Specification.* Englewood Cliffs, NJ: Prentice-Hall, 1979.

5. Gane, C., and T. Sarson, *Structured Systems Analysis: Tools and Techniques.* Englewood Cliffs, NJ: Prentice-Hall, 1979.

6. Lurcott, E., "F²D², A System Mangement Tool," *Defense Systems Management Review*, I, no. 4.

7. IBM HIPO *A Design Aid and Documentation Technique.* GC20-185D, New York: IBM Corporation, 1974.

8. Warnier, J., *Logical Construction of Systems*. New York: Van Nostrand Reinhold, 1981.

9. Orr, K., *Structured Requirements Definition*. Topeka, KS: Ken Orr and Associates, Inc., 1981.

10. Jackson, M. A., *Principles of Program Design*. New York: Academic Press, 1975.

11. Jackson, M. A., *System Development*. Englewood Cliffs, NJ: Prentice-Hall, 1983.

12. Martin, J., and C. McClure, *Action Diagrams, Clearly Structured Program Design*. Englewood Cliffs, NJ: Prentice-Hall, 1985.

13. Eisner, H., *The Assessment of Military Systems' Performance Using Parameter Dependency Diagram Techniques*. The Third Annual Israel Conference on Operations Research, July 1969.

14. Nassi, I., and B. Shneiderman, "Flowchart Techniques for Structured Programming," *ACM SIGPLAN Notices*, 8, no. 6, August 1973.

15. Eisner, H., "A Generalized Network Approach to the Planning and Scheduling of a Research Project," *Operations Research*, 10, no. 1, February 1962.

16. Elmaghraby, S. F., "An Algebra for the Analysis of Generalized Activity Networks," *Management Science*, 10, no. 3, 1964; "On Generalized Activity Networks," *Journal of Industrial Engineering*, 18, no. 11, November 1966.

17. Whitehouse, G. E., *Systems Analysis and Design Using Network Techniques*. Englewood Cliffs, NJ: Prentice-Hall, 1973.

18. Pritsker, A. A. B., and C. E. Sigal, *Management Decision Making: A Network Simulation Approach*. Englewood Cliffs, NJ: Prentice-Hall, 1983.

19. *System Engineering Management Guide*. Defense Systems Management College, Fort Belvoir, VA, October 3, 1983.

20. Peterson, J. L., *Petri Net Theory and the Modeling of Systems*. Englewood Cliffs, NJ: Prentice-Hall, 1980.

PROBLEMS

P4.1. Construct a simple process flow chart describing all the steps you through in:
    a. studying for an examination
    b. a typical day

P4.2. Write and explain the blocks in a signal flow block diagram for an
    a. AM radio
    b. FM radio

P4.3. Develop a functional hierarchy chart for a typical stereo system using the following diagramming techniques:
    a. an HDD
    b. an F^2D^2 Diagram
    c. a HIPO Diagram
    d. a WOD
    e. a MIJAC
    f. an AD

P4.4. Develop a functional hierarchy chart for a typical command, control, and communications system using the diagramming techniques cited in Problem P4.3.

P4.5. Develop a functional hierarchy chart for our national aviation system using the diagramming techniques cited in Problem 4.3.

P4.6. Formulate a data flow diagram for the processing of information through the various departments of a bank.

P4.7. Develop a hierarchy (organization) chart for a typical systems engineering project. Formulate a data flow diagram for the processing of information through the various organizational units of the project.

P4.8. Construct an STD for an automobile engine's internal combustion cycle, showing as a minimum the spark-plug firing and valve positions for a four-cylinder configuration.

P4.9. Formulate a PDD for a typical communications system. Discuss the functional relationships represented by the blocks of the diagram.

P4.10. Formulate a PDD for our national aviation system with principal output parameters being airspace capacity (number of aircraft handled per hour) and airport capacity (number of takeoffs and landings per hour). Discuss the functional relationships represented by the blocks of the diagram.

P4.11. Construct a flow chart and a Nassi–Shneiderman chart for a typical configuration management function for a systems engineering project.

P4.12. Construct a flow chart and a Nassi–Shneiderman chart for the synthesis function within a systems engineering process.

P4.13. Develop a DND for the purchase of a personal computer and its associated starter software packages.

P4.14. Develop a DND for a project of your own selection where your objective is to write and then market your own software package.

*We sail within a vast sphere, ever drifting
in uncertainty, driven from end to end*

Pascal

OVERVIEW OF PROBABILITY CONCEPTS

5.1 INTRODUCTION

This chapter presents the rudiments of probability theory and relates them to systems engineering issues and problem areas. The intent is to provide the reader with the tools necessary to address both these systems engineering issues as well as the computer software packages that utilize probability theory concepts.

There is no attempt to provide either a complete or a rigorous discourse on the subject of probability theory. Only selected topics are presented and the more formal foundations of the theory are left to the reader to explore in other texts [1,2].

Some of the systems engineering areas in which probability theory applications are found include

- risk analysis
- decision analysis
- error analysis and allocation
- requirements analysis
- simulation and modeling
- RMA assessments
- test and evaluation

- scheduling
- cost analysis
- performance analysis

Examples will be drawn from these areas so that we can better understand the role of probability concepts in systems engineering.

5.2 DISCRETE EVENTS AND PROBABILITIES

First we consider the world of discrete events and the assessment of the probabilities of occurrence of these events. Examples of discrete events include the throwing of the number five on a die or the receipt of a bit (e.g., positive electronic pulse) of data designating a *one* in a communications system. These are each discrete and observable events and it is of interest to examine the likelihoods or probabilities of their occurrence.

We often refer to a graphic known as a *Venn diagram* to show the population of events with which we are concerned. This diagram, illustrated in Figure 5.1, shows the events as contained within a closed boundary and the associated probabilities of occurrence can be thought of as a line or vector drawn perpendicular to the page and out of the page. The length of such a line is visualized as a measure of the numerical value of the probability.

To further illustrate the essential notions of events and probability, we show two situations in Figure 5.1. In the first we have two cubes, one containing the conventional numbers one through six, and the second containing the numbers four through nine. In the second situation we also have two cubes, one with the numbers one through six and the second shows numbers 7 through 12.

Events that we may wish to explore might include the following in either situation in regard to the throw of a single cube:

- the throw of a 3
- the throw of a 6
- the throw of a 9
- the throw of a 12

We may likewise wish to examine the same events in relation to the toss of both cubes where we then add together the two numbers shown on the two cubes. The two tables in Figure 5.1 show the resultant probabilities in relation to the above four events.

5.2.1 Examples

Situation one. Here the individual events relative to cube one are the numbers one through six, as depicted in the Venn diagram of Figure 5.1(a). For cube two the events are the numbers four through nine. The common set of events, namely the numbers four through six, represent an *intersection* of the individual sets

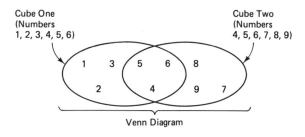

		Probability	
Event	Cube One	Cube Two	Both Cubes Thrown
Throw of a 3	$\frac{1}{6}$	0	0
Throw of a 6	$\frac{1}{6}$	$\frac{1}{6}$	$\frac{2}{36} = \frac{1}{18}$
Throw of a 9	0	$\frac{1}{6}$	$\frac{5}{36}$
Throw of a 12	0	0	$\frac{4}{36} = \frac{1}{9}$

(a)

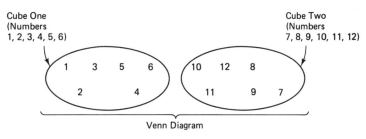

		Probability	
Event	Cube One	Cube Two	Both Cubes Thrown
Throw of a 3	$\frac{1}{6}$	0	0
Throw of a 6	$\frac{1}{6}$	0	0
Throw of a 9	0	$\frac{1}{6}$	$\frac{2}{36} = \frac{1}{18}$
Throw of a 12	0	$\frac{1}{6}$	$\frac{5}{36}$

(b)

Figure 5.1 Venn diagrams, events, and probabilities: (a) situation 1, (b) situation 2.

of events. For the toss of cube one, the probability of a three is $\frac{1}{6}$ as is the probability of a six. This reflects the so-called *frequency* interpretation of probability in that for a large number of tosses each of the six sides will come up approaching one sixth of the time. The same is true for cube two so that the probability of a six $= P(6) = 1/6 = P(9)$. Also, since cube two does not have a 3 or a 12, for that cube $P(3) = 0 = P(12)$.

When both cubes are thrown and the numbers added, we can look at all the possible outcomes as shown in Figure 5.2. Clearly there are 36 possibilities and in no case can the resultant sum come to a three so that in this case $P(3) = 0$. However, there are two ways (of the 36) that a six can be thrown so that $P(6) = 2/36 =$

Cube One (Numbers 1, 2, 3, 4, 5, 6)
Cube Two (Numbers 4, 5, 6, 7, 8, 9)

Cube Two Possibilities

	4	5	6	7	8	9
1	5	6	7	8	9	10
2	6	7	8	9	10	11
3	7	8	9	10	11	12
4	8	9	10	11	12	13
5	9	10	11	12	13	14
6	10	11	12	13	14	15

Cube One Possibilities

(a)

Cube One (Numbers 1, 2, 3, 4, 5, 6)
Cube Two (Numbers 7, 8, 9, 10, 11, 12)

Cube Two Possibilities

	7	8	9	10	11	12
1	8	9	10	11	12	13
2	9	10	11	12	13	14
3	10	11	12	13	14	15
4	11	12	13	14	15	16
5	12	13	14	15	16	17
6	13	14	15	16	17	18

Cube One Possibilities

(b)

Figure 5.2 Illustrative outcomes for toss of two cubes: (a) situation 1, (b) situation 2. (Note that entries represent the sum of the toss of both cubes.)

1/18. Similarly, a nine can occur in five ways and a 12 can occur in four ways, leading to $P(9) = 5/36$, and $P(12) = 4/36 = 1/9$.

Situation two. Here the possible events represented by the two cubes do not intersect and are therefore said to constitute *disjoint* or *mutually exclusive* sets (see Venn diagram of Figure 5.1(b)). The construction of probabilities when both cubes are thrown is shown in Figure 5.2(b) leading to the results: $P(3) = 0 = P(6)$; $P(9) = 2/36 = 1/18$; $P(12) = 5/36$.

5.2.2 Generalization

The examples in Figures 5.1 and 5.2 actually illustrate a variety of points, three of which can be considered axiomatic for purposes of this presentation. These three points are expressible as

1. $P(E) \geqq 0$ for every event E
2. $P(T) = 1$ for the certain event T
3. $P(E \cup F) = P(E) + P(F)$ if $EF = \varnothing$ (null set)

The first point simply expresses the fact that a probability is defined as a nonnegative number. The second point indicates that for an event that is certain to occur, the probability is defined as unity. For example, in the cube-tossing situations, if a particular cube had only the number six on all faces, then the probability of a six would obviously be one. The third point relates to the *union* of sets, designated by the symbol $\cup$ and associated with the *or* operator. The point is read as "the probability of event E or F is equal to the sum of the probabilities, if events E and F are mutually exclusive." To illustrate the point, we refer again to Figure 5.1(b), situation two, in which the Venn diagram shows the two cube events to be mutually exclusive since they do not intersect, in other words, they have no events in common. Thus if we define E as a five on cube one and F as 11 on cube two, we have

$$P(E \cup F) = P(5, \text{ cube } 1) + P(11, \text{ cube } 2)$$

$$= \frac{1}{6} + \frac{1}{6}$$

$$= \frac{1}{3}$$

If the events are not mutually exclusive, it means that they have a mutual intersection, as in situation one in Figure 5.1. The union (*or*) relationship then becomes

$$P(E \cup F) = P(E) + P(F) - P(EF)$$

where EF is read as E *and* F. In order not to double count, we need to subtract out the probability of both E and F occurring. In situation one, if E is the event of a five for cube one, and F is the event of a six for cube two, then

$$P(E \cup F) = \frac{1}{6} + \frac{1}{6} - \left(\frac{1}{6}\right)\left(\frac{1}{6}\right)$$

$$= \frac{1}{3} - \frac{1}{36} = \frac{11}{36}$$

5.2.3 Joint Probabilities and Independence

We have already alluded to the notion of a joint probability, namely the probability of E and F. This may be written as

$$P(E \cap F) = P(E \text{ and } F) = P(E|F)P(F) = P(F|E)P(E)$$

This relationship contains $(E|F)$, which is read as E *given* F and allows for the possibility that events E and F might be dependent upon one another. In the examples in

Figure 5.1, the events do not depend upon one another, that is, the two cube tosses can be performed independently and the results of one have no effect whatever upon the results of the other. In the more general case in which there might be some kind of dependence, we calculate the probability of E given F and then multiply it by the unconditional probability that F occurs. In the case of independence, $P(E|F) = P(E)$ and the relationship becomes

$$P(E \text{ and } F) = P(EF) = P(E)P(F)$$

for independence between E and F.

We can explore the notion of dependence by posing a problem in which the events in fact depend upon one another. We will resort to a simple urn problem as follows. We postulate an urn that has three balls in it: two black balls and one white ball. We will select two balls in succession from the urn, without replacement, as shown in Figure 5.3. We ask for the probability that the two balls selected will both be black. This can be written as

$$P(BB) = P(B \text{ and } B)$$

$$= P(B \text{ on second draw given } B \text{ on first draw})P(B \text{ on first draw})$$

The $P(B$ on the first draw) is simply $\frac{2}{3}$ since when we start there are two black balls and one white ball. The $P(B$ on second draw given B on first draw) is $\frac{1}{2}$ since the urn at that point contains a black ball and a white ball. Therefore the result becomes

$$P(BB) = \left(\frac{2}{3}\right)\left(\frac{1}{2}\right) = \frac{1}{3}$$

Since the second event depends upon the result of the first event, we have a case of dependence which we must account for.

We also observe situations in which a probability is conditional upon other events. We illustrate such a case by using two urns and placing three white balls in one. As shown in Figure 5.4 we now have two urns (X and Y), where urn X has two black balls and a white ball and urn Y has three white balls. We now ask for the probability that we will select a white ball on the first draw, given that we select an urn at random, that is, with equal likelihood, from which to make the selection. Clearly, if we select urn X, the chances of drawing a white ball are $\frac{1}{3}$ since there are two black balls and one white ball. However, if we have selected urn Y then the probability of drawing a white ball is one since there are only white balls in that urn. We form a general relationship to address problems of this type as follows:

$$P(E) = P(E|F)P(F) + P(E|\text{not } F)P(\text{not } F) \quad \text{where } F + (\text{not } F) = F + \overline{F} = T$$

Note the dependence upon event F and the fact that we account for all the possibilities with respect to F, namely, both F occurring and *not* F occurring (i.e., F not occurring). The event F in this situation is interpreted as the selection of an urn and the event *not* F is the selection of the other urn. Since we select the urns at random, both of these probabilities in this situation are $\frac{1}{2}$. Thus the relationship becomes

$$P(W) = P(W|\text{urn } X)P(\text{urn } X) + P(W|\text{urn } Y)P(\text{urn } Y) = \left(\frac{1}{3}\right)\left(\frac{1}{2}\right) + (1)\left(\frac{1}{2}\right) = \frac{2}{3}$$

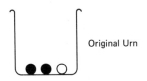
Original Urn

First-Draw Possibilities

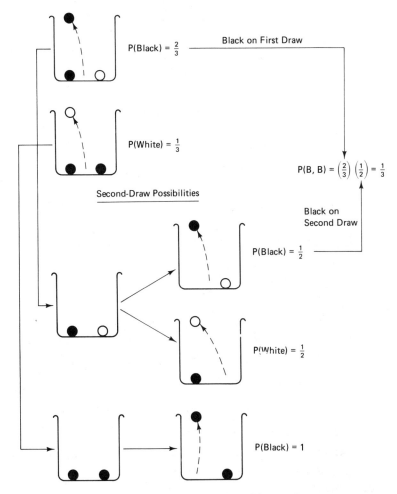

$P(\text{Black}) = \frac{2}{3}$ Black on First Draw

$P(\text{White}) = \frac{1}{3}$

Second-Draw Possibilities

$P(\text{Black}) = \frac{1}{2}$ Black on Second Draw

$P(\text{White}) = \frac{1}{2}$

$P(\text{Black}) = 1$

$P(B, B) = \left(\frac{2}{3}\right)\left(\frac{1}{2}\right) = \frac{1}{3}$

Figure 5.3 Dependence by selection without replacement.

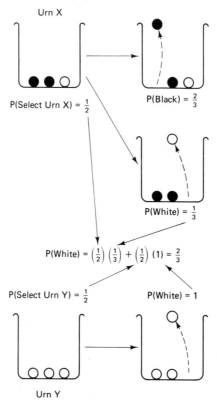

P(Select Urn X) = $\frac{1}{2}$

P(Black) = $\frac{2}{3}$

P(White) = $\frac{1}{3}$

P(White) = $\left(\frac{1}{2}\right)\left(\frac{1}{3}\right) + \left(\frac{1}{2}\right)(1) = \frac{2}{3}$

P(Select Urn Y) = $\frac{1}{2}$

P(White) = 1

Urn Y

Figure 5.4 Selections at random from two urns.

We summarize the key relationships we have explored in the tabulation here

- $P(E) \geqq 0$ for every event E (5.1)

- $P(T) = 1$ for the certain event T (5.2)

- $P(E \cup F) = P(E + F)$

$$= P(E) + P(F), \text{ if } EF = \varnothing \text{ (null set)} \qquad (5.3)$$

- $P(E \cap F) = P(EF) = P(E|F)P(F) = P(F|E)P(E)$ (5.4)

- $P(E \cap F) = P(EF)$

$$= P(E)P(F) \text{ for independence between } E \text{ and } F \qquad (5.5)$$

- $P(E) = P(E|F)P(F) + P(E|\bar{F})P(\bar{F}), F + \bar{F} = T$ (5.6)

We explore one additional example that involves a combination of Equations 5.4 and 5.6. For this example, we add a third urn to the previous problems and arrange them as follows:

- two white balls in urn X
- a white ball and a black ball in urn Y
- two black balls in urn Z

        Overview of Probability Concepts   Chap. 5

We select an urn at random and draw a ball from it. We find it to be a black ball. We then ask the question: What is the probability that the next draw from the same urn will also be a black ball?

We are looking for $P(B2|B1)$, which will be called the probability that we draw a black ball on the second draw, given that we have drawn a black ball on the first draw. From Equation 5.4, we have

$$P(B2|B1) = P(B2, B1)/P(B1) \qquad (5.7)$$

The joint probability $P(B2, B1)$ is simply $\frac{1}{3}$ since we are choosing the urns at random. The denominator of the right side of Equation (5.7) is derivable from Equation (5.6) as follows:

$$P(B1) = P(B1|\text{urn } X)P(\text{urn } X) + P(B1|\text{urn } Y)P(\text{urn } Y) + P(B1|\text{urn } Z)P(\text{urn } Z) \qquad (5.8)$$

The three terms in Equation 5.8 can readily be calculated as

$$P(B1) = (0)(1/3) + (1/2)(1/3) + (1)(1/3)$$

which has a value of $\frac{1}{2}$. Thus the final result numerically is

$$P(B2|B1) = (1/3)/(1/2) = 2/3$$

This result may be counter intuitive since one might argue that having drawn a black ball on the first draw, we can then reject urn X as the urn from which we are drawing. Since there are two remaining urns (Y and Z) that we may be drawing from, the probability of being in either is equal and thus is $\frac{1}{2}$. This turns out to be incorrect as shown by the previous computation. In effect, the way to look at this problem, having drawn the first ball, is to pool the remaining three balls which must be two black balls and a white ball. From this correct point of view it is easily seen that the chance of drawing another black ball is $\frac{2}{3}$.

5.2.4 Risk-Analysis Example

Using the relationships discussed, we can begin to explore a simple example of a risk analysis. Safeguarding nuclear materials and facilities is a major problem that continues to face the nuclear regulatory community. They have framed this issue in terms of a risk assessment with the following basic relationship [3]:

$$R = \sum_{i, j} F(i, j)P(i, j)C(i) \qquad (5.9)$$

where

R is the societal risk associated with a failure to maintain safeguards

F is the frequency of an attempt to penetrate safeguards, P is the probability of successful penetration, given an attempt to do so

C is the consequence to society of a successful breach of safeguards

The indices i and j refer to an event sought by an adversary and an adversary-action sequence, respectively. This relationship demonstrates that

a. risk assessments in the real world can be and often are placed in a probabilistic context

b. risk assessments can be and often are structured in terms of a string of conditional probabilities

c. there is often a great deal of analysis required in order to calculate the components of a simple probability expression such as Equation 5.9 shows

The Nuclear Regulatory Commission has for years had a series of programs underway in order to understand and be able to compute the various elements of the Equation 5.9 (see Reference 3). Chapter 13 provides a further discussion of the subject of risk analysis.

5.2.5 Reliability Example

A simple reliability example will also demonstrate the use of probability theory in a systems engineering context. We will take two pieces of equipment, shown in Figure 5.5, characterized by their reliabilities $R(1)$ and $R(2)$. These reliabilities can be interpreted as the probability that the respective equipment is operating properly. In order for the function that they are performing to be operative, both equipments need to be working. If the equipments are independent, then the probability that both are working properly can be expressed simply as a joint probability, namely

$$P(E1, E2) = P(E1)P(E2) = R(1)R(2) \tag{5.9}$$

If we have a configuration in which one equipment *or* the other needs to be operating (rather than both), we concern ourselves with the *union* situation (see aforemen-

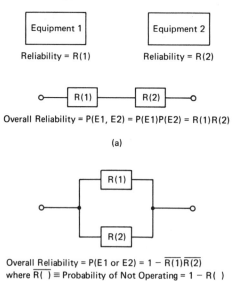

Overall Reliability = P(E1, E2) = P(E1)P(E2) = R(1)R(2)

(a)

Overall Reliability = P(E1 or E2) = 1 − $\overline{R(1)}\,\overline{R(2)}$
where $\overline{R(\ )}$ ≡ Probability of Not Operating = 1 − R( )

(b)

Figure 5.5 Elementary reliability calculations: (a) reliability block diagram for both equipments necessary, (b) reliability block diagram for one of the two equipments necessary.

tioned discussion) rather than the *intersection* situation. Then the only case that is harmful is one where both equipments are simultaneously inoperative. The probability of having both equipments down is simply

$$[1 - R(1)][1 - R(2)] = [(\text{not } R(1)][\text{not } R(2)] = \overline{R(1)} \; \overline{R(2)}$$

All other events have at least one of the equipments up so that the probability of this situation is

$$P(E1 \text{ or } E2) = 1 - \overline{R(1)} \; \overline{R(2)} \qquad (5.10)$$

Additional examples of reliability computations are explored later in this chapter and also in Chapter 15.

5.3 DISCRETE DISTRIBUTIONS

5.3.1 Uniform Distribution

The uniform discrete distribution has already been shown by example in the discussion regarding the tossing of cubes. In all of those cases, the probability was the same and equal to one sixth for each of the sides of the cube. When all the probabilities deal with discrete events and are equal, we are dealing with a discrete uniform distribution. We will represent this distribution as shown in Figure 5.6 with specific reference to the cube-tossing situation. A similar distribution with probabilities of one half derives from the matter of flipping an unbiased coin.

We often wish to examine the probability of obtaining a result greater than or less than some amount. For example, in the toss of the conventional cube in Figure 5.1 we may ask about the probability of throwing a three or greater, or a five or less. This is the *or* or *conjunctive* statement that we examined earlier where we found that for mutually exclusive events we were able to simply add the individual probability components. This process has been formalized so as to produce a cumulative-probability distribution function (CDF). The CDF for the cube tossing situation is likewise shown in Figure 5.6. Note that, in general, it displays the probability of having a result equal to or less than the abcissa value.

5.3.2 Binomial Distribution

The binomial distribution is a discrete distribution that is concerned with repeated trials in which each trial is independent and has a probability of success equal to p and a probability of failure equal to q, such that $(p + q) = 1$, as expected. The binomial can therefore be seen to apply directly to the earlier examples of tossing a cube. The discrete probability function for the binomial is given as

$$p(x) = \binom{n}{x} p^x q^{n-x} \qquad \text{for } x = 0, 1, 2, \ldots, n$$
$$= 0 \qquad \text{otherwise} \qquad (5.11)$$

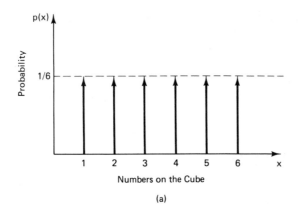

(a)

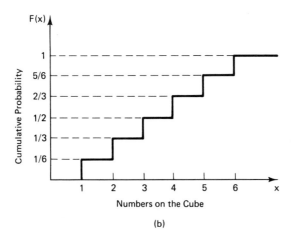

(b)

Figure 5.6 Uniform distributions for the throwing of a cube: (a) discrete probability distribution function, (b) discrete cumulative distribution function.

The CDF is represented as

$$F(x) = \sum \binom{n}{x} p^x q^{n-x} \tag{5.12}$$

and we note that when we sum over all discrete values we obtain the result

$$\sum_{x=0}^{n} \binom{n}{x} p^x q^{n-x} = (p + q)^n = 1 \tag{5.13}$$

With the discrete binomial we are able to examine directly the probability of obtaining exactly x successes in n trials. With its CDF we compute the probability of x or fewer successes in n trials.

The binomial distribution is often used in connection with a variety of engineering problems. As an example, we may be receiving a message of some number of bits of digital information and, assuming independence, wish to calculate the probability of error or of no error. To illustrate, assume that we are receiving characters of eight data bits and the probability of error for each of the bits is estimated at

0.01 (which would be high for a normal communications channel, but will be useful for this example). The binomial distribution and its tabulations can then be used to determine the probability that the entire character is received without error. Formally, this would be expressed as

$$p(8) = \binom{8}{8}(0.99)^8(0.01)^0 = (0.99)^8 = 0.923$$

If we use parity bits to detect and correct single errors, we can use the binomial to see what the chances are that we might have more than one error, which would not be found or corrected by the parity bits. In the same way, we can also combine characters to make words and combine the words to constitute typical messages and then calculate the probabilities of error-free message transmission and receipt.

5.3.3 Poisson Distribution

This distribution derives from situations, among others, in which there is a demand for and a response to service. It has been applied to such situations as

- airplanes arriving at a certain rate at an airport
- providing service lines at a telephone exchange
- electrons being discharged from a cathode in a vacuum tube
- traffic movements through a toll point
- the appearance of defective parts being produced by an assembly line

The Poisson distribution specifically allows one to compute the probability that exactly x (discrete) events occur within a defined period of time or space by means of the formula

$$p(x) = \frac{(ut)^x \exp(-ut)}{x!} \tag{5.14}$$

where $x = 0, 1, 2, \ldots$, and u is the rate in time or space at which events are occurring. The t parameter is the time (or space) one is considering for the problem at hand.

The Poisson can be shown to be a limiting case of the binomial, in particular when the binomial is very far from being bell-shaped, which will occur when the probability of success parameter p of the binomial is no greater than, for example, one tenth. We also would require that n be relatively large, say $n > 50$, and that the parameter represented by the product (ut) be less than of the order of ten.

To illustrate the possible use of the Poisson distribution, let us assume that in this book of roughly 550 pages, somehow a total of 275 typographical errors were not found and remain in this text. We now pose the question: what is the probability that two pages selected at random will not contain any typographical errors? Under the consideration of the use of the poisson distribution, we first calculate the rate of error, $u = 275/550$, which yields the average value of $\frac{1}{2}$ error per page. For two pages, we then have $t = 2$. Thus the Poisson becomes

$$p(0) = \frac{(1)^0 \exp(-1)}{0!} = \frac{1}{e}$$

(Note that 0! is defined as unity.)

It is left as a challenge to the reader to find the typographical errors, if indeed they exist!

5.4 CONTINUOUS DISTRIBUTIONS

5.4.1 Density and Cumulative Distribution Functions

Continuous distributions deal with probabilities in a manner somewhat different from discrete distributions. The fact of the infinite density of numbers in any continuous interval leads us to define probability elements by looking at a normalization and limiting process. First we consider the intuitive meaning of a probability as a ratio of successful events to total possible events, as with the discrete case. We can interpret this as a frequency count or ratio and plot the result as a bar graph, as shown in Figure 5.7. We can also simultaneously shrink the width of the interval (Δx) under consideration. In constructing what we call a *density* function for the continuous case, however, we also divide or normalize the frequency ratio by the size of the interval, Δx. The final step is to take the limit of this ratio as Δx becomes dx. The resultant construction is defined as the probability-density function, and is illustrated in Figure 5.7.

The calculation of probabilities then becomes a matter of integrating the probability-density function between certain limits. The multiplication of the probability-density function by the element dx under the integral then allows us to remove the prior normalization and add up probability elements for the application at hand. The integration procedure yields the CDF as shown in Figure 5.7. We explore now a variety of continuous distributions and examples of their use.

5.4.2 Uniform Distribution

The uniform distribution has the same essential form in the continuous case as it does in the discrete case. It can be expressed by the formula

$$p(x) = \frac{1}{b - a} \qquad \text{for } a < x < b$$
$$= 0 \text{ elsewhere} \tag{5.15}$$

Its CDF will take the form of a ramp function as the area under the uniform density function is integrated. A uniform distribution would be generated, for example, by spinning a pointer so that it is equally likely to stop at any point along the circumference of the circle. Random variables expressed in polar coordinates often define the angle in terms of a uniform distribution.

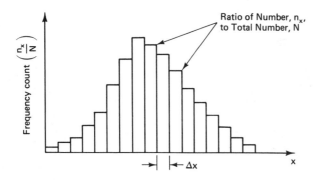

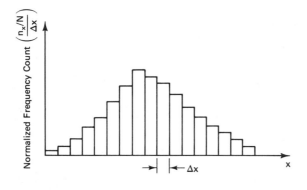

$$p(x) = \lim_{\substack{\Delta x \to 0 \\ N \to \infty}} \frac{n_x/\Delta x}{N}$$

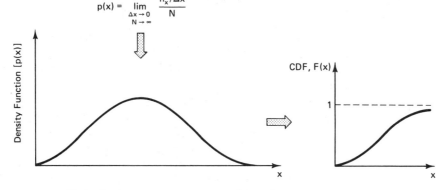

Figure 5.7 Continuous-distribution-(density) function construction.

5.4.3 Normal Distribution

The normal or Gaussian distribution is the most commonly used in the fields of probability and statistics. The graph of the normal distribution is a symmetric bell-shaped curve, as shown in Figure 5.8, and it is given by the relationship

$$p(x) = \frac{1}{\sigma\sqrt{2\pi}} \exp\left(\frac{-x^2}{2\sigma^2}\right) \tag{5.16}$$

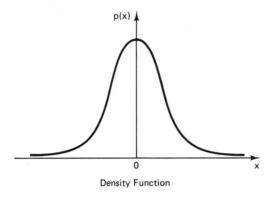

p(x)

0   x

Density Function

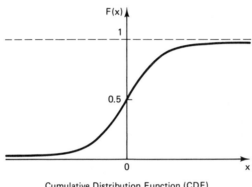

F(x)

1

0.5

0   x

Cumulative Distribution Function (CDF)

$$F(x) = \int p(x)\,dx$$

Figure 5.8 The normal (Gaussian) distribution.

The CDF is illustrated in Figure 5.8 and is obtained by integrating the density function. The parameter σ is called the standard deviation of the distribution and will be explained further in Section 5.5.4.

Experience with the normal distribution is likely in terms of the educational procedure of "grading on a curve" where the expectation is that the grades for a large population of students will follow this type of random distribution. The normal has also been found to represent a limiting distribution for a variety of processes and it is used extensively, as such. More importantly, it has been shown that if we are taking the sum of a large number of random variables, whatever their individual distributions, the sum will tend to be normally distributed. This is called the *central limit theorem*.

For the moment, we can examine the portion of the distribution that lies between ± 1 and ± 2 sigma values, that is, one and two standard deviations from the line of symmetry of the distribution, otherwise known as the mean value of the distribution. In order to do this we integrate the density function between those limits, as illustrated in Figure 5.9, and obtain the following results:

$$P(-\sigma < x < \sigma) = \int_{-\sigma}^{+\sigma} p(x)\,dx = 0.6827$$

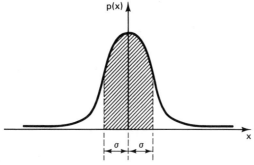

Area Between ±σ for Normal Density Function

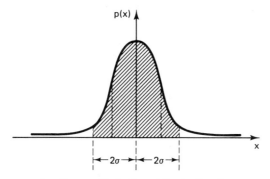

Area Between ±2σ for Normal Density Function

Figure 5.9 Integration of the normal distribution.

$$P(-2\sigma < x < 2\sigma) = \int_{-2\sigma}^{+2\sigma} p(x)\,dx = 0.9545$$

Thus we find that 68% and 95% of the normal distribution lie between the plus and minus one and two sigma values, respectively. We note further that the normal distribution is not directly integrable so that in order to obtain the values for the integrations we resort to widely published tables [4] for this distribution. Additional examples involving this important distribution are provided later in this chapter.

5.4.4 Exponential Distribution

It has been shown [2] that if we are examining a series of events that occur in a way that satisfies the Poisson distribution, then the waiting time to any selected event is distributed according to the gamma probability law. A special case of that law is the exponential, where we are concerned with the waiting time to the first event. We will therefore interpret the variable x as time-oriented and write the exponential distribution (density) as

$$p(x) = u \exp(-ux) \qquad \text{for } x \geq 0$$
$$= 0 \qquad\qquad \text{for } x < 0$$
(5.17)

A plot of the exponential density function is shown in Figure 5.10(a).

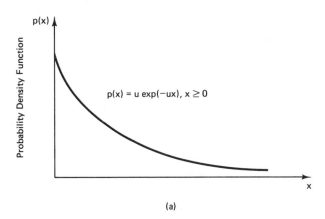

(a)

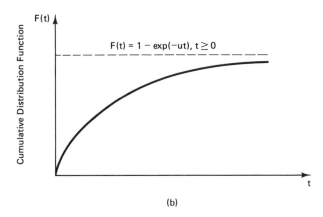

(b)

Figure 5.10 The exponential distribution.

The exponential distribution is widely used in reliability analysis in the aforementioned context. If the parameter u in Equation 5.17 is interpreted as the failure rate for a certain piece of equipment, and we can consider this failure rate to be a constant, then the exponential distribution allows us to calculate the probability of failure as a function of time. The reciprocal of the failure rate u is the mean time between failures, sometimes written as MTBF.

As a density function, this distribution must be integrated to obtain the CDF, as shown in Figure 5.10(b), which in turn will determine the desired probability. This integration process yields

$$F(t) = \int_0^t u \exp(-ux)\, dx$$

$$F(t) = 1 - \exp(-ut) \tag{5.18}$$

and the reliability, $R(t)$, is

$$R(t) = 1 - F(t) = \exp(-ut)$$

In terms of reliability theory, the previous relationship is the probability that a piece of equipment will survive without failure to time t, given that it exhibits a constant failure rate equal to the value u. With this specific characterization, then, we can re-iterate the series and parallel reliability block diagrams of Figure 5.5 as developed in Figure 5.11. Here we see the reliabilities of two equipments. Figure 5.11(a) shows a series arrangement and the overall reliability as the product of the two exponentials. From Figure 5.5(b) we recall that the overall reliability of a parallel reliability block diagram yielded the result

$$\text{Reliability} = 1 - \overline{R(1)}\,\overline{R(2)} \qquad (5.19)$$

With the exponential this becomes

$$
\begin{aligned}
R &= 1 - [1 - \exp(-u_1 t)][1 - \exp(-u_2 t)] \\
&= \exp(-u_1 t) + \exp(-u_2 t) - \exp[-(u_1 + u_2)t]
\end{aligned} \qquad (5.20)
$$

This can be generalized to a series of blocks in both series and parallel arrangements (see problems at end of this chapter). The latter are also called redundant or cross-strapped configurations and are examined further in Chapter 15.

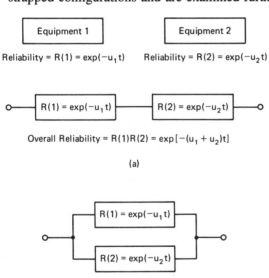

(a)

(b)

Figure 5.11 Reliability computations with the exponential distribution: (a) series reliability block diagram (both equipments necessary), (b) parallel reliability block diagram (only one equipment necessary).

5.4.5 Weibull Distribution

The Weibull distribution is included here because it represents a generalization of the exponential distribution and it is therefore widely used in dealing with reliability problems. Note that the exponential is characterized by a constant failure rate. This is not true for the Weibull so it is possible to us the Weibull where there is wearout and the failure rate follows the so-called *bathtub* type of curve [5,6]. This refers to a high early-failure rate as quality control defects are observed followed by a lower relatively constant failure rate, followed by an increase due to wearout. The density function for the Weibull distribution [6] takes the following form:

$$p(x) = \alpha \lambda x^{\alpha-1} \exp(-\lambda x^{\alpha}) \tag{5.21}$$

Clearly when $\alpha = 1$, this reduces to the exponential case. As a two-parameter distribution (α and λ), the failure behavior of the equipment under consideration can be approximated by selecting values of these two parameters. Whereas the constant failure rate characterization has been appropriate for wide classes of purely electronic equipment, it has not been sufficient for electromechanical equipment that exhibits wearout. In these cases the Weibull distribution has been used for reliability analyses.

5.4.6 Rayleigh Distribution

The Rayleigh distribution is a single-parameter distribution given by the relationship

$$\begin{aligned} p(x) &= \frac{x}{\sigma^2} \exp\left(\frac{-x^2}{2\sigma^2}\right) &&\text{for } x > 0 \\ &= 0 &&\text{for } x < 0 \end{aligned} \tag{5.22}$$

The Rayleigh distribution is found, for example, in analyzing the clutter observed in the use of radar systems. Clutter is represented by unwanted echo-signal returns that come from a large number of individual scatterers of about the same size. These scatterers produce individual echoes whose phase relationship is random and in the aggregate they produce an amplitude distribution that is very well represented by the Rayleigh distribution.

Unlike the normal distribution, the Rayleigh is directly integrable. The CDF is found directly by integration as

$$F(x) = 1 - \exp\left(\frac{-x^2}{2\sigma^2}\right) \tag{5.23}$$

It will also be shown more formally later in this chapter that a Rayleigh distribution is obtained when transforming from rectangular to polar coordinates wherein the rectangular variable distributions are independent and normally distributed. This representation is the basis for the clutter model described qualitatively in this subsection.

5.5 AVERAGES

Average values in the context of probability theory bear a very close relationship to one's intuitive sense of calculating an average. This can be demonstrated by means of a simple example. Returning to the case of throwing a cube, now a conventional die with the numbers one through six, we can imagine the tossing of such a die a large number of times. Since the die is assumed to be unbiased, we may expect that each of the sides will come up an equal number of times and approximately equal to

       Overview of Probability Concepts   Chap. 5

one sixth of the total number of throws. This reflects the case of the uniform discrete probability function discussed in Section 5.2.

If each time we throw the die we record the value observed and finally add all these values together and divide by the number of throws, we would be calculating an average in the conventional sense. The same result is obtained when we weight the probability function by the (discrete) value of all possible observed values, as given by the formula

$$\bar{x} = \sum_{all\,x} xP(x) \tag{5.24}$$

This is called the average value of x and is sometimes also called the expected value of x, written as $E(x)$, or the *first moment* of the given distribution, or its *mean value*. In the example selected, we find this average value as

$$\bar{x} = 1\left(\frac{1}{6}\right) + 2\left(\frac{1}{6}\right) + 3\left(\frac{1}{6}\right) + 4\left(\frac{1}{6}\right) + 5\left(\frac{1}{6}\right) + 6\left(\frac{1}{6}\right) = 3.5$$

We note that there are only six possible values that x can take but that the average or expectd value of x is none of these values and in fact lies in the center of the distribution, that is, midway between the lowest value, one, and the highest value, six. This is characteristic of symmetric distributions, but the point is that for discrete distributions the average or mean value need not be any of the admissable discrete values of the abcissa.

5.5.1 Mean Value for the Binomial

The average or mean value normally can be expressed in terms of the parameters of a given distribution. This can be demonstrated by the mean value of the binomial distribution, as below

$$\bar{x} = \sum_{all\,x} x\binom{n}{x}p^x q^{n-x} = np \tag{5.25}$$

Thus we find that for the binomial the mean value is simply the product of the number of trials and the probability of success on each trial, also interpretable as the expected number of successes in n trials. If we then toss a coin 1000 times and call a head a success, then we can expect a total of $(1000)(\frac{1}{2}) = 500$ successes, which is in consonance with our intuition in the matter. Using the basic definition, we can also find the average or mean value for the Poisson distribution, namely (ut), which is left as a problem at the end of the chapter.

5.5.2 Mean Value for the Normal Distribution

Finding averages for the continuous case follows the same basic principle as for discrete variables. In this situation we weight the probability density function by the value of x yielding the relationship

$$\bar{x} = \int_{-\infty}^{\infty} xp(x)\,dx \tag{5.26}$$

The general form of the density function for the normal distribution includes the mean value, designated by the parameter m

$$p(x) = \frac{1}{\sigma\sqrt{2\pi}} \exp\left[\frac{-(x-m)^2}{2\sigma^2}\right]$$ (5.27)

In this more general form, shown in Figure 5.12, the distribution is displaced to the right such that it is symmetric about the mean value, m. We find the mean value formally by substitution into Equation 5.26. Proof of this result is left as a problem cited at the end of this chapter.

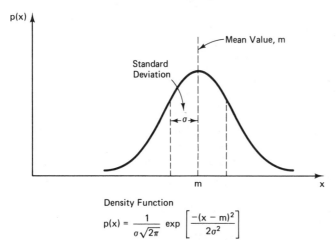

Density Function

$$p(x) = \frac{1}{\sigma\sqrt{2\pi}} \exp\left[\frac{-(x-m)^2}{2\sigma^2}\right]$$

Figure 5.12 Generalized normal distribution.

5.5.3 Higher Order Moments

Given the description in Section 5.5.2 of the average value or the first moment of a distribution, we may find second and higher order moments by the following relationships:

$$\left.\begin{aligned}
m_2 &= E(x^2) = \int x^2 p(x)\, dx \\
m_3 &= E(x^3) = \int x^3 p(x)\, dx \\
&\;\;\vdots \\
m_n &= E(x^n) = \int x^n p(x)\, dx
\end{aligned}\right\}$$ (5.28)

Note that these are expressed in Equation 5.28 as expected values. They can also be viewed as analogous to finding moments of bodies in the field of physics.

5.5.4 Variance and Standard Deviation

We used the parameter sigma, σ, in expressing several of the previous distributions, including the normal distribution. This parameter is known as the *standard deviation* of the distribution and its square is the *variance*. It is a measure of the spread of a distribution about its mean value.

Formally, in order to calculate a variance, we take the difference between values of the distribution and its mean value, square it, and then find the expected value. The general formula for the variance in the continuous case is thus given as

$$\sigma^2 = E[(x - m)^2] = \int (x - m)^2 p(x) \, dx \tag{5.29}$$

An analogous form is apparent for the discrete case.

The variance can also be expressed in terms of the first and second moments (note problems at end of chapter) of a distribution by the relationship

$$\sigma^2 = m_2 - m_1^2 = E(x^2) - E^2(x) \tag{5.30}$$

Thus at least one use of the general relationship for the moments of a distribution is to be able to calculate more simply its variance and standard deviation.

The standard deviation can be thought of as a measure of the error represented by a given distribution. When the term *root-mean-square* is used it likewise is a representation of the standard deviation parameter. Therefore, it is extremely important in systems engineering problems and issues.

5.5.5 Signal-to-Noise Example

As an example of signal-to-noise analysis, assume that we are sending a digital signal as a series of *on–off* pulses, and the *on* pulse has a magnitude V. This signal may be corrupted by noise, which we will assume to be additive Gaussian noise with variance equal to $\sigma^2 = N_o$. In such a situation, we represent the distribution of signal plus noise as

$$p_{S+N}(x) = \frac{1}{\sqrt{2\pi N_o}} \exp\left(\frac{-(x - V)^2}{2N_o}\right) \tag{5.31}$$

and the signal-to-noise (power) ratio as V^2/N_o. This becomes the crucial connection between the signal-to-noise ratio concept in engineering and the notions of mean values and variances in probability theory. It also has immediate application to all situations in which we are attempting to detect signals in the presence of noise. Thus if the above *on* pulse is a signal in a radar system, we can immediately compute the detection and false-alarm probabilities for a simple threshold-detection scheme.

The situation is represented in the two parts of Figure 5.13. The probability density function of the Gaussian noise is shown both with and without the presence of a signal. If the threshold for detection is set at the value T, then the probabilities of detection and false alarm, are respectively

$$P(det) = \int_T^\infty \frac{1}{\sqrt{2\pi N_o}} \exp\left(\frac{-(x - V)^2}{2N_o}\right) dx \tag{5.32}$$

$$P(fa) = \int_T^\infty \frac{1}{\sqrt{2\pi N_o}} \exp\left(\frac{-x^2}{2N_o}\right) dx \tag{5.33}$$

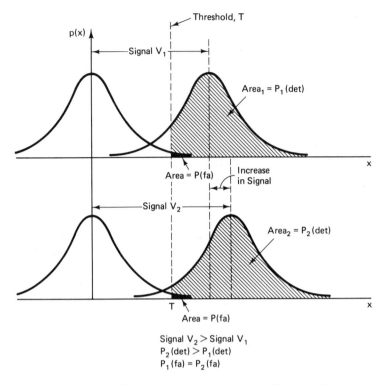

Figure 5.13 Detection and false-alarm probability examples.

In the case of detection, we calculate the probability of the signal plus noise exceeding threshold (Equation 5.32). For false alarms, it is necessary that the noise alone exceed the threshold (Equation 5.33). In the top portion of Figure 5.13, the areas under the curves show the detection and false alarm probabilities. In the bottom portion of the figure, the signal is increased thereby increasing the signal-to-noise ratio. It can be seen from the diagram that there is no effect on the false-alarm probability. However, the distribution of signal-plus-noise moves to the right as the signal is increased. Thus the area under the curve to the right of the threshold increases, which represents an increase in the detection probability. This is how an increase in signal-to-noise manifests itself in terms of a typical pulse-detection problem.

5.5.6 Table of Means and Variances

In Table 5.1 we list the means and variances for some of the various distributions covered in this chapter as a summary of material covered with respect to averages for discrete and continuous distributions. Other texts [2] provide more complete tabulations of these and other parameters.

TABLE 5.1 SELECTED DISTRIBUTIONS AND THEIR MEANS AND VARIANCES

Probability Law	Distribution Function	Mean $m = E(x)$	Variance σ^2
Discrete cases—mass function			
Bernoulli	$p(x) = p \quad x = 1$ $ = q \quad x = 0$ $ = 0 \quad \text{otherwise}$	p	pq
Binomial	$p(x) = \binom{n}{x} p^x q^{n-x}$ for $x = 0, 1, 2, \ldots, n$	np	npq
Poisson	$p(x) = \dfrac{e^{-ut}(ut)^x}{x!}$ for $x = 1, 2, \ldots$	ut	ut
Continuous cases—density function			
Uniform	$p(x) = \dfrac{1}{b-a}$ $a < x < b$	$\dfrac{a+b}{2}$	$\dfrac{(b-a)^2}{12}$
Generalized Normal (Gaussian)	$p(x) = \dfrac{1}{\sigma\sqrt{2\pi}} \exp\left[-\dfrac{(x-m)^2}{2\sigma^2}\right]$	m	σ^2
Standard Normal	$p(x) = \dfrac{1}{\sqrt{2\pi}} \exp\left[-\dfrac{x^2}{2}\right]$	0	1
Exponential	$p(x) = ue^{-ux}$ $x > 0$	$\dfrac{1}{u}$	$\dfrac{1}{u^2}$
Rayleigh	$p(x) = \dfrac{x}{\sigma^2} \exp\left[-\dfrac{x^2}{2\sigma^2}\right]$ $x > 0$	$\sigma\sqrt{\dfrac{\pi}{2}}$	$\dfrac{\sigma^2(4-\pi)}{2}$

5.6 SUMS OF RANDOM VARIABLES

For sums of random variables, we state here three important relationships for independent random variables

1. the mean value of a sum is equal to the sum of the mean values
2. the variance of a sum is equal to the sum of the variances, and
3. the distribution of a sum is equal to the convolution of the distributions.

We demonstrate (but do not prove) these relationships by means of an example.

We return to the previously discussed situation of throwing a pair of dice. The throw of each die is considered a random process and when we add the values of the results we are concerned with the sum of two independent random variables. The probability distribution for each of the dice is shown in Figure 5.14(a). Figure 5.14(b) shows all possible values of the sum of these two random variables. The probability of obtaining the value in any one cell of this two-way table is 1/36 by the product rule. However, since several of the cells yield the same sum, the discrete probability distribution is not uniform but is as tabulated next and plotted in Figure 5.14(c)

$$P(2) = \frac{1}{36}$$

$$P(3) = \frac{2}{36} = \frac{1}{18}$$

$$P(4) = \frac{3}{36} = \frac{1}{12}$$

$$P(5) = \frac{4}{36} = \frac{1}{9}$$

$$P(6) = \frac{5}{36}$$

$$P(7) = \frac{6}{36} = \frac{1}{6} \quad \text{(mean value point)}$$

$$P(8) = \frac{5}{36}$$

$$P(9) = \frac{4}{36} = \frac{1}{9}$$

$$P(10) = \frac{3}{36} = \frac{1}{12}$$

$$P(11) = \frac{2}{36} = \frac{1}{18}$$

$$P(12) = \frac{1}{36}$$

We have seen before that the mean value for an individual die is simply 3.5. Using the aforementioned distribution of the sum we see that the mean value of the sum is clearly the sum of the individual mean values, or 7. The variance for each separate die is calculated as

$$\sigma^2 = \frac{1}{6}(-2.5)^2 + \frac{1}{6}(-1.5)^2 + \frac{1}{6}(-0.5)^2 + \frac{1}{6}(0.5)^2 + \frac{1}{6}(1.5)^2 + \frac{1}{6}(2.5)^2 = 2.9$$

The variance of the sum using this distribution can be shown to be the sum of each variance or 5.8 (see end-of-chapter problems). By direct observation, the distribution of the sum, as developed in Figure 5.14, is triangular. This results from the convolution of the two uniform distributions. This process also provides some insight into the central limit theorem previously mentioned. As additional distributions are convolved we can imagine how the resultant distribution might approach the normal distribution.

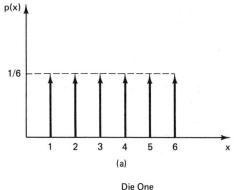

(a)

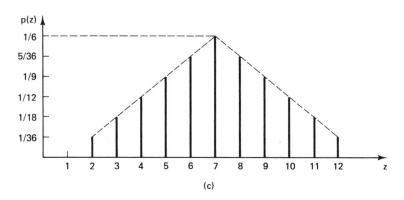

(b)

(c)

Figure 5.14 Distribution of the sum of two uniform distributions.

In order to convolve two distributions, one can enumerate all possibilities, as shown in Figure 5.14, and then plot the resultant distribution. A more formal way to do this is to use the convolution relationship which is written as Equation 5.34 for the continuous case.

$$p(x + y) = p(z) = \int_{-\infty}^{+\infty} g(x)h(z - x)dx \qquad (5.34)$$

It is noted that this is precisely the same relationship used in engineering to calculate the output signal of a linear system when both the input signal and transfer function of the system are known [7]. This is a very powerful and often-used relationship in the fields of probability theory and engineering.

5.6.1 Error Analysis Example

We have previously alluded to the matter of requirements analysis and allocation as activities of systems engineering. For a large-scale system we often have a need to consider the requirement that the overall error be no greater than some specified amount. This sets an upper bound for the total error and the issue of requirements allocation is to suballocate maximum errors to phenomena that contribute to the overall error. In the absence of a better model, it is sometimes assumed that the overall error is a combination of the component errors, given as

$$\sigma_T^2 = a_1^2 \sigma_1^2 + a_2^2 \sigma_2^2 + \ldots + a_n^2 \sigma_n^2 \tag{5.35}$$

If we know the overall error bound and can estimate the coefficient values, then bounds can be set for each of the individual errors. As this error equation shows, this process does not provide unique results unless additional constraints are placed upon the relationship. This is often done using engineering judgment and knowledge as to the expected hardware and software implementation of the individual error sources. In any case, the individual error sources are often expressed as *one sigma* standard deviation values and for independent additive error sources, Equation 5.35 expresses the overall error.

5.7 FUNCTIONS OF RANDOM VARIABLES

A very important situation is reflected by the processing of a random variable in which some transformation is effected. In mathematical terms, this may be viewed as a transformation of variables through some functional relationship $y = f(x)$, where x is the original variable and y is the transformed variable. If the variable x has some distribution, the question then is: given that the function $f()$ is known, what is the resulting distribution of the random variable y? The situation is depicted in Figure 5.15. A random variable is fed into a black box that processes this variable through the functional equation $y = f(x)$. In this case, the distribution of y can be found by the following formula:

$$g(y) = p[f^{-1}(y)] \left| \frac{df^{-1}(y)}{dy} \right| = p[f^{-1}(y)] \left| J\left(\frac{Y}{X}\right) \right|^{-1} \tag{5.36}$$

The notation, $J(Y/X)$, represents the Jacobian of the transformation, which is a determinant of derivatives as shown in Figure 5.15. In short, it is not sufficient to simply make the transformation by substitution; the Jacobian of the transformation must be taken into account as reflected in Equation 5.36.

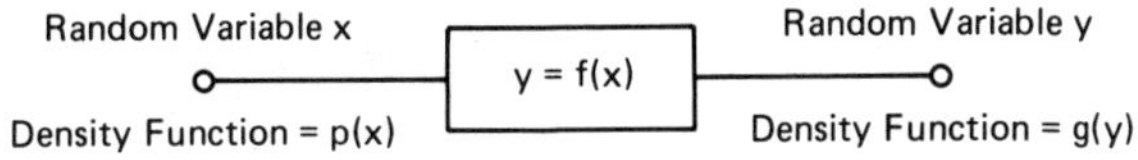

$$g(y) = p[f^{-1}(y)] \left| \frac{df^{-1}(y)}{dy} \right|$$

For
$$y_1 = f_1(x_1, x_2, \ldots x_n)$$
$$y_2 = f_2(x_1, x_2, \ldots x_n)$$
$$\vdots$$
$$y_n = f_n(x_1, x_2, \ldots x_n)$$

$$\left| \frac{df^{-1}(y)}{dy} \right| \quad \text{becomes} \quad \left| J\left(\frac{Y}{X}\right) \right|^{-1} = \begin{vmatrix} \frac{\partial f_1}{\partial x_1} & \frac{\partial f_1}{\partial x_2} & \cdots & \frac{\partial f_1}{\partial x_n} \\ \vdots & & & \\ \frac{\partial f_n}{\partial x_1} & \frac{\partial f_n}{\partial x_2} & \cdots & \frac{\partial f_n}{\partial x_n} \end{vmatrix}^{-1}$$

and $J\left(\frac{Y}{X}\right) \equiv$ Jacobian of the Transformation

Figure 5.15 Transformations of random variables.

5.7.1 Standard Normal Distribution

The matter of transformations or functions of random variables may be illustrated by converting from the generalized normal distribution to the so-called standard normal. This is done by defining the functional transformation $y = f(x)$ as

$$y = \frac{x - m}{\sigma} \tag{5.37}$$

By substitution into Equation 5.36 we have

$$g(y) = \frac{1}{\sqrt{2\pi}} \exp\left(\frac{-y^2}{2}\right) \tag{5.38}$$

This distribution is normal but has a mean value of zero and a standard deviation of unity, otherwise known as the standard normal distribution. It is this distribution that is found in the look-up tables for the normal distribution [2,4].

The transformation of variables shown by Equation 5.37 will lead to a mean value of zero and a standard deviation of unity even if we are not dealing with the normal distribution. This can be seen in a more general sense by considering the transformation

$$y = \frac{x - E(x)}{\sigma(x)}$$

in which case

$$E(y) = \frac{1}{\sigma(x)}[E(x) - E(x)] = 0$$

and

$$\sigma^2(y) = \frac{1}{\sigma^2(x)}[\sigma^2(x) + \text{variance of } E(x)]$$

$$\sigma^2 = 1 + 0 = 1$$

Thus it is noted that this transformation produces a zero mean and unity variance no matter what the distributions in question are.

5.8 JOINT DISTRIBUTIONS

Joint distributions can be thought of in terms of two or more random variables that may or may not be related in some fashion to one another. In that sense we are concerned with the distribution of X and Y (or additional variables) in much the same way that we considered Prob(X,Y), or Prob(XY), or Prob(X and Y) previously. We can now describe the formal distributions of the variables X and Y and how they might interrelate. This can be illustrated by means of a simple example.

If we are shooting at a bull's-eye target, we can imagine a distribution of results in two dimensions. If we draw rectangular coordinates on the bull's-eye we can see each shot as having both an x and a y position. The aggregate positions of the shots represent some joint distribution viewed in x–y space. In the same way, we can see the concentric rings of the bull's-eye as a polar coordinate system such that each shot has a radius from the center and an angle from some zero-angle position (normally the positive x-axis). The latter can be called an *r–theta polar coordinate system* and the random variables would be designated as r and θ. The distribution represented by the x–y system is generally different from that shown by the r–θ system, even for precisely the same positions of the shots.

5.8.1 Rectangular and Polar Coordinate Systems

Let us assume that we are in fact shooting at a bull's-eye and that we can represent the x and y distributions as normal and independent with equal variances and zero mean values. This type of joint distribution is illustrated in Figure 5.16 by a three-dimensional plot. The joint distribution (density) in rectangular coordinates would thus be represented as

$$p(x, y) = \frac{1}{2\pi\sigma(x)\sigma(y)} \exp\left[-\frac{x^2}{2\sigma^2(x)} - \frac{y^2}{2\sigma^2(y)}\right] \tag{5.39}$$

Since the distributions have been postulated as independent, the joint distribution

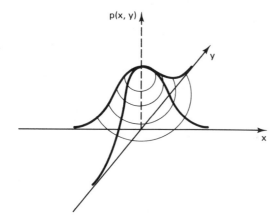

Joint Normal-Distribution (Density) Function

⬇

Transformation To Polar Coordinates

$$g(r, \theta) = \frac{r}{2\pi\sigma^2} \exp\left(\frac{-r^2}{2\sigma^2}\right) \qquad \text{(Rayleigh)}$$

⬇

Marginal Distribution of Radius, r

$$h(r) = \frac{r}{\sigma^2} \exp\left(\frac{-r^2}{2\sigma^2}\right) \qquad \text{(Rayleigh)}$$

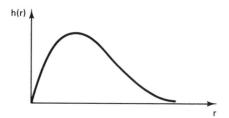

Figure 5.16 Transformation from normal in rectangular coordinates to Rayleigh in polar coordinates.

can be written as a product, namely

$$p(x, y) = p_1(x)p_2(y)$$

$$= \frac{1}{\sigma(x)\sqrt{2\pi}} \exp\left(\frac{-x^2}{2\sigma^2(x)}\right) \frac{1}{\sigma(y)\sqrt{2\pi}} \exp\left(\frac{-y^2}{2\sigma^2(y)}\right)$$

In order to find the corresponding form of these distributions in polar coordinates, we have to make a change of variables in exactly the manner described in Section 5.7 and in Figure 5.15. The transformation equations are the familiar ones

$$\left.\begin{array}{l} r = (x^2 + y^2)^{1/2} \\[2mm] \theta = \tan^{-1}\dfrac{y}{x} \end{array}\right\} \tag{5.40}$$

From Figure 5.15 we can define and calculate the Jacobian of the transformation as

$$J = \begin{vmatrix} \dfrac{\partial r}{\partial x} & \dfrac{\partial r}{\partial y} \\ \dfrac{\partial \theta}{\partial x} & \dfrac{\partial \theta}{\partial y} \end{vmatrix} = (x^2 + y^2)^{-1/2} \tag{5.41}$$

Thus the joint distribution in polar coordinates can be expressed as

$$g(r, \theta) = (x^2 + y^2)^{1/2} p(x, y)]_{r,\theta}$$

By substitution this becomes (assuming $\sigma(x) = \sigma(y) = \sigma$) simply

$$g(r, \theta) = \frac{r}{2\pi\sigma^2} \exp\left(\frac{-r^2}{2\sigma^2}\right) \tag{5.42}$$

This can be seen as the Rayleigh distribution as previously discussed. It can also be shown that the distribution of θ is uniform over the interval from zero to 2π. This accounts for the 2π factor in the denominator. If we integrate Equation 5.42 with respect to θ over the interval from zero to 2π, this factor will be eliminated and we will obtain the Rayleigh distribution in one variable as shown in Section 5.4.6.

Thus we find two interesting results: a transformation of coordinates yields different distributions, and if we start with independent normal distributions in rectangular coordinates we have a Rayleigh distribution in r and a uniform distribution in θ. All of these results are illustrated in Figure 5.16.

In dealing with the aforementioned joint Rayleigh distribution, the notion of integrating out the distribution of θ was considered. This can be made more formal by examining the joint distribution $p(x, y)$. The variables for a general distribution may or may not be independent but it is always possible to determine the unconditional distribution of one of the variables by integrating out the other variable. The remaining distribution is sometimes called the marginal distribution and can be calculated generally as

$$g(x) = \int_{-\infty}^{+\infty} p(x, y) \, dy \tag{5.43}$$

The integration should be carried out in Equation 5.43 over all values of the y variable.

5.8.2 Averages for Joint Distributions

In general terms, we may find the expected value of x and y for joint distributions as

$$E(xy) = \int\int xy \, p(x, y) \, dx \, dy \tag{5.44}$$

This is a simple generalization of the previous result for a single variable distribution and deals with a mean value concept. There is a generalization of the notion of a variance that applies as well for joint distributions that is called the *covariance* and that is given as

$$\text{Cov } (xy) = E\{[x - E(x)][y - E(y)]\}$$
$$= E(xy) - E(x)E(y) \tag{5.45}$$

Further, the variance of the sum of two random variables that are jointly distributed is

$$\text{Var } (x + y) = \text{Var } (x) + \text{Var } (y) + 2\text{Cov } (xy) \tag{5.46}$$

If there is a covariance term other than zero we say that the two jointly distributed random variables are *correlated* and we compute a correlation coefficient, ρ, as

$$\rho(xy) = \frac{\text{Cov } (xy)}{\sigma(x)\sigma(y)} \tag{5.47}$$

This correlation coefficient lies between the values plus and minus one since it is normalized with respect to the product of the individual standard deviations.

Some discussion is in order relative to the concepts of independence and correlation. If two random variables are independent, then the joint distribution (density function) can be partitioned as

$$p(x, y) = g(x)h(y) \tag{5.48}$$

and the expected value is likewise partitionable as

$$E(xy) = E(x)E(y) \tag{5.49}$$

Examining Equation 5.45, we see that this implies that the covariance term becomes zero. Thus if the two variables are independent, the covariance is zero and the two variables are said to be uncorrelated. However, the converse is not necessarily true. That is, if two random variables are uncorrelated, it is not necessarily the case that they are independent (although it is true for the normal distribution). Also, in the special case for which the correlation coefficient is either plus or minus one, the two variables are said to be linearly dependent or correlated.

5.8.3 Correlation Examples

A simple example will illustrate the matter of correlation and independence. Assume that $y = 1/x$ and that $E(x) = E(y) = 1$. In such a case we have

$$\text{Cov } (xy) = E(xy) - E(x)E(y)$$
$$= E(1) - (1)(1) = 1 - 1 = 0$$

Thus in this situation we have uncorrelated variables (covariance is zero) that are dependent upon one another through the functional relationship $y = 1/x$.

We consider here two additional numerical examples that will illustrate the computation of a covariance or correlation coefficient. We will calculate the correlation coefficient for the following two cases in which the joint density functions are described as:

$$\text{Case I: } p(x, y) = \frac{x + y}{12}$$

$$\text{Case II: } p(x, y) = \frac{xy}{4}$$

For both cases, assume that the values of x and y are in the domains

$$0 \leq x \leq 2, \qquad 0 \leq y \leq 2.$$

The basic idea in both situations is to first find the (marginal) distributions x and y, then their first moments, the joint first moment, the covariance, the second moments and standard deviations, and finally the correlation coefficient.

For case I, we therefore find

$$g(x) = \int_0^2 p(x, y) \, dy = \frac{x + 1}{6}$$

and

$$h(y) = \int_0^2 p(x, y) \, dx = \frac{y + 1}{6}$$

By symmetry, it is evident as well that $g(x)$ and $h(y)$ have the same form. We then calculate

$$E(x) = \int_0^2 x \, g(x) \, dx = \frac{7}{9}$$

and

$$E(y) = \int_0^2 y \, h(y) \, dy = \frac{7}{9}$$

The joint moment is found as

$$E(xy) = \int_0^2 \int_0^2 xy \, p(x, y) \, dx \, dy = \frac{8}{9}$$

and the covariance therefore is

$$\text{Cov }(xy) = E(xy) - E(x)E(y) = \frac{23}{81}$$

The individual variances can be found by first finding the second moments for the marginal distributions

$$E(x^2) = \int_0^2 x^2 g(x) \, dx = \frac{10}{9}$$

and the same value is found for $E(y^2)$. Thus

$$\sigma^2(x) = \sigma^2(y) = m_2 - m_1^2 = \frac{41}{81}$$

The correlation coefficient for this case therefore is

$$\frac{\text{Cov }(xy)}{\sigma(x)\sigma(x)} = \frac{23/81}{41/81} = \frac{23}{41}$$

In case II in which $p(x, y) = xy/4$, we note that the joint distribution can be expressed as a product so that we can expect that the two marginal distributions are independent and thus the correlation coefficient (and covariance) is zero. To show this more formally, we will compute the covariance by the same procedure as in case I. The marginal distributions of x and y are found as

$$g(x) = \int_0^2 p(x, y)\, dy = \frac{x}{2}$$

and

$$h(y) = \int_0^2 p(x, y)\, dx = \frac{y}{2}$$

$$E(x) = E(y) = \frac{4}{3}$$

$$E(xy) = \int_0^2 \int_0^2 xy\, p(x, y)\, dx\, dy = \frac{16}{9}$$

$$\text{Cov }(xy) = E(xy) - E(x)E(y) = \frac{16}{9} - \frac{16}{9} = 0$$

In case I and case II it should be verified that the double integral of the joint distribution over the full range of the variables indeed yields the value unity and is thus a proper probability function (see problems P5.10 and P5.11).

The notion of correlation is extremely important in a large number of engineering problems. In general, we are able to detect signals in the presence of noise by means of correlation. This may be done by autocorrelation (i.e., correlating a signal-plus-noise function with itself) or cross-correlation (i.e., correlating a signal-plus-noise function with a locally generated new signal with the same periodicity as the original signal). Several texts are available to explore in detail why and how this works [8, 9].

There is also a correlation function that is defined entirely in the time domain [8, 9] that, under certain conditions (i.e., ergodicity), is equal to (or proportional to) the covariance function. It can also be shown that the autocorrelation function and the power spectrum of a signal are Fourier transforms of one another (the Wiener–Khintchine theorem). Thus we have a linkage that allows one to examine the frequency spectral content of a signal with noise, or noise alone, by starting first with the statistical properties of the function under examination, namely the covariance or correlation function.

5.9 SUMMARY

We noted in Chapter 3, Table 3.1, that various functions have been built into spreadsheet software. Most such software will allow computations of mean values and standard deviations, as dealt with more formally in this chapter. In addition, there are a large number of software packages available that specialize in handling probability and statistics computations. These are examined in Chapter 7 on specialized software, which can hopefully be better understood with the results of this chapter as background.

Throughout this chapter we have alluded to areas of systems engineering in which probability theory may be applied. Several examples were drawn from situations that present themselves in a systems engineering context. In Table 5.2 we cite selected areas of the potential application of probability theory to systems engineering.

TABLE 5.2 SELECTED AREAS OF APPLICATION OF PROBABILITY THEORY
TO SYSTEMS ENGINEERING

- Risk Analysis: Risk expressed as series of conditional probability statements
- Decision Analysis: Decisions framed in form of decision tree with conditional probabilities representing alternative choices along various branches of trees
- Error Analysis and Allocation: Errors described as standard deviations; overall errors computed as weighted sums of either variances or transformations, or both, of variables that describe how errors are propagated
- Requirements Analysis: Some requirements described in terms of errors (see *Error Analysis* entry); some requirements expressed in terms of probabilistic statements such as the bit error rate (BER) of a communications channel, the probability of detection and false-alarm probability of a radar system; the waiting-time distribution for a system of customer service
- Simulation and Modeling: Very often involves Monte Carlo techniques, which require sampling from random distributions of various types; computations and outputs expressed as probabilistic statements such as kill probabilities, probability of waiting less than or equal to some number of minutes; probability of responding to a request for service equal to or less than some number of seconds
- Reliability-Maintainability-Availability (RMA) Assessments: Reliability calculations using the exponential and Weibull distributions
- Test and Evaluation: Testing to verify requirement values that are expressed as probabilistic statements; sampling plans using probability and statistics to verify performance levels in specifications
- Scheduling: PERT charts that include distributions of activity times; stochastic scheduling includes conditional probabilities of selecting various activity paths
- Cost Analysis: Estimation of mean values and variances for aggregated costs for systems; development of cost-estimating relationships (CERs) utilizing regression analysis and principles of probability theory
- Performance Analysis: Performance often expressed in terms of probability specifications (as in Simulation and Modeling entry)

A summary of important distributions together with their mean values and variances has already been given in Table 5.1. In Table 5.3 we list some of the other significant relationships covered in this chapter. These may be used as a point of departure in addressing the problems that follow.

TABLE 5.3 SUMMARY OF SELECTED IMPORTANT RELATIONSHIPS

Probability Concept	Mathematical Relationship
Inclusive Or Probability	$P(E \cup F) = P(E + F) = P(E) + P(F) - P(EF)$ $P(E \cup F) = P(E + F) = P(E) + P(F) \quad$ if $EF = \varnothing$
And Probability	$P(E \cap F) = P(EF) = P(E\mid F)P(F) = P(F\mid E)P(E)$ $P(E \cap F) = P(EF) = P(E)P(F) \quad$ if E and F independent
Unconditional Dependence	$P(E) = P(E\mid F)P(F) + P(E\mid \overline{F})P(\overline{F}), F + \overline{F} = T$
Density Function	$p(x) = \lim_{\substack{\Delta x \to 0 \\ N \to \infty}} \dfrac{n_x/\Delta x}{N}$
Cumulative Distribution Function	$F(x) = \int p(x)\, dx$
First Moment, Mean, m	$\bar{x} = E(x) = \int x\, p(x)\, dx$
Second Moment	$\overline{x^2} = E(x^2) = \int x^2 p(x)\, dx$
nth Moment	$\overline{x^n} = E(x^n) = \int x^n\, p(x)\, dx$
Variance	$\sigma^2 = E[(x - m)^2] = \int (x - m)^2 p(x)\, dx$ $\sigma^2 = E(x^2) - E^2(x)$
Convolution, Density of Sum	$p(x + y) = p(z) = \int g(x)h(z - x)\, dx, \quad z = x + y$
Transformation of Variables	$g(y) = p[f^{-1}(y)]\,\lvert J(Y/X)\rvert^{-1}, \quad y = f(x)$
Marginal Density Function	$g(x) = \int p(x, y)\, dy$
Mean Value of Sum	$E(z) = E(x + y) = E(x) + E(y), \quad z = x + y$
Expected Value, Joint Variables	$E(xy) = \iint xy\, p(x, y)\, dx\, dy$
Covariance, Joint Variables	$\mathrm{Cov}\,(xy) = E\{[x - E(x)][y - E(y)]\} = E(xy) - E(x)E(y)$
Variance of Sum	$\mathrm{Var}\,(x + y) = \mathrm{Var}\,(x) + \mathrm{Var}(y) + 2\,\mathrm{Cov}\,(xy)$ $\left.\begin{array}{l} \mathrm{Var}\,(x + y) = \mathrm{Var}\,(x) + \mathrm{Var}\,(y) \\ \sigma^2(x + y) = \sigma^2(x) + \sigma^2(y) \end{array}\right\}$ if x and y independent
Correlation Coefficient	$\rho = \dfrac{\mathrm{Cov}\,(xy)}{\sigma(x)\sigma(y)}$

REFERENCES

1. Feller, W., *An Introduction To Probability Theory and Its Applications*. Vol. I, New York: John Wiley, 1950.
2. Parzen, E., *Modern Probability Theory and Its Applications*. New York: John Wiley, 1960.
3. Bennett, C. A., W. M. Murphey, and T. S. Sherr, *Societal Risk Approach to Safeguards Design and Evaluation*. ERDA-7, Washington, D.C.: Energy Research and Development Administration, June 1975.
4. *Tables of the Normal Probability Functions*. Applied Mathematics Series 23, National Bureau of Standards, Department of Commerce, Washington, D.C., 1953. See also *Handbook of Mathematical Functions*, Applied Mathematics Series 55, National Bureau of Standards, Department of Commerce, Washington, D.C., June 1964.
5. Bazofsky, I., *Reliability Theory and Practice*. Englewood Cliffs, NJ: Prentice-Hall, 1961.
6. Lloyd, D. K., and M. Lipow, *Reliability: Management, Methods, and Mathematics*. Englewood Cliffs, NJ: Prentice-Hall, 1962.
7. Gardner, M. F., and J. L. Barnes, *Transients In Linear Systems*. Vol. I, New York: John Wiley, 1942.
8. Schwartz, M., *Information Transmission, Modulation and Noise*. New York: McGraw-Hill, 1959.
9. Lee, Y. W., *Statistical Theory of Communication*. New York: John Wiley, 1960.
*10. Cramer, H., *Mathematical Methods of Statistics*. Princeton: Princeton University Press, 1946.
*11. Bowker, A. H., and G. J. Lieberman, *Engineering Statistics*. Englewood Cliffs, NJ: Prentice-Hall, 1972.
*12. Mood, A. M., *Introduction to the Theory of Statistics*. New York: McGraw-Hill, 1950.

PROBLEMS

P5.1. Prove that the mean or expected value for the binomial distribution is equal to np, where n is the total number of trials and p is the probability of success on each trial. Discuss the meaning of this result.

P5.2. Prove that the mean or expected value for the Poisson distribution is equal to ut, where u is the rate of event occurrence and t is the total time (or space) under consideration. Discuss the meaning of this result.

P5.3. Starting with the general form of the normal distribution, show that the mean value is equal to m (see Figure 5.12).

P5.4. Prove that the variance of any distribution can be found as the second moment minus the square of the first moment.

P5.5. Use the relationship in P5.4 to find the variance for the exponential distribution.

P5.6. Prove that for independent random variables

---

*References are relevant to the text but are not specifically cited.

a. the mean value of the sum is equal to the sum of the mean values

b. the variance of the sum is equal to the sum of the variances

P5.7. By using the definition of a variance, compute the numerical variance of the distribution obtained in the throwing of two dice. Confirm that this value is numerically equal to the sum of the variances for the individual dice.

P5.8. Find a table of the standard normal distribution in another text. Find the area under this distribution between

    a. minus and plus 0.5 sigma

    b. minus and plus 1.0 sigma

    c. minus and plus 1.5 sigma

    d. minus and plus 2.0 sigma

    e. minus and plus 2.5 sigma

    f. minus and plus 3.0 sigma

P5.9. Prove that the standard normal distribution is given as

$$p(x) = \frac{1}{\sqrt{2\pi}} \exp\left(\frac{-x^2}{2}\right)$$

starting with the generalized normal distribution, that is, prove Equation 5.38 starting with the generalized normal and the transformation suggested in the text.

P5.10. In the chapter's discussion of correlation, two probability density functions are examined, namely $p(x, y) = (x + y)/12$ and $p(x, y) = xy/4$. Show that these are properly defined density functions by demonstrating that they integrate to the value 1. Fill in all the missing steps in the text for these examples in terms of calculating the correlation coefficients.

P5.11. Assume a probability density function of the form $p(x, y) = x + y$, where both x and y range over values from 0 to 1. Compute the correlation coefficient for this joint distribution. Confirm that this is a properly defined density function by integration to obtain the value 1 over the full range of the two variables.

P5.12. Develop the reliability equations for the exponential distribution for the following series and parallel configurations:

    a. N series equipments of equal failure rate

    b. a series of N banks of parallel equipments where each parallel bank has M equipments in parallel, all of which have equal failure rates.

P5.13. For the cubes in Figure 5.1(a) and 5.2(a), find

    a. the mean value for each cube

    b. the variance and standard deviation for each cube

    c. the distribution of the sum of the two cubes, and show its plot

    d. the mean value of the sum

    e. the variance and standard deviation of the sum

Do step 1 through 5 above for Figures 5.1(b) and 5.2(b).

P5.14. Assume that you have four urns with white balls (W) and black balls (B) in them as follows:

    1. Urn W: BBB

    2. Urn X: BBW

    3. Urn Y: BWW

    4. Urn Z: WWW

    a. An urn is selected at random. What is the probability that a ball selected from it will be a white ball?

b. An urn is selected at random. A ball is drawn from it which turns out to be white. What is the probability that the next draw will also be a white ball? What is the probability that it will be a black ball?

c. An urn is selected at random and two white balls are drawn at random from it. What is the probability that the next draw will be (a) a white ball (b) a black ball?

P5.15. If you have two pieces of equipment with mean times between failure of 500 hr. and 1000 hr., find the overall reliability when these equipments are in

a. a series reliability configuration

b. a parallel reliability configuration

P5.16. Show that the variance of a uniform distribution defined over the range from a to b is given by $(b - a)^2/12$.

P5.17. Find the detection and false-alarm probabilities in reference to Figure 5.13 and its associated text for the following values:

a. $N_o = 4, V = 4, T = 1$

b. $N_o = 4, V = 8, T = 1$

c. $N_o = 4, V = 10, T = 1$

P5.18. For the three pulse-detection cases in P5.17, find the probability of detecting at least one pulse if a stream of

a. 4 pulses are sent

b. 8 pulses are sent

P5.19. Aircraft are arriving at an airport at a rate of 30/hr. What is the probability that the following numbers of aircraft will arrive in a 6-hr. period

a. 90 aircraft

b. 180 aircraft

c. 210 aircraft

P5.20. You are shooting at a bull's-eye as described in Figure 5.16 and its related text. The standard deviation in x–y coordinates is 4 in. What is the probability that a shot will lie no further from the center than

a. 2 in.

b. 4 in.

c. 8 in.

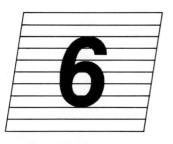

SPECIALIZED COMPUTER TOOLS

In this chapter we introduce a variety of computer-based tools applicable to systems engineering tasks that differ from the basic computer tools shown in Chapter 3. These tools are defined in the following broad categories:

- multitask applications managers
- project management aids
- alternatives and preference evaluators
- decision support systems
- toolchests desk/managers
- idea processors
- management evaluators/aids

A variety of software packages for these categories are also shown in Appendix A. These packages are all available for the p-c, which is the main orientation of this book. Selection of the particular packages in Appendix A is illustrative and is not meant to be a definitive or complete representation of what may be available to the user. Indeed, new packages are continuously under development that may become popular and successful. Thus we discuss the capabilities represented by the classes of software in a generic sense rather than any single package's features.

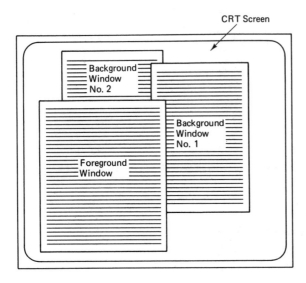

CRT Screen

Background Window No. 2

Background Window No. 1

Foreground Window

Figure 6.1 Split-screen windowing.

6.1 MULTITASK APPLICATIONS MANAGERS

After the introduction of integrated software packages discussed in Chapter 3, the so-called multitask applications manager came upon the scene. These packages allow the simultaneous use of both single- and multiple-function software. In both cases this is done by using a split-screen or windowing approach, as illustrated in Figure 6.1.

In the single function case, through the use of two or more subscreens or *windows* on one screen, the user can work back and forth between two aspects of the same function. This allows visibility into the overall function as with a manual operation along with instant access and simultaneous viewing. A simple example is that of a word-processing function in which the user may be updating an old draft copy and wishes to view both the new and old copy at the same time. For a spreadsheet it would then be possible to view two or more portions of the spreadsheet at the same time in different windows where in the normal single-function case or integrated package it would be necessary to scroll back and forth between the various portions of the spreadsheet that are of interest. This is the operating environment provided by this class of multitask applications software.

In addition to the multiple and concurrent use of the same function, these multitask applications managers can provide for the viewing and transfer of information from one function to another. Thus we can view a spreadsheet, database management system, and word processor simultaneously in three windows on the same screen. While we're viewing this, we can modify or transfer data from one to another by command. Note that the multitask software does not include the basic single- or multifunction software that appears in the windows; rather the multitask software "overlays" the single- or multifunction software and must be purchased separately and entered into the machine. This also differs from windowing that may be available on the more sophisticated integrated packages.

Some of the characteristics that can be found in these multifunction applications managers include

- on-line help menus
- scrolling within a window
- opening, expanding, shrinking, and moving of windows
- programmable function keys

In some cases, separate packages are available that overlay the multitask application manager that may be viewed as programming packages (sometimes called tool kits), and allow the user to write additional software to perform special functions. These multiple overlays are sure to become more extensive and capable with time in order to achieve higher levels of productivity with the large number of existing software packages.

6.2 PROJECT MANAGEMENT AIDS

These software tools are in abundance and provide a wide range of capabilities for managing projects. Therefore, they are available for application to the program planning and control function within the process of systems engineering management (see Chapter 2). As with all software packages, there are built-in functions and screens that provide the user with the tools considered necessary to set up the application. The various packages differ in method of presentation, graphics capability, ease of use, and other features that appeal to various potential users in different ways.

An important part of essentially all project management aids is the scheduler. This portion of the software contains one or more devices for defining and viewing a schedule for a project. Two of the most prevalent schedule mechanisms used today are the Gantt chart and the PERT chart, which are illustrated in Figure 6.2. The Gantt chart is a horizontal bar chart that depicts the time period over which an activity is scheduled. The activity is listed on the left side of the chart and the bars are constructed horizontally along a time scale. For most of the software packages available, the time scale is arbitrary and can be set to correspond to days, weeks, months, or even years. Along the bar it is possible to set up a series of milestones, shown by the triangles, that indicate when a particular activity begins or ends. An open triangle may be set up to represent a future scheduled milestone and when that milestone is achieved, the triangle may be blackened to indicate the successful accomplishment. Other annotations may be placed on the Gantt chart to code the activities to such things as the name of the project, the originator of the chart, and other features that are to be tracked.

The PERT chart is normally set up with a series of events and activities. The events are discrete points in time at which an activity begins or ends. The events are shown as boxes or bubbles on the chart. The activities imply the expenditure of resources in order to achieve the events and are depicted by lines on the chart that con-

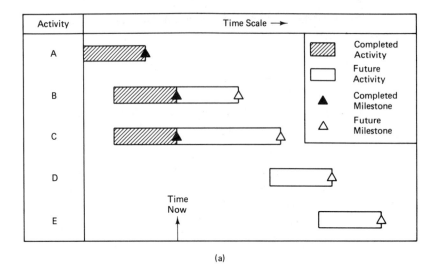

(a)

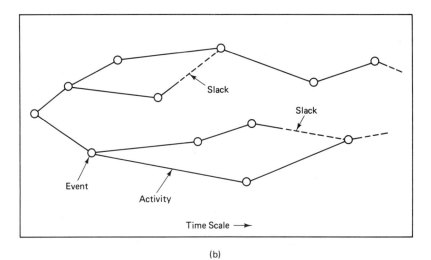

(b)

Figure 6.2 (a) Gantt chart, (b) PERT chart.

nect the events. The main feature embodied in the PERT chart that is not on the Gantt chart is the set of interdependencies between events that can be shown by activity lines that may interconnect any set of events. Through this mechanism, the PERT chart becomes a network of events and activities and leads to the concept and calculation of a *critical path* through the network. The precise nature and meaning of this critical path is dealt with in Chapter 11, Scheduling and Costing; in the context of this chapter we simply note that the PERT-chart software will usually calculate the critical path for the user once the essential times are entered for the various activities and the overall network structure is defined.

Another feature of this class of software packages is that of cost estimation and tracking. Developers of these kinds of software have recognized that most project managers wish to allocate resources to the project activities and then determine the cost of the project or portions thereof. This leads to the capability to assign resources in the form of people and equipment, and in the determination of the costs of these resources. Once this is done and the results aggregated, both schedule and cost tracking are possible.

A variety of other features are built in to the various software that facilitate its use in differing degrees. Examples include the availability of multiple windows for viewing several aspects of the project at the same time, sorting capability to aggregate resources by categories (as with a DBMS), help screens, automatic scaling (as in the case of redefining the time axis from days to weeks), control at the keyboard or through a mouse, and scrolling. In selecting an appropriate package for a particular application, the user is interested in all of these as well as the number of elements that can be simultaneously handled on the screen (e.g., number of horizontal bars on a Gantt chart), the speed of operation, other graphics capabilities (e.g., organization charting, trend lining, pie charting) and the overall eye appeal of the results. The user may also wish to explore the features in a package that allow the specific exploration of "What if?" questions in the simplest manner. In any case, the packages within this category of software are plentiful and powerful and are available at costs under $100.

6.3 ALTERNATIVES AND PREFERENCE EVALUATORS

This class of software can be construed as contained within a broader category of decision support systems. However, the latter definition is somewhat vague and arbitrary and we wish here to identify and discuss the particular kinds of software that focus upon the process of decision making in a rather broad context. Several developers of software have recognized that decisions often involve the incorporation of the user's value system as well as subjective judgments as to the importance of the various alternatives that may be present. This type of package is emphasized here since it is considered to be of growing significance in the field of top-level systems engineering and the synthesis of large-scale systems for which many (often subjective) factors and criteria must be taken into account. This class of software is designed specifically to assist the user in coming to terms with these kinds of considerations. It is also believed that these packages will continue to develop in scope and utility and will increasingly be applied to a variety of fields.

The first matter, explored by example in Chapter 3, is that of establishing the user's preferences or criteria for evaluation in regard to assessment of the merits of a given system or alternative systems. In short the question is—what does the user consider to be the elements of the system, the elements of the process of evaluating the system, and their relative importance? These will be lumped together in the concept of *weighting* the various alternatives, whether they are system elements or criteria elements. The second matter is that of rating the system or its subsystems according to the various weighted elements or criteria, and thereby obtaining an overall score or evaluation rating. Both of these are viewed as important aspects of this class of software. However, not all the packages have both aspects and the user should examine the package features before making a purchase.

The issue of criteria for evaluation of alternatives has a central role in large-scale systems engineering and therefore appears again and again. In the context of software, the package does not care whether one is dealing with systems engineering or the purchase of a house, a car, or a stereo system. The software is useful in all of these as well as in other contexts. Much attention has been given to the study of subjective evaluations and their attributes and some of those considerations will be dealt with in Chapter 13. With respect to software, it is important for the student or practitioner of systems engineering to be aware of the existence of this category of available programs and to experiment with those that appear to be most suitable in meeting particular systems engineering needs. Essentially all decisions that contain subjective aspects can be aided by the use of this type of software package.

A hierarchy of criteria may be structured by the user, using the same software package, to constitute a decision tree. A *decision tree* defines the hierarchy of a complex decision process and will be analyzed in greater detail in Chapter 13. Suffice it to say here that much of the software for this category also facilitates the construction and analysis of decision trees, which constitute a formal mechanism in the field of decision analysis.

The final point to be made is in regard to rating the various alternatives as to their relative merits, given the relative weights as to their importance in the evaluation process. This was previously illustrated in Chapter 3. Here we note that some of the software packages provide for both weighting and rating and some only for weighting. These procedures lay out quite simply in a spreadsheet format, as shown previously in Chapter 3. This fact does not negate the importance of alternatives and the preference-evaluation software packages that are designed specifically to aid the user in both structuring the decision process and then assessing the alternative decisions that may be available. Indeed, as these packages become more capable, more computational routines will be built in that will add a stronger quantitative set of factors. In addition, with the use of (the first category of) multitask applications managers, the capabilities of multiple-program packages can be brought together under the overall direction and control of the operating environment set up by the multitask applications package. Possibly this will suggest to the entrepreneurial reader some avenues for development of new multitask applications managers using some of the suggested techniques for synthesis (e.g., Zwicky's morphological box) described in Chapter 2.

6.4 DECISION SUPPORT SYSTEMS

Many producers of software packages claim that their product is a decision support system, including many of those who provide the category of software discussed above as alternatives and preference evaluators. This is certainly the case, but there is little precision as to exactly what constitutes a decision support system. Essentially all of the packages discussed in this book (including spreadsheets, DBMSs, project management aids) can be used to support a decision process of one sort or another. For that reason we will not linger upon this category nor attempt to provide a taxonomy of decision support systems. We simply wish to note here that there is such a class of software with that particular title and also that the user should examine systems advertised in this manner with extreme care to determine exactly what the capabilities are of the packages under consideration. They differ widely in the specifics and often are oriented to certain kinds of applications, for example, financial analysis and forecasting.

It appears that the central core of most of the systems designated as decision support systems is a database management system of one type or another, all with the capability of sorting and performing some level of quantitative analysis and graphics after each sort has been performed. Thus the fields of data can be aggregated and displayed in a variety of ways, depending upon what the user needs.

The kinds of broad capabilities represented by this class of software include the following:

- ad hoc query
- graphics
- modeling
- statistical analysis
- a menu generator
- spreadsheet interfaces
- DBMS interfaces
- word processing interfaces
- messaging
- data communications
- forms processing and reporting
- text processing
- internal language/command structure
- data security

6.5 TOOLCHESTS/DESK MANAGERS

This category of software is extensive, versatile, and very useful for the person who spends a lot of time at the computer at home or has the computer at or near the desk at work. It is essentially a large number of special functions that can be separately accessed and viewed at the same time. By so doing, data can be pulled from different sources simultaneously by opening and closing windows. In some implementations, these packages will also serve to overlay other packages and in that sense they subsume the same capability as the multitask applications manager.

A list of selected functions that appear in various toolchests/desk managers is provided here as Table 6.1. The operating environment allows multiple screens so that several of the features in the table can be overlaid and observed together on the screen. For example, a note pad can be in the foreground with several screens in the background such as: this month's calendar, a calculator, and a spreadsheet. The automatic dialer simply calls for putting the cursor next to a telephone number on the

TABLE 6.1 SELECTED FEATURES OF TOOLCHESTS/DESK MANAGERS

Alarm	DOS services	Phone message pad
Appointment calendar	Electronic mail	Programmable special keys
ASCII table	File cards	Special stored tables
Automatic dialer	Games	Template maker
Calculator	Keyboard macros	Time and expense diary
Card printer	Mailing label printer	To-do list
Cut and paste	Notepad	Windows
Day calendar	Phone directory	

screen and with a keystroke the system will dial that number, given that your p-c is equipped with a modem. Finally, one can purchase these types of packages for as little as $50.

6.6 IDEA PROCESSORS

This group of software packages basically helps the user define and organize ideas and thoughts. They assist in writing by bringing up an automated tool to move ideas around, sort them, prioritize them, and put them into categories and subcategories.

One can think about this type of software first as a means of developing an outline of a report or paper that is to be written. The user starts by developing an outline of headings and subheadings, usually in a hierarchical structure. On-screen menus help to access the command set. At some point the user has formulated an outline of headings and subheadings, and for convenience these can be thought of as numbered in a conventional hierarchical manner. If the user wishes to see only the first-level headings, this may be done by the stroke of a key and only these headings will be displayed. If a top-level number is to be changed, the program will automatically change all the subordinate numbers to conform. Additional commands are available to reorganize material, expand it, collapse it, and manipulate it in ways that help the user explore various combinations that are more pleasing or logical.

Some of the packages have a keyword search capability that will help the user look for inconsistencies or overlaps in ideas and thoughts—multiple portions of the text in which essentially the same idea is expressed. In one package in particular, the command structure contains a set of modes that correspond to ways of thinking. For example, eight ways of thinking are identified, one of which is associated with the notion of prioritizing. Therefore, one of the commands is to go into a *prioritize* mode that allows the user to rank order the various ideas or headings. After prioritization is complete, the headings are in a new order that refects their importance and the user can reorganize the report along these lines. This "brainstorming " feature is a clever way to both identify and provide a rationale for the imbedded command structure of the package.

An important capability in these packages is that of providing file formats that are compatible with the transfer to other programs, especially word processors. In general, this class of software is not amenable to writing long reports within its own structure, mainly due to storage limitations. However, with the ability to transfer results in a compatible format to a more powerful single-function word processor, for example, a vital link is established between the idea processor and the production-oriented word processor. Examples in this area include the ability to store and transfer files in ASCII format or in the format of a particular word processor.

Finally, some of these packages contain their own imbedded language that allows for the creation of customized subroutines, templates. and utility programs. One such language has been called a *thought processing language* and likened to LISP (see Chapter 16) in its structure and syntax. Although these so-called languages are rather simple and unsophisticated they still provide additional capability to allow the user to develop new functions, which is an important job. This capability can also be thought of as a feature to create simple macros to perform certain specific user-defined functions.

6.7 MANAGEMENT EVALUATORS/AIDS

These software packages are filling what appears to be an endless need to help people become better managers. Despite the very long list of management books and face-to-face management training seminars that are available, software packages have now entered the scene and they too have been successful in terms of public acceptance.

The software in this area is quite novel and in most cases has considerable eye appeal due to the use of embedded graphics. Extensive attention has been paid to user interfaces and friendliness so that the user can easily move through the software. On the other hand, the material for the most part is structured linearly, moving the user through the subject areas as one would normally read a book. The added ingredient, however, is user interaction, which helps maintain user interest and continually tests and reinforces the subject matter. In some cases there is a simulated role-playing exercise, which is a powerful technique for the user to test skills in the various subject areas.

The material in several of the available software packages is quite extensive. As an example, listed next are the various modules offered by Thoughtware (see Appendix A)

- assessing personal management skills
- evaluating organizational effectiveness
- understanding personal interaction styles
- leading effectively
- motivating to achieve results
- defining goals and objectives
- improving employee performance
- performance appraisal
- managing time effectively
- conducting successful meetings
- managing by exception

This is indeed a comprehensive "course" in management training.

Another innovative software producer provides packages dealing with communications, negotiations, management, and sales. An interesting approach is to ask the user to take a test by answering a series of questions—agreeing or disagreeing with statements posed by the program. Based upon the answers, the computer then produces a profile of the user that is developed from the combination of answers. This has also been applied on a third-party basis, that is, attempting to profile another person (such as a boss or a customer) in terms of trying to better understand how to deal with that person. In these cases we now have developers of software that are actually performing a type of psychological profiling within the context of management training.

As indicated, these management aids are filling a marketplace need that was previously handled by books, seminars, tapes, and television. With the ever-increasing presence and power of the p-c it can be safely predicted that they will flourish and find their way more and more in systems engineering applications.

6.8 SUMMARY

In this chapter we have provided a broad overview of various types of computer tools that can be used by the student and practicing engineer to assist in the process of systems engineering. Not one of these classes of software has specifically been designed to apply to systems engineering. It is up to the user to see the potential areas of application and to experiment with the available software to explore its applicability. In Table 6.2 we assist the reader by listing some of the systems engineering

---

TABLE 6.2 POTENTIAL APPLICATIONS OF TOOLS TO SYSTEMS ENGINEERING

Multitask Applications Managers
  • Requirements tracking and traceability by overlay on spreadsheet and DBMS packages
  • Systems modeling and trade-off analyses
Project-Management Aids
  • Project management and control aspects of systems engineering management (see Chapter 2)
  • Cost, schedule, and (TPM) tracking
Alternatives and Preference Evaluators
  • Definition and weighting of criteria for systems' evaluation
  • Decisions as to most cost-effective system and subsystem alternatives
Decision Support Systems
  • Life-cycle cost analysis and trending for system
  • Personnel assignments, tracking, and program costing
Toolchests/Desk Managers
  • Auto-dialing for customers and suppliers
  • Keeping of appointments' calendar
Idea Processors
  • Develop logic for system synthesis, analysis, and evaluation
  • Write required systems engineering reports (e.g., SEMP)
Management Evaluators/Aids
  • Assessment and improvement of personal skills for performing job
  • Improvements in defining and achieving goals and objectives

---

areas that may be addressed by the seven categories of software tools discussed in this chapter.

The costs of these packages can easily vary within about one order of magnitude, ranging generally from about $50 to $500. As software has become more capable, the trend has been to provide more capability at the same cost. This means that manufacturers that normally sell packages between $50 and $100 tend to stay in this price range even as improvements are made in the software they offer. The same trend appears to hold for producers in the higher price range.

The problems at the end of this chapter are designed to encourage the reader to explore options and applications in the seven categories of software tools that have been introduced. Chapter 7 continues in the same vein, but deals with new categories of tools drawn generally from the fields of mathematics and engineering.

REFERENCES

The software tools discussed in this chapter are part of an ever-evolving and improving set of available packages that are represented most adequately by the magazine article literature. Therefore, rather than attempting to cite particular articles, the references for this chapter are of two types: (1) magazines that consistently contain relevant articles on this vast array of software, and (2) catalogs of software packages whose latest edition may be reviewed in order to locate particular packages and determine their characteristics and features.

MAGAZINES

1. *Business Computer Systems,* Cahners Publishing Company, Boston, MA.
2. *Business Software,* M&T Publishing, Inc., Redwood City, CA.
3. *BYTE,* The Small Systems Journal—Byte Publications Inc., Peterborough, NH.
4. *Computer Business News,* C. W. Communications, Inc., Framingham, MA.
5. *Compute!* Compute Publications Inc., Greensboro, NC.
6. *Computer Decisions,* Hayden Publishing Company, Inc., Riverton, NJ.
7. *Computer Design,* Computer Design Publishing Co., Litteton, MA.
8. *Computerworld,* C. W. Communications Inc., Framingham, MA.
9. *Creative Computing,* Ahl Computing, Inc., Los Angeles, CA.
10. *Datamation,* Technical Publishing Company, New York, NY.
11. *Dr. Dobbs' Journal of Software Tools,* M&T Publishing, Inc., Redwood City, CA.
12. *Family Computing,* Scholastic, Inc., New York, NY.
13. *Infosystems,* Hitchcock Publishing Company, Wheaton, IL.
14. *Mini-Micro Systems,* Cahners Publishing Company, Inc., Denver, CO.
15. *Output,* Technical Publishing, Barrington, IL.
16. *PC Magazine,* PC Communications Corp., New York, NY.
17. *Personal Computing,* Hayden Publishing Company, Inc., Hasbrouck Heights, NJ.
18. *Popular Computing,* McGraw-Hill Inc., Peterborough, NH.
19. *Software News,* Sentry Database Publishing, Hudson, MA.
20. *Systems & Software,* Hayden Publishing Company, Inc., Hasbrouck Heights, NJ.

CATALOGS

1. *Introduction to Software Packages,* National Bureau of Standards (NBS) Special Publication 500-114, U. S. Department of Commerce, April 1984.

2. *Software Development Tools,* National Bureau of Standards (NBS) Special Publication 500-88, U. S. Department of Commerce, March 1982.

3. *Software Tools Survey,* Office of Software Development, Federal Software Testing Center, Report OSD/FSTC-83/015, Office of Information Resources Management, GSA, June 1983.

4. Brand, S., ed., *Whole Earth Software Catalog,* Garden City, New York: Quantam Press/Doubleday, 1985.

5. Kelly, B. W., and D. J. Grimes, *IBM PC Compatible Computer Directory,* New York: John Wiley, 1985.

6. Mellin, M., and N. Hays, ed., *The Book of IBM Software 1985,* Los Angeles, CA: Arrays, Inc., 1985.

7. *IBM PC Expansion & Software Guide,* Indianapolis, IN: Que Corporation, 1985.

8. *Directory of Computer Software, 1986,* National Technical Information Service (NTIS), PB86-135357, U. S. Department of Commerce, 1986.

PROBLEMS

P6.1. Develop ten system-evaluation criteria for a system that you select and at least three subcriteria for each of the ten major criteria. Formulate a normalized weighting system for these criteria. Develop and apply a rating scheme of your choice for three hypothetical systems to be evaluated.

P6.2. Define two systems engineering application areas for each of the seven categories of software tools presented in this chapter, other than those shown in Table 6.2.

P6.3. Describe ten additional features, other than those listed in Table 6.1, that you might design into a toolchest/desk manager.

P6.4. Lay out a Gantt chart for all the things that you would like to do tomorrow, listing the time axis in hourly increments. Tomorrow evening, check off the items that you accomplished.

P6.5. Design an idea processor with ten separate and distinct features. Draw a logic chart showing the command structure and the conditions under which the user moves from mode to mode.

P6.6. Consider having a multitask applications manager at your disposal. Develop a systems engineering application example that would require the simultaneous use of a spreadsheet, DBMS, and graphics package and illustrate how this would work.

P6.7. By consulting appropriate computer software magazines, find examples of specific packages, other than those listed in Appendix A, for each of the categories of specialized computer tools. List and explain five major features of each.

P6.8. Construct a morphological box (see Chapter 2) whose elements are the five software types discussed in Chapter 3 and the seven software categories discussed in this chapter. Examine the various cells in such a box and rate the cells according to your judgment as to what cells would make (a) the most useful software package, and (b) the most saleable software package. For the five cells with the highest rating in (a) and (b), explain the reasons for your choices.

When we deliberate it is about means and not ends

Aristotle

MATHEMATICAL AND ENGINEERING COMPUTER TOOLS

In the previous chapter we explored a variety of specialized software tools that may be used as computer aids in the systems engineering process. In this chapter we continue to examine such tools, but we deal with software that tends to have direct or indirect elements that are part of the fields of mathematics or engineering. The categories of interest for this chapter are

- statistical
- mathematical
- engineering
- reliability
- mathematical programming
- simulation
- curves and plotting
- CAD/CAM/graphics
- workbenches
- other special purpose packages

As before, the emphasis is on the p-c and therefore on software packages that are designed to run on the microcomputer. In some cases, larger software capabilities are considered since these capabilities are likely to be available on the next generation of microcomputers. The reader is also referred to Appendix A in which the categories of interest are listed together with specific packages and the manufacturers of these packages. A summary of some of the systems engineering application areas with respect to the selected software capabilities is provided at the end of this chapter.

7.1 STATISTICAL TOOLS

There is a large number of statistical software packages available for the p-c user. In general, in contemplating such use one starts with a set of data elements that are collected through some empirical process. Examples of such data elements include

- reliability data that is taken for a set of components in field operation to determine the potential failure characteristics of these components
- training data taken through testing of either operations or maintenance personnel, or both, to ascertain levels of proficiency
- system- and subsystem-level data taken during the test and evaluation (T&E) phase of a system to see if the system satisfies the requirements as reflected by the test plans and procedures

We can envision starting the statistical-analysis process with a relatively large amount of data and then examining the nature of this data by means of formal statistical procedures.

Most statistical analysis packages will provide the capability of calculating standard types of sample data characteristics such as

- mean
- standard deviation and variance
- maximum value
- minimum value
- range
- skewness
- kurtosis

In addition to these standard types of parameters, specialized statistical tests and analyses are available such as those listed in Table 7.1.

Many of the statistical software packages that are available are command driven in distinction to being menu driven. This design approach is highly debatable; manufacturers make a virtue out of their particular design and each has advantages and disadvantages. For the novice, it can be a great advantage to have a menu-driven package that will walk the user through its various capabilities. As proficiency is gained, however, the user generally finds the menu-driven package somewhat laborious and prefers a command-driven approach. The more powerful packages tend to be command driven since they assume long-term need and application of the package such that the user will gain considerable proficiency. This has evolved to the

TABLE 7.1 SELECTED STATISTICAL SOFTWARE FEATURES

- Analysis of variance (ANOVA)
- Binomial test
- Box–Jenkins methodology
- Chi-square tests
- Cochran Q test
- Confidence limits
- Contingency coefficient *C*
- Covariance matrix calculations
- Cramer's *V* or phi
- Durbin–Watson statistic
- Extended median test
- Factor analysis
- Fisher exact probability test
- Fisher–Tocher test
- Friedman 2-Way ANOVA
- F-tests
- Gamma test
- Kendall coefficient of concordance
- Kendall rank/partial rank correlation coefficient
- Kolmogorov–Smirnov one/two sample test
- Kruskal–Wallis ANOVA

- Likelihood ratio tests
- Mann–Whitney U test
- McNemar test
- Median test
- Moses test
- Multidimensional scaling
- Nonparametric procedures
- Normal distribution tests
- Pearson's *r*
- Probability density & cumulative distribution functions
- Randomization test for matched pairs
- Regression analysis
- Sampling plans and designs
- Somers' *d*
- Spearman rank correlation coefficient
- T tests
- Uncertainty coefficient
- Wald–Wolfowitz runs test
- Walsh test
- Wilcoxon matched pairs signed ranks test
- Yates 2 by 2 chi

point where, as with DBMSs, the command structure is considered to be essentially a new language to by mastered by the user.

Virtually all statistical software packages have the capability of merging, joining, and developing subsets of multiple files. Three characteristics to be considered and compared with respect to these packages include speed, accuracy, and ability to exchange data with other packages. Speed is important since in most cases one is dealing with large amounts of data. A coprocessor often is recommended in order to speed up calculations with large data sets. Accuracy is clearly important in statistical computations and the various manufacturers often show comparisons of the accuracy they obtain in relation to their competitors and other standards. With respect to interchangeability with other packages, one can look for the transferability of data with spreadsheets, DBMS's, and languages such as BASIC (see Chapter 9). In addition, another key feature in this area is that of transferring files between the p-c and a mainframe (mainframe to micro linking). An example of the advantages of this capability is when the p-c is used to write the needed programs using the package's command language, the program is uploaded to a mainframe for rapid execution, and then the program is downloaded so the results can be displayed and printed out by the p-c. At least one manufacturer advertises the use of KERMIT, a terminal emulator and file-transfer utility, at no additional cost, for mainframe to micro linking. KERMIT, developed at Columbia University, is considered a public-domain file-transfer system that embodies error detection and correction to assure data and computational accuracy.

As one might expect, essentially all of the available and better-known statistical packages have extensive graphics capabilities, including such outputs as

- histograms
- scattergraphs
- probability plots
- bar, pie, and star plots (2- and 3-D)
- maps

Due to the large number of available statistical packages and the broad range of prices (from $50 to $500), the potential user should examine the application area in some detail before making a selection. Many of the packages, however, provide extraordinary features at affordable prices.

7.2 MATHEMATICAL TOOLS

The mathematics packages available for the p-c are not as extensive as the statistical packages. Further, some of the packages can be viewed as bordering on engineering types of tools. At the simplest level, there are rather inexpensive packages (around $50) that will provide solutions to linear, quadratic, and higher order simultaneous equations. Usually, these equations have to be entered in a standard form but in some packages provision is made for free-form entry at the keyboard with the program recognizing the terms and the form of the input equations. Some packages allow for an "educated guess" as to the roots of an equation and will automatically search for and converge upon the solutions. Other packages have built-in conversions (such as from the English to the Metric system) and built-in functions such as square, square root, logarithm, exponentials, and trigonometric functions. These types of packages, in general, may be considered one step up from a calculator, providing some additional capability and elementary forms of programming languages.

At a next higher level, packages are available that will move from algebra into calculus as well as matrix and vector analysis. In that sense such packages extend one's ability from scalar to vector or state variable computations. Finally, a third level of tools, still reasonably priced (under $100), will move one into forecasting algorithms, logical operations, Fourier and other types of series expansions, differential equations, numerical integration, various sorting techniques, and fast Fourier transforms (FFTs).

It would appear that unless one has a very particular and specialized calculation in mind, it is more cost effective to purchase a programming language package such as Pascal, BASIC, or Fortran to carry out a wide variety of mathematical computations. The price differential is negligible and the computing power can be much more flexible and varied. Chapter 9 deals more explicitly with the capabilities of two programming languages, namely BASIC and Pascal.

7.3 ENGINEERING TOOLS

The engineering tools that are of interest here tend to be focused on detailed conventional engineering subjects in distinction to the broader subject of systems engineering. As discussed earlier in this text, the latter deals primarily with top-down engineering at the systems level. After systems engineering tasks and activities are performed, they are normally followed by more detailed "black box" conventional engineering, which is the emphasis with respect to the packages discussed here and the subject of computer-aided engineering (CAE). The latter is also a highly creative process requiring considerable talent, expertise, and ability to synthesize. Fortunately, the state of the art is such that many of the lower indenture level engineering-analysis tasks have been captured in software and can be used in later more detailed stages of subsystem and component design.

Recognition of capability as well as more extensive interest in p-c engineering packages is showing itself in the literature. For example, some articles argue for and demonstrate the availability of such packages for detailed engineering design [1]. Other articles tout the p-c moreso than the software, for example [2], discussing the use of the microcomputer as an effective tool for the design of aircraft. In the latter case as well as others of a similar type, a combination of user-developed software and commercially available packages provide the type of capability most useful for the application area in question. Micros are also being used as on-line and support equipment in a variety of research applications, many of which can be classified as engineering disciplines [3]. Specialized "problem solving" software that offer a variety of applications, including engineering, are also being recognized, developed, and used [4]. We see then that engineering packages for the p-c have become extremely plentiful. They also span a rather complete range of engineering subjects with the greatest emphasis on electrical and electronics engineering. Table 7.2 provides a list of some of the capabilities represented by a variety of such packages.

TABLE 7.2 SELECTED ELECTRICAL AND ELECTRONICS ENGINEERING CAPABILITIES AVAILABLE IN SOFTWARE PACKAGES

- Active and passive circuit analysis
- Analog computer simulation
- Antenna gain plots
- Bode plots
- Cascaded networks
- Computerized circuits, schematics, and parts lists
- Control system simulation
- Convolution
- Digital filter design
- Digital image processing
- Fast Fourier transforms
- Filter analysis
- Frequency and phase spectra
- Harmonic analysis
- Logic circuit design
- Matrix manipulations
- Microcomputer design
- Polynomial roots
- Transfer functions
- Transient and steady state analyses
- State variable calculations

In addition, engineering packages can be found for the p-c that will provide routines that deal with

- fluid flow
- heat transfer
- beam loading and deflection
- hydraulic systems
- vibrational analysis
- building design
- chemical reactions
- pipe-line networks
- kinematics
- reactor simulation
- topographic and field surveys
- geometric design

Computer-aided design and manufacturing (CAD/CAM) application packages are also available and are referred to later in this chapter.

7.4 RELIABILITY

Reliability analysis is a specific subject within the context of systems engineering and as such is a specialized engineering- and mathematics-based area. There are apparently a limited number of available reliability packages. In addition, the p-c packages appear to handle only the exponential failure distribution. This means that the mathematics is relatively simple since the failure rates can be thought of as additive, unlike the situation where wearout is accounted for by using the Weibull or other more complex distributions (see Chapters 5 and 15).

A typical reliability package bears a strong relationship to Military Handbook 217 (MIL-HDBK-217), which describes a more-or-less standardized method for calculating the reliability of a system. A base failure rate for each type of system component is entered from what amounts to a catalog of failure rates for elementary components. These component types include

- resistors
- capacitors
- inductors
- transistors
- diodes
- microcircuits
- relays
- switches
- connectors
- printed circuit boards

Starting with a base failure rate, adjustments are made to these rates based upon environmental factors (ground versus airborne use, etc.), ambient temperature and stress level (e.g., power, current, and voltage) in relation to rated values. This leads to modified failure rates, which are then aggregated to obtain composite failure rates, which are then aggregated to obtain composite failure rates for user-defined black boxes. The composite failure rates are then entered as the "lambdas" for the exponential distribution, thus yielding an immediate reliability prediction. The reciprocal of the failure rate is, of course, the MTBF.

These packages have the same options that are present with other packages in terms of menus, graphics, and other user aids. Some have manual worksheets that are presumed to be helpful to the user in preparing input data. Such aids will likely be incorporated into software formats in later more sophisticated versions of this class of software. It can also be anticipated that nonexponential distributions will eventually be incorporated into such software although this modification is not a simple one. Many reliability models have been built and run on mini and mainframe computers that do account for nonexponential failure characteristics. Further discussion of the matter of reliability analysis and modeling is discussed in Chapter 15 of this text.

7.5 MATHEMATICAL PROGRAMMING

Mathematical programming is a broad topic in mathematical analysis that includes linear programming, quadratic and higher order programming, as well as a subject known as dynamic programming. Specifically excluded from this area is computer language programming, a subject considered in some detail in Chapter 9.

Most of the available software packages in this category deal with linear programming, in all likelihood as a result of the relative simplicity of this technique. Being linear in nature, even large-scale linear programming problems can be handled in a p-c environment. The essence of a linear programming problem takes the form of a series of inequalities of the form

$$A(1, i)X(1) + A(2, i)X(2) + \cdots + A(m, i)X(m) \gtrless C(i)$$

where $(i = 1, 2, \ldots, n)$ and both $A(\ )$ and $C(i)$ are a series of constants. Together with these equations is a typical cost function of the variables $X(i)$ which is usually expected to be minimized. Alternatively, other objective functions can be used (in place of the cost function), which are to be maximized, such as the mix of available channels in a communications network or the expected defensive mix of capabilities of a military system [5]. Whichever the case, the series of linear inequalities can be thought of as representing a boundary in n-dimensional space. An example is shown in Figure 7.1 in which three inequalities are plotted on an x–y axis. It is normally the case that the solution (e.g., minimum cost x–y point) lies at one of the vertices of the intersecting lines. A way of solving such a problem is to calculate the cost at each vertex and then select the solution as the x–y point at which the cost is a minimum (or the objective function is a maximum). Several texts can be consulted to obtain additional information on the specific subject of linear programming [5, 6] and

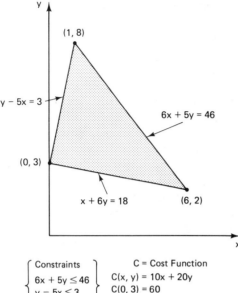

Constraints	C = Cost Function
$6x + 5y \leq 46$	$C(x, y) = 10x + 20y$
$y - 5x \leq 3$	$C(0, 3) = 60$
$x + 6y \geq 18$	$C(1, 8) = 170$
	$C(6, 2) = 100$

Figure 7.1 Linear programming example.

its duality with game theory. Software packages can be found for less than $50 that provide more than an adequate capability.

As we move into more complex linear and higher-order programming problems we enter the world of mathematical programming in general. Specific solution algorithms such as the simplex algorithm, Lagrange multipliers, and branch and bound can be used as solution techniques. Both continuous and integer variables are also considered and embodied in several of the available software packages. As we move into more complex problem areas the cost of the software increases into the $500 and greater price range.

The same program features that we have discussed before are to be examined in selecting an appropriate package for systems engineering applications, such as

- numbers of variables that can be handled (e.g., 100 × 100)
- speed of computation
- editing capabilities
- menu versus command-driven (user friendliness)
- available report generators

Some of the packages have the feature of interacting with the files and capabilities of the more popular spreadsheet software.

7.6 SIMULATION

Simulation can be a powerful systems engineering tool for evaluation and trade-off analyses for large-scale systems. As such, it is dealt with separately in Chapter 12, which focuses on performance analysis and modeling. *Simulation* is a special class of modeling techniques that, in effect, mirrors or represents the detailed behavior of a system in regard to a set of attributes that are to be studied in a quantitative manner.

A relatively large number of simulation languages have been written. Some of these languages, such as SIMSCRIPT, are available for the p-c; others require greater computing power and storage capacity to be useful for larger scale problems. Typical areas in which simulation has been used extensively include

- the analysis of routing for a packet-switched telecommunications network
- the operation of a rapid-transit system
- the possible outcomes of a two-sided war game (air, ground, sea based)
- the dynamics of air/ground transportation routing and scheduling

The reader is referred to Chapter 12 for further specific information on the subject of simulation.

7.7 CURVES AND PLOTTING

In this category we refer mainly to matters of curve fitting and generating graphs of results. In this sense it can be viewed as a subset of statistical analysis techniques. However, fitting curves to available data comes up in many areas of engineering and the user may wish to purchase a package that is specialized to that purpose.

The technique most applicable to curve fitting is to find a "least squares" fit and perhaps the simplest problem area is that of finding the appropriate linear fit to a series of data points. The mathematics of such an approach is straightforward and is described briefly next.

We start out with a series of data points or observations $X(i)$, $Y(i)$ and we wish to find a least squares line of the form $y = mx + b$ that will fit these data points. The least squares concept is one of minimizing the mean square error between the line and the aggregate data points. We assume that the $X(i)$ in the series of observations represent true values and that the corresponding $Y(i)$ have some error associated with them. We define the error as the point-by-point difference between the observed $Y(i)$ value and the y-value of the line, that is, error $= Y(i) - y(i) = Y(i) - mx(i) - b$. This error value is then squared and a solution found by using conventional calculus techniques (see Problem P7.4). The solution provides a means of solving for the slope, m, and y-intercept, b, of the least squares line, using the following equations:

$$\left.\begin{aligned}
m \sum X^2(i) + b \sum X(i) &= \sum X(i)Y(i)\\
m \sum X(i) + bn &= \sum Y(i)
\end{aligned}\right\} \tag{7.1}$$

It is clear that with the known observations $X(i)$ and $Y(i)$ the unknown values of m and b can be found.

The same linear least squares fit approach is often used in the matter of forecasting or extrapolation. For this approach one starts with a series of data points for a range of x and y variables. Such variables may represent

1. the power rating of aircraft engines
2. the data rates and frequencies of communications channels
3. the number of lines of code of a computer program
4. the cost of a system, as a function of items 1 to 3

By performing a least squares fit to existing data, one often attempts to extrapolate to new ranges of the variables in question. Such an approach has its limitations, particularly where some nonlinearity or effective discontinuity (breakthrough) in a process may be experienced. However, this technique is used extensively with considerable success when tempered with good sense and real-world judgments.

A variety of curves can be used (assumed) as appropriate fits to the given data, with the criterion for fitting these curves remaining the same, that is, a least squares fit. The selection of a linear curve versus some particular nonlinear curve is usually a matter of blending engineering experience with some insight into the physical behavior of the phenomenon. Blind application of a linear fit can lead to much difficulty and violation of the dynamics of the behavior of large-scale systems. In most of the curve-fitting packages, the user can call for a plot of the resultant curve, which can be "eye-balled" to see if the results make sense physically. This is particularly helpful when attempting to extrapolate to regimes for which there is little or no real-world data. The reader may be treated to some wonderful examples of inappropriate and humorous applications in the arena of developing large-scale military systems [7].

7.8 CAD/CAM/GRAPHICS

Computer-aided design (CAD), manufacturing (CAM), and graphics capabilities are present in many low-cost software packages for the p-c. Drawing on the screen requires high resolution and processing speed, and considerable storage capacity. The reason is simple: A picture on the screen contains a great amount of information.

In terms of resolution, the unit of concern is the number of pixels on the screen. A pixel represents the smallest discernible point of light that can be placed on the screen. A so-called standard now is of the order of 300×200 pixels with a professional level at least as great as 600×400 pixels. In achieving higher levels of resolution, users must consider the inherent capability of the system that they have, the available graphics boards, and the graphics software capabilities. The user can also choose between monochrome monitors and color monitors and their associated resolution capabilities. All of these choices depend upon the requirements that the user has and it is important that the system be set up compatibly for a good CAD or CAM capability.

There are also several different ways in which to input information to a CAD/CAM system. Among the devices that can be used are the standard keyboard, a light or touch pen, a mouse, a trackball, and a joystick. Also available are various types of digitized tablets. These tablets are pads that can run up to four foot squares by which one can enter data into the p-c using a stylus. The process is simply to lay a chart or drawing upon the tablet and then trace the figures using the stylus. The information is digitized and entered as data through the software for the particular program that is being used. Often the pad or tablet is backlit to facilitate this process.

Another way to place images on the screen is to access a library that is part of the software. The packages contain certain primitives that include lines, curves, circles, borders, fonts, initials, faces, figures, maps, and various types of hatching. By means of menus and command structures, the various input devices can be employed to bring these primitives to the screen and then proceed to manipulate them. Techniques for the latter include movement across the screen, expansion, contraction, and replication. Of particular importance in a CAD/CAM context is zooming so that one can work on a portion of a figure in great detail and place it into a larger drawing when completed. Conversely, from the overall drawing one may wish to zoom in on a particular detailed feature. Dimensioning capabilities are often provided in an attempt to replicate as closely as possible the information usually found on a conventional mechanical drawing. Proportional scale changing is also a feature found in most of the better packages.

The CAD/CAM/graphics packages can also be divided into 2-D and 3-D categories and the user must make this fundamental choice based upon need. A 2-D package may be very capable in all respects other than to show a true three dimensional representation of an object. In 3-D packages, isometric and other projections can be obtained and the more sophisticated ones allow for the rotation of objects in space. As expected, the 3-D packages generally are more expensive than the 2-D ones. Since these capabilities require more complexity, greater speed, and increased storage capacity, the user should be prepared to spend somewhere between $200 and $2000 to obtain the features that are desired.

7.9 WORKBENCHES

There is a close relationship between CAD/CAM/graphics and workbenches. The former is often thought of as being implemented on a workbench; the latter almost always is thought of as containing some type of diagramming and graphics features. In addition, the notion of CAE is often added so that practitioners in the field may couple CAD/CAM/CAE with the implementing facility of a workbench. As an example, the yearly international conference and exposition on CAE workstation technology typically brings these concepts and capabilities together in proceedings that deal with such subjects as

- workstation hardware trends and CAD implementation
- overview of CAE technology
- workstation purchasing and pitfalls
- workstations—a solution for the VLSI designer
- simulation speed and VLSI modeling

- the p-c and CAE
- CAE software links
- analog CAE workstations
- workstations for electromagnetic CAD
- design management and communications in CAE and CAD/CAM

A workbench can be viewed as a computer terminal facility that contains many tools and has an integrated capability to perform a variety of tasks. In our context, these tasks are primarily CASE, but can also include CAD, CAM, and CAE. Another capability for workstations is that of supporting the software development process [8]. In this context, a workstation can be thought of as establishing a *software engineering environment*—a facility that will support such functions as

- project management
- requirements analysis
- software design
- coding and testing
- configuration management
- documentation

In summary, we see that workstations of various types can be configured to support CASE, CAE, CAD, CAM, and software development.

Notwithstanding the aforementioned, we also see the following topics as the structure for an IEEE conference on workstation technology and systems:

- workstation hardware technology and architecture
    performance hardware
    adaptive control
    VLSI engines
    geometry hardware
    networking hardware
- workstation software technology and hardware
    networking and communications
    concurrency control
    performance metrics
    user interfaces
    databases
    data interface formats
- workstation subsystems
    input devices
    graphics
    image and signal processing
    array processors

- special purpose workstations
  - design strategies for workstations
  - expert systems for workstations
  - product assurance and quality control through workstations
  - robotic control through workstations

Two points are notable with respect to the previous list: (1) there is no explicit mention of CAD, CAM, CAE, CASE, or many aspects of software development, and (2) other interesting subjects are highlighted, such as image, signal, and array processing. Clearly there is much to be done to define and organize the topical content under the subject of workstations.

The workbench packages that we focus on here are those available for the standalone p-c, keeping our emphasis on the personal computer. The p-c, when outfitted with workbench capabilities, constitutes a workstation in our context. The various specific software workbench packages listed in Appendix A are generally oriented to support of the software development process. However, many of these capabilities will also support CASE, but were not specifically designed to do so.

The workbench features for the p-c appear to be centered around a diagramming and graphics capability. The diagramming software allows the user to formulate almost any kind of chart (see Chapter 4), including structure charts, data flow diagrams, and functional flow diagrams. These are viewed as generally supportive of the various methodologies of structured analysis. In other words, the workbench allows the user to perform a variety of structured analyses, with emphasis on the diagramming techniques that are part of these analyses.

The diagrams are supported by a variety of data analysis, report generation, and documentation capabilities. This includes a data dictionary and an imbedded or associated word processor. It also generally includes at least some type of DBMS capability. Of course, all of these features are accessed through screens that are called up by reference to a hierarchy of menus.

The workbench design graphics is manipulated also from top level to lower and more refined detail. Thus a particular diagram can be "exploded" to call for the data associated with a portion of a larger diagram. The data dictionary ties all the data structures together and is the central information source of all design data. The emphasis on diagramming and graphics requires storage capacity and resolution on the screen. A typical package, with the proper graphics board, will have of the order of 700×350 pixels and will depend upon at least a half megabyte of storage.

In general, these p-c packages will not be designed to support a particular design language, although there are workbenches to do so, for example, to support COBOL on the p-c. Apparently, most designers felt that this would limit the appeal of the software. This was a good choice especially when considering the cost of these workbenches, which run in the range of $2000 to $9000. That is a considerable amount of investment for an individual (in distinction to a company) and can be compared with alternative methods of achieving similar capabilities. Ten other packages at $500 each plus a multitask manager can cost less than the integrated workbench—an attractive alternative for the individual user.

7.10 OTHER SPECIAL PURPOSE PACKAGES

There are a large number of other types of specialized software packages for the p-c user in a mathematical and engineering context. Three other notable types include

1. survey/questionnaire instruments
2. utilities
3. artificial intelligence

Several packages have been developed for number 1 that embody generalized questionnaire structures and scoring systems [9]. All the user has to do is enter the specific questions and the remainder of the scoring and tabulation of results is built into the software. In broad terms, utilities interact with a system's operating system and allow the user greater flexibility in such areas as file transfer and recovery, partitioning, copying, preventing crashes and other loss of data, and interpreting operating system messages. Finally, there are a variety of p-c packages that have appeared in the area of artificial intelligence (AI). AI applications normally require a considerable amount of storage so that the boundaries of p-c capabilities are being stretched for this application area. Since AI represents an important field of the future, we reserve further discussion of it for Chapter 16 in which we consider forecasting and AI, with an emphasis on expert systems.

Whereas this book emphasizes the personal computer, it should be recognized that software packages for the minicomputer and mainframe can be obtained. As indicated in Chapter 1, the government has taken on an important role in cataloguing and disseminating information regarding such packages. For example, one such catalog [10] cites software in the following categories that directly relate to the subject areas of this chapter:

- aerodynamics and fluid dynamics
- business and economics
- chemistry
- civil and structural engineering
- communications
- computer science
- energy
- environmental pollution and control
- industrial and mechanical engineering
- library and information sciences
- mathematics and statistics
- nuclear science and technology
- physics
- transportation

The systems engineer with access to a larger computer thus has these and many other sources regarding available software. This point is reiterated with respect to specific elements of systems engineering in Part III of this book.

7.11 SUMMARY

In this chapter we have continued our broad examination of some of the classes of software tools that are available to help the user carry out the tasks and activities of systems engineering. Emphasis has been placed upon generic types of mathematical and engineering tools rather than particular packages that are being sold today, although a representative list of the latter is provided in Appendix A. Table 7.3 shows some of the systems engineering areas that may be addressed by the types of tools cited in this chapter.

TABLE 7.3 POTENTIAL APPLICATIONS OF TOOLS TO SYSTEMS ENGINEERING

Statistical
- Analysis of schedule deviations for a city-wide bus system over a sample time period
- Evaluation of mean-time-to-repair for various classes of operational field equipment from on-site log data

Mathematical
- Maximizing the flow through a fuel-transfer system using classical calculus optimization techniques
- computation of trajectories using spherical trigonometric relationships

Engineering
- Analysis of the composite transfer function of an analog process control system
- Design of a digital logic processor subsystem

Reliability
- Reliability assessment of a complex assemblage of electronic communications equipment
- Reliability requirements allocation for a hierarchy of systems, subsystems, and components

Mathematical Programming
- Scheduling of operators and maintainers of a rapid-transit system
- Formulation of an optimum routing scheme for intercity pickup and delivery of parcels

Simulation
- Evaluation of the defensive capabilities of a postulated "star wars" system architecture
- Assessment of the airspace capacity of an advanced automated air-traffic-control system

Curves and Plotting
- Formulation of a least squares fit curve to the tracking data obtained from a position location system
- Ray tracing of underwater signal propagation

CAD/CAM/Graphics
- Layout of a major systems engineering design facility and associated test laboratories
- Layout of a printed circuit board

Workbenches
- Flow charting the systems engineering process in a large-scale R&D program
- Formulation of structured analysis diagrams for a software development project

---

Looking at this chapter as well as Chapters 3 and 6 we see about two dozen classes of software tools that can be used to perform systems engineering tasks. Further software tools will also be considered in later chapters (e.g., simulation, AI) in relation to specific systems engineering subjects. If this is a bewildering situation, it is also one in which change is occurring rapidly and new tools as well as machines are coming out on the market every year. In addition, prices are dropping as some manufacturers have adopted a low-cost market-penetration strategy. Thus the user is in an increasingly advantageous position. The challenge is to clearly focus upon the application areas that are of greatest importance and to select, within the categories of packages set forth here and others that may emerge in the future, those that are most cost effective.

REFERENCES

1. Kull, D., "Personal Computers in Engineering," *Computer Decisions,* January 1984.
2. Lerner, E. J., "Designing Aircraft on Small Computers," *Aerospace America,* Journal of the American Institute of Aeronautics and Astronautics, Inc., June 1985.
3. Stewart, D., "Micros in Research," *Popular Computing,* April 1985.
4. McMullen, B. E., and J. F. McMullen, "Specialized Software," *Popular Computing,* April 1985.
5. Saaty, T. L., *Mathematical Methods of Operations Research.* New York: McGraw-Hill, 1959.
6. Best, M. J., and K. Ritter, *Linear Programming: Active Set Analysis and Computer Programs.* Englewood Cliffs, NJ: Prentice-Hall, 1985.
7. Augustine, N. R., *Augustine's Laws.* © Norman R. Augustine; all rights reserved, New York: American Institute of Aeronautics and Astronautics, Inc., 1982.
8. Verity, J. W., "Empowering Programmers," *Datamation,* October 15, 1985.
9. Smithy-Willis, D., J. Willis, and M. Miller, *How to Use Telofacts.* Beaverton, OR: Dilithium Press, Ltd., 1983.
10. *Directory of Computer Software, 1986,* PB86-135357, National Technical Information Service (NTIS), U.S. Departmnt of Commerce, 1986.

MAGAZINES

See References for Chapter 6.

CATALOGS

See References for Chapter 6.

PROBLEMS

P7.1. Define two systems engineering application areas for each of the nine categories of software tools presented in this chapter, other than those shown in Table 7.3.

P7.2. By consulting appropriate computer software magazines and other source materials find two examples of specific packages, other than those listed in Appendix A, for each of the categories of mathematical and engineering tools. List and explain five major features of each.

P7.3. Throw a pair of dice ten times. Calculate the mean and variance for the samples obtained. Compare with the mean and variance of the a priori known distribution and explain why they are different (if they are).

P7.4. Prove the least squares fit relationship shown in Equation 7.1.

P7.5. Find a least squares fit line to the following (X, Y) data points: $(1, 2)$, $(2, 5)$, $(3, 7)$, $(4, 9)$, $(5, 9)$.

P7.6. Identify five features that you would build into a mathematics software package for each of the following areas:
  a. algebra
  b. calculus
  c. differential equations
  d. vector analysis

P7.7. You have a personal computer and $5000 to spend on software. By referring to catalogs and magazines, develop a list of software tools (by name and manufacturer) from the categories in Chapters 3, 6, and 7 that you would purchase to develop your own workstation. Explain your rationale for each selection. Develop a similar list and rationale for a workstation devoted entirely to systems engineering applications.

P7.8. List and discuss ten attributes of software packages that are common to all of the tool categories presented in this chapter.

There is one thing certain, namely, that we can have nothing certain; therefore it is not certain that we can have nothing certain

Samuel Butler

INFORMATION AND SEARCH THEORY

This chapter deals with selected aspects of information theory and search theory, two subjects not usually found in systems engineering texts. The motivation for including information theory here is that we are in an information age, one likely to be with us for some time to come, implying that we must better understand the nature and theory of information—how to define it, how to measure it, how to process it, how to display it, and how human beings assimilate it. We are at the very early stages of linking information theory with systems engineering and computer aids, and hopefully this will set the stage for some further developments in this linkage.

With respect to search theory, two observations are noteworthy. The first, and more important for purposes of this text, is that the central core of system synthesis is to "search" for so-called optimal configurations from among a large array of possibilities. As discussed in earlier chapters, a systematic approach to design or synthesis often leads to combinatorics, which represent numerous options—too numerous to be treated entirely through engineering judgment. Therefore, we must increasingly use computer-aided techniques to search for the most desirable solutions. Secondly, we find that certain aspects of search theory relate directly to both concepts and algorithms that come from the field of information theory. Specifically, various information theory principles lead directly to efficient search procedures. These areas of information and search theory are explored, on a selected and largely heuristic basis, in this chapter.

8.1 BASIC INFORMATION THEORY CONCEPTS

The beginnings of what is now called information theory are traceable to the work of Claude Shannon [1], who defined the basic concepts and measures of information theory and related them to the fields of communication theory and engineering. A central concept in the theory is the notion of uncertainty and that if there is no inherent existence and resolution of uncertainty, then information is not and cannot be transmitted. Uncertainty is dealt with in a very formal way, requiring the definition of a probability scheme for both discrete and continuous situations. To illustrate with a simple case, we consider the matter of tossing a coin. The probabilities of a head or a tail are both one half and before we run the "experiment" of tossing a coin we have uncertainty as to which of these results will be observed, known as a priori probabilities. We will define the *self information* as to one or the other of these two possibilities as

$$I = -\log p(i) \tag{8.1}$$

where $p(i)$ is the a priori probability of a head, which of course is the same as the a priori probability of a tail. This self-information notion is tied to a particular result (a head, for example). The self information is thus the same for the *head* result as it is for the *tail* result and is numerically calculated as

$$I = -\log\left(\frac{1}{2}\right) = \log 2 = 1 \text{ bit of information}$$

We note two aspects of this last computation. First, we have taken the logarithm to the base two in which case the unit of information is measured in *bits* (*b* for binary). Had we taken the logarithm to the base ten the unit of information would have been measured in *dits* (*d* for decimal). This choice is arbitrary and simply connects the choice of base directly to the units in which information is measured. Second, we observe the fact that the essential measure is logarithmic. There are rather fundamental reasons for this selection, which will not be explored here. The reader is referred to Shannon's work, as well as other writers on this subject, to understand this rationale.

Thus, if we run an experiment that consists of tossing a coin many times, we will obtain one bit of information each time the result of a toss is observed. In ten tosses we will have obtained ten bits of information since all tosses are independent. In each case there was initial uncertainty as to the result of a toss and when the result was observed the uncertainty was resolved. In general, we can distinguish between the information obtained on each trial of an experiment and the average amount of information obtained per trial. It is a conventional mean or expected value concept, as developed earlier in Chapter 5, and it recognizes that the average value may differ from any and all values obtained in a series of trials. Using formal probability relationships we find this mean or average value as

$$H = E[-\log p(i)] = -\sum p(i) \log p(i) \tag{8.2}$$

and call this the entropy of the probability scheme. The entropy value in the coin toss situation is therefore

$$H = -\left(\frac{1}{2}\right) \log \left(\frac{1}{2}\right) - \left(\frac{1}{2}\right) \log \left(\frac{1}{2}\right) = 1 \text{ bit}$$

Clearly, in this case, the information obtained on each trial is always one bit, which is the average information obtained per trial. This average value per trial, or *entropy*, is the central concept of information theory and in fact is related to the notion of thermodynamic entropy in physics [2].

The entropy measure is immediately applicable to other probability distributions. To illustrate, consider two cases in the toss of a die—the first is a normal die whose faces are numbered one through six, and the second is a special die that has four faces with the letter A and two faces with the letter B. For the normal die the self information in all cases is log 6 as is the entropy. We note that this value is 2.6 bits, which is larger than the entropy for the coin toss case. The reason is that in the normal die situation the prior probability was smaller (1/6 vs. 1/2) and therefore a greater uncertainty was resolved when the result was observed, all based upon the particular logarithmic definition of entropy. In the case of the special die, we calculate the prior probabilities as

$$p(A) = \frac{4}{6} = \frac{2}{3} \text{ and } p(B) = \frac{2}{6} = \frac{1}{3}$$

The self information in each case is

$$I(A) = 0.585 \text{ bits and } I(B) = 1.585 \text{ bits}$$

and the entropy is computed as

$$H = -\left(\frac{2}{3}\right) \log \left(\frac{2}{3}\right) - \left(\frac{1}{3}\right) \log \left(\frac{1}{3}\right)$$

$$= -\left(\frac{2}{3}\right) \log 2 + \left(\frac{2}{3}\right) \log 3 - \left(\frac{1}{3}\right) \log 1 + \left(\frac{1}{3}\right) \log 3$$

$$= \log 3 - \left(\frac{2}{3}\right) \log 2 = 0.918 \text{ bits}$$

Thus in this special case we can see that each self-information measure is different and both are different from the entropy or average information per trial measure.

Some investigators of information theory have generalized the notion of self information to account for situations in which the uncertainty is not completely resolved after an experiment is run. For such a case, the probability after the experiment, known as the a posteriori probability, is other than zero and the self information may be defined as

$$I = \log \frac{p(\text{a posteriori})}{p(\text{a priori})} \tag{8.3}$$

In the case for which there is no residual uncertainty after the experiment (the observation instrument is assumed to be perfect, completely resolving the uncertainty, which is not always the situation), p (a posteriori) $= 1$, and the generalization in Equation 8.3 reduces to the form previously defined, namely,

$$I = -\log p \text{ (a priori)} \tag{8.4}$$

Two properties of the entropy function are especially noteworthy. First, it is seen that in general as the number of possible outcomes, n, increases, the entropy increases. This means that as greater differentiation occurs in terms of increased probability states, there is greater uncertainty. It makes intuitive sense that this should be the case. For example, as we have seen in this section, greater overall uncertainty exists in a die toss situation in distinction to a coin toss. This concept is extended in moving to a roulette wheel from the toss of a die. Second, for a given number of outcomes, n, the entropy becomes a maximum when $p(i) = p(j)$, that is, when all probabilities are equal. This means that no event has preference over the others, which intuitively represents a situation of maximum uncertainty. The value of the entropy under this circumstance is found simply as $H = \log n$.

We pause for a moment to comment upon the entropy notion and its relationship to systems engineering and computer applications. Since entropy is a measure of uncertainty we first observe that linkages exist only when uncertainty can be said to be present in systems engineering. This is clearly the case in a large number of systems engineering areas, for example, risk analysis, schedule uncertainty, and probabilistic modeling and analysis (see problems at end of chapter). To be able to utilize the information-theoretic notion of entropy, as can be seen from the simple examples in this section, it is necessary that we estimate a priori probabilities (and a posteriori probabilities, when the post-trial observation may have errors in it). From a computer-aid point of view, we now have the ability to generate massive amounts of data—pages and pages of rows and columns of numbers. The *resolution of uncertainty* concept of information tells us that there is no information conveyed when we know in advance the results of a given trial or experiment. Therefore if we receive a data tabulation (a spreadsheet, DBMS output, a Gantt chart, etc.) in a systems engineering activity, and 90% is what we have seen (and remember) from earlier data outputs, then at least 90% conveys no real information to us. This represents another context in which to view the relationship between information theory and systems engineering. The former may provide some guidance in regard to cutting down on redundancy, attempting to convey real information and sparing the systems engineer as well as the customer from having to sort through endless amounts of data containing only sparse amounts of information. Here again, from an information theory point of view, conveying what one already knows is to convey no information at all.

8.2 REDUNDANCY AND CODING

The issues referred to in the previous discussion can be placed in a more rigorous and quantitative context, which will be discussed in this section. There is considerable redundancy in the English language and likely even more redundancy in the way that alphanumeric data is organized and presented within the tasks and activities of systems engineering. However, redundancy can be a two-edged issue. Data conveyed with a great deal of redundancy helps users when they are learning—when contact first occurs and the data format as well as content is being assimilated. An example might be the matter of menu-versus command-driven software packages. When the user is unfamiliar with the package, the menu approach is often preferred since it is highly structured, sequential, and repetitive. It is also highly redundant, that is, at each level, the commands are well defined and explained, usually in English phrases. However, when the user becomes proficient in the use of a package it is often laborious to have to track forward and back through the various menus. Shortcuts are desired by the well-trained user and the original redundancy is viewed as an obstacle to speed and efficiency. All of a sudden a nonmenu command structure is more desirable, as is the benefit of operating with considerably less redundancy. Thus, at times redundancy is useful and at other times it can reduce productivity.

Redundancy has a rather particular meaning within the realm of information theory. If we consider a source that is generating a sequence of symbols, such as the letters of the alphabet (and a space), we have a total of 27 possible symbols that can be developed. If each of these symbols had an equal likelihood of being selected and transmitted, then we would have a situation of maximum entropy which is easily calculated as $H(\text{max}) = \log n = \log 27 = 4.75$ bits of information per symbol. This clearly can be expanded if we allow additional symbols as provided on a conventional keyboard such as the numbers zero through nine and symbols such as $, %, &. To the extent that the probabilities of symbol selection and transmission differ from being precisely equal, then we say that there is some built-in redundancy in the source. If we define the relative entropy as the ratio of the actual source entropy to the maximum entropy, that is,

$$E(R) = \frac{H(\text{actual})}{H(\text{max})} = \frac{H(\text{actual})}{\log n} \qquad (8.5)$$

then the redundancy is given as

$$R = 1 - E(R). \qquad (8.6)$$

Clearly, if $H(\text{actual}) = H(\text{max}) = \log n$, then $E(R)$ is unity and the redundancy is zero.

Looking at some specific data with respect to the English language, with a 27 letter-and-space representation, we have a set of estimated probabilities as provided in Table 8.1 [3]. Analyses of the English language have concluded that printed texts are of the order of 75% redundant. Thus the true information content is about one fourth of the raw text, implying that with proper coding one could send the same information with only one quarter of the original number of transmitted symbols. For a given transmission time per symbol this translates into a time saving in the same ratio.

Coding, as referred to previously, is a means by which one can achieve higher information-transmission rates. As shown in Figure 8.1, the encoder is placed between the source and the transmission channel. Likewise, the decoder is at the output of the channel, translating the coded symbols back into a data stream recognizable by the user. The essential purpose of the coding process is to remove

TABLE 8.1 PROBABILITIES
FOR ENGLISH LANGUAGE
LETTERS

Letter	Estimated Probability
A	0.0642
B	0.0127
C	0.0218
D	0.0317
E	0.1031
F	0.0208
G	0.0152
H	0.0467
I	0.0575
J	0.0008
K	0.0049
L	0.0321
M	0.0198
N	0.0574
O	0.0632
P	0.0152
Q	0.0008
R	0.0484
S	0.0514
T	0.0796
U	0.0228
V	0.0083
W	0.0175
X	0.0013
Y	0.0164
Z	0.0005
Space	0.1859

redundancy (thus increasing true information-transmission efficiency). Other purposes, in a more sophisticated context, include error detection and correction as the channel introduces errors, and encryption of the source message so as to prevent interception and understanding of the message by a possible interloper. We will focus here on the matter of improving efficiency and removing redundancy.

We illustrate some coding procedures by means of a simple example. The symbols used for coding will be only zeros and ones, corresponding, for example, to the dots and dashes of the Morse code. Strictly speaking, any number of symbols can be used to represent this *coding alphabet* but limiting ourselves to just the two symbols will demonstrate the major points of coding. In addition, to avoid confu-

Figure 8.1 Coding and decoding an information transmission system.

sion, we will call the output of the source a series of *messages* instead of symbols. Thus, if the source is sending the letters of the alphabet, each such letter (and a space) will be referred to here as a message. The encoding process will then assign a sequence of zeros and ones to each of the messages to be sent by the source (i.e., the 27 alphabetic letters).

In Table 8.2 we show three possible zero–one codes for the 27 alphabetic letters now being interpreted more generally as 27 individual messages. Different numbers as well as sequences of zeros and ones have been assigned to the alphabetic messages and it is of interest to calculate the average number of symbols per message as the efficiency of transmission is directly related to this measure. Since the source messages have different likelihoods of being selected, the average number of code symbols per message is

$$L = \sum p[m(i)] \, n(i) \tag{8.7}$$

where $p[m(i)]$ is the probability that message i is selected and $n(i)$ is the number of coded symbols assigned to message i. For the three codes in Table 8.2 and the prob-

---

TABLE 8.2 ILLUSTRATIVE CODING PROCEDURES

Letter	Binary Code	Alphabetic Code	Huffman Code
A	00000	0100	0100
B	00001	010100	0111111
C	00010	010101	11111
D	00011	01011	01011
E	00100	0110	101
F	00101	011100	001100
G	00110	011101	011101
H	00111	01111	1110
I	01000	1000	1000
J	01001	1001000	0111001110
K	01010	1001001	01110010
L	01011	100101	01010
M	01100	10011	001101
N	01101	1010	1001
O	01110	1011	0110
P	01111	110000	011110
Q	10000	110001	0111001101
R	10001	11001	1101
S	10010	1101	1100
T	10011	1110	0010
U	10100	111100	11110
V	10101	111101	0111000
W	10110	111110	001110
X	10111	1111110	0111001100
Y	11000	11111110	001111
Z	11001	11111111	0111001111
Space	11010	00	000

abilities for the English letters in Table 8.1, the average numbers of code symbols per message are listed below

- binary code: 5.0 symbols/message
- alphabetic code: 4.1978 symbols/message
- Huffman code: 4.1195 symbols/message

The fact that the so-called Huffman code has the shortest code length means that it is the most efficient. That is, we can send the same message sequence by means of a shorter overall coded sequence of symbols, thereby either sending more information in a given time or a given amount of information in a shorter period of time. The efficiency of a coding procedure can be computed by the relationship

$$E(C) = \frac{H}{L \log D} \qquad (8.8)$$

where H is the entropy of the message source, D is the number of symbols in the coding alphabet, and L is the average number of code symbols per message as defined in Equation 8.7. For a given source entropy, H, and number of symbols in the coding process ($D = 2$ for coding with zeros and ones), we verify that coding efficiency is inversely related to the average length of the code (L), as alluded to in the previous discussion.

Another simple example will illustrate the notions of coding. We will consider simple binary coding versus the Huffman code for four messages, as shown in the table.

Message	Message Probability	Binary Code	Huffman Code
A	$\frac{1}{2}$	00	0
B	$\frac{1}{4}$	01	10
C	$\frac{1}{8}$	10	110
D	$\frac{1}{8}$	11	111

The entropy, H, of the message source does not change with the coding procedure and is calculated directly as

$$H = -\left(\frac{1}{2}\right) \log \left(\frac{1}{2}\right) - \left(\frac{1}{4}\right) \log \left(\frac{1}{4}\right) - \left(\frac{1}{8}\right) \log \left(\frac{1}{8}\right) - \left(\frac{1}{8}\right) \log \left(\frac{1}{8}\right)$$

$H = 1.75$ bits per message

where bits are interpreted as before as a measure of information content since the logarithm is taken to the base two. Similarly, we use the same logarithm base for the $\log D$ term so that $\log D = \log 2 = 1$. The average code length for the binary code is numerically equal to 2, by inspection, that is, each coded message is of length 2.

For the Huffman code the average length is computed as

$$L = \frac{1}{2}(1) + \frac{1}{4}(2) + \frac{1}{8}(3) + \frac{1}{8}(3)$$

$$L = 1.75 \text{ symbols per message}$$

Thus the coding efficiencies are seven eighths or 87.5% for the binary code and 100% for the Huffman code. Clearly one cannot do better than the Huffman code. We note that the latter is ideally matched to the source message probabilities. Its construction may be viewed in the following light. We partition the source message set into two equally likely groups since we are using two symbols for the coding (a zero and a one). This grouping puts message A into one group and messages B, C, and D into the other group such that each group has a probability of one half. We then encode the first digit of the code as a zero for the first group (message A) and a one for the second group (messages B, C, and D). We now partition the B, C, D group into two equally likely subgroups and encode the second digit in an analogous manner. The assigns a second digit of 0 to message B and 1 to messages C and D. This process continues for messages C and D, yielding the final Huffman code. This process is optimal for this type of coding as seen by the 100% efficiency. It is also related to the process of binary search through a file as discussed later in this chapter.

There is a great deal of literature available on the matter of coding and properties of various codes, in relation to which selected references are listed at the end of this chapter. It is important to note that the coding considerations discussed here apply to the efficient coding of a source with known message probabilities and no noise. Other considerations are necessary for coding so as to combat noise that is introduced by the medium of transmission, that is, the channel, for purposes of error detection and correction. Such codes are examined in detail in the extensive literature available on this subject.

8.3 CAPACITY

A central concept embodied in the information theory is that of capacity of a channel. A channel is often thought of in terms of communications but it can be more broadly interpreted as any conduit or processor of information, including a computer or a human being. In the latter context, researchers in the field of man–machine interactions and experimental psychology have been examining and attempting to assess the maximum rate at which people can process information. Using the basic entropy measure, the issue is posed in terms of a human's ability to perceive, assess, and select among alternatives. As these alternatives increase in powers of 2, such as 2^1, 2^2, 2^3, selections are made so as to process 1, 2, and 3, and so on, bits of information. Unfortunately, results can be strongly dependent upon the particular manner in which these alternatives are presented, the degree of human processing necessary, and other factors. G. Raisbeck [4] considers some rough bounds by examining reading at one extreme and chess playing at the other. The former is calculated in terms of 500 words per minute, 5 letters per word, and 1 bit per letter (accounting for the 75% English-language redundancy), yielding an estimate of some 42 bits per second. With respect to chess, his rough estimate is of the order of 0.2 bits/sec. Attneave's book [5], which deals with applications of information theory to psychology, cites several experiments and generally appears to confirm rates within these two bounds, depending upon the experimental designs and conditions. A ham radio operator receiving and decoding the Morse code can operate at 20–25 words/minute, also fitting within the above rate estimates. Whatever the specific rate, it is seen that the ability of a human is quite limited in terms of assessing and selecting among even simple alternatives in a short period of time. Certainly machines can be programmed to carry out such selections much more quickly, as long as the assessment process is well understood and can be appropriately converted into rules.

For some of the aforementioned reasons we have difficulty, especially in real-time or near real-time systems, in finding a proper place for the "person in the loop." At the same time we know that it is crucially important to have the person in the loop for systems requiring considerable human judgment. An example is the strategic defense or "star wars" scenario. The time lines for the boost phase can be viewed to have an upper bound of some five minutes (300 sec). This is almost no time at all for a decision maker to decide if the threat is real and then to direct defensive forces at such a threat. At the same time it is undesirable to bypass the human decision maker and have machines essentially making this kind of decision on their own. Sensors and computers can provide and process enormous amounts of data in just a couple of minutes but the human decision maker is mismatched to this high data rate environment. Thus we have capacity and time line mismatches in some of our most important large-scale systems—essentially all such systems with near real-time requirements, multimegabits of data flowing and the human in the loop.

An example of a system with a relatively good inherent man–machine capacity match is the modern rapid-transit system for a large metropolitan area. Surely it is possible to run such a system strictly by computers. However, for safety as well as other institutional reasons, we normally have train operators that do not appear to be near overload in terms of information handling. Moving to air traffic control, however, the situation appears to be somewhat more stressful for the air-traffic controller, although this individual is not the actual vehicle operator. Improvements in air-traffic-control computer systems are designed, at least in part, to improve controller productivity and reduce stress by offloading appropriate processing tasks to the computer where this can be achieved safely. In broad terms, all of these considerations are attempts at matching the capacity capabilities of system elements while maintaining acceptably low error rates.

Information theory has a set of rather formal constructs dealing with channel capacity. They all have as a basis the entropy concept developed earlier. Briefly demonstrating these relationships, we first define probability functions $p(x)$ as representing the inputs to a channel and $p(y)$ as representing the outputs. We define the *transinformation* with respect to the channel and these input–output probability functions as

$$I(X; Y) = E\left[\log \frac{p(x, y)}{p(x)\,p(y)}\right]$$

$$= \sum\sum p[x(i), y(j)] \log \frac{p[x(i), y(j)]}{p[x(i)]\,p[y(j)]} \qquad (8.9)$$

We see that the essential relationship is logarithmic and the ratio of the argument is the joint distribution divided by the product of the distributions—such a ratio interpretable as the extent to which the joint distribution is *close to* or *far from* (statistical) independence. The capacity of a channel, then, is the maximum value of this transinformation, expressed as

$$C = \max I(X; Y) \qquad (8.10)$$

In order to understand both the notions of capacity, C, as well as transinformation, $I(X; Y)$, we define the following entropy relationships:

$$H(X) = -\sum p[x(i)] \log p[x(i)] \qquad (8.11)$$

$$H(Y) = -\sum p[y(j)] \log p[y(j)] \qquad (8.12)$$

$$H(X, Y) = -\sum\sum p[x(i), y(j)] \log p[x(i), y(j)] \qquad (8.13)$$

$$H(X|Y) = -\sum\sum p[x(i), y(j)] \log p[x(i)|y(j)] \qquad (8.14)$$

$$H(Y|X) = -\sum\sum p[x(i), y(j)] \log p[y(j)|x(i)] \qquad (8.15)$$

These definitions and related concepts can be clarified by reference to an example. Figure 8.2 shows a simple binary channel such that the inputs $x(i)$ can take on the values 0 and 1 as can the outputs $y(j)$. The channel, in general, may introduce

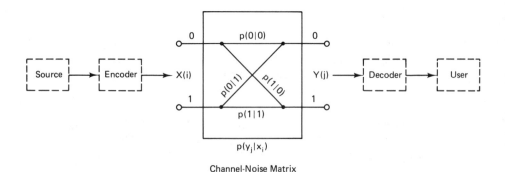

Channel-Noise Matrix

Figure 8.2 A binary channel.

Information and Search Theory   Chap. 8

noise, converting an input 0 into an output 1 or an input 1 into an output 0 with probabilities $p[y(j)|x(i)]$. These probabilities are called the *channel-noise matrix,* an example of which might be

$$Y = \text{output}$$

$$p(y_j|x_i) = \begin{array}{c} \\ \\ \text{X = input} \end{array} \begin{array}{c} \quad 0 \quad\quad 1 \\ 0 \\ 1 \end{array} \begin{bmatrix} \dfrac{2}{3} & \dfrac{1}{3} \\ \dfrac{1}{4} & \dfrac{3}{4} \end{bmatrix}$$

If the probability of converting an input 0 into an output 1 is 1/3, then $(1 - 1/3) =$ 2/3 is the *no error* probability for a 0 input. Similarly, the error probability for a 1 input is 1/4 and the probability of no error is 3/4.

We will now assume that the input zeros and ones $[x(i)]$ are both transmitted with probabilities 1/2. We note that in general

$$p[x(i), y(j)] = p[y(j)|x(i)]\,p[x(i)] \tag{8.16}$$

so that

$$p(0, 0) = p(0|0)p(0) = \left(\frac{2}{3}\right)\left(\frac{1}{2}\right) = \frac{1}{3}$$

$$p(0, 1) = p(1|0)p(0) = \left(\frac{1}{3}\right)\left(\frac{1}{2}\right) = \frac{1}{6}$$

$$p(1, 0) = p(0|1)p(1) = \left(\frac{1}{4}\right)\left(\frac{1}{2}\right) = \frac{1}{8}$$

$$p(1, 1) = p(1|1)p(1) = \left(\frac{3}{4}\right)\left(\frac{1}{2}\right) = \frac{3}{8}$$

The matrix of probabilities $p[x(i), y(j)]$ is called the *system matrix* and contains all joint probabilities of input and output, as shown in the previous equations. We interpret the various entropies in Equations 8.11 through 8.15 as follows.

Entropy	Uncertainty with Respect to
$H(X)$	The input messages
$H(Y)$	The output messages
$H(X, Y)$	The joint input-output message scheme
$H(X|Y)$	The input, given the output
$H(Y|X)$	The output, given the input

We compute their values as

$$H(X) = H\left(\frac{1}{2}, \frac{1}{2}\right) = 1$$

$$H(Y) = H\left(\frac{11}{24}, \frac{13}{24}\right) = 0.996$$

$$H(X, Y) = H\left(\frac{1}{3}, \frac{1}{6}, \frac{1}{8}, \frac{1}{8}\right) = 1.865$$

$$H(X|Y) = 0.869$$

$$H(Y|X) = 0.865$$

It can be shown (problem P8.7) that the previous transinformation and entropies are related as follows:

$$H(X, Y) = H(X|Y) + H(Y) = H(Y|X) + H(X) \tag{8.17}$$

$$I(X; Y) = H(X) + H(Y) - H(X, Y) \tag{8.18}$$

$$I(X; Y) = H(X) - H(X|Y) \tag{8.19}$$

$$I(X; Y) = H(Y) - H(Y|X) \tag{8.20}$$

yielding the numerical value $I(X; Y) = 0.131$ for this example. From Equations 8.19 and 8.20 we observe that to maximize $I(X; Y)$ and thereby find the channel capacity we need to look for

$$C = \max \left[H(X) - H(X|Y)\right] \tag{8.21}$$

or

$$C = \max \left[H(Y) - H(Y|X)\right] \tag{8.22}$$

noting that $H(Y|X)$ is given and fixed as the channel noise matrix. A formal procedure [3] for finding this capacity is to determine $Q(i)$ such that

$$C = \ln \sum \exp \left[-Q(i)\right] \tag{8.23}$$

where the $Q(i)$ can be found from

$$\sum_j p[y(j)|x(i)] Q(j) = H(i) \tag{8.24}$$

The particular set of input probabilities $p^*[x(i)]$ (in general different from the 1/2, 1/2 illustrative example) that will achieve a maximum transinformation can be

Information and Search Theory   Chap. 8

found by the relationship

$$\sum_i p[y(j)|x(i)] \, p^*[x(i)] = \frac{\exp[-Q(j)]}{\sum \exp[-Q(j)]} \qquad (8.25)$$

Two special cases are now cited to examine bounds on channel entropy and capacity relationships. The first is the so-called noiseless channel for which the noise-matrix cross elements are zero, that is, $p(1|0) = p(0|1) = 0$ and $p(0|0) = p(1|1) = 1$. Clearly in this case $H(Y|X) = H(X|Y) = 0$ and

$$I(X; Y) = H(X) = H(Y) = H(X, Y)$$

This means that whatever uncertainty there is on the input side is not further confused by the channel. We also must transmit the input messages with equal probability (1/2 each) to maximize the above transmission and the channel capacity is max $H(X) = \log n$. In the second limiting case the inputs and outputs are independent, meaning $p[x(i), y(j)] = p[x(i)] \, p[y(j)]$. In such a situation, from the definition in Equation 8.9, the transinformation goes to 0 and no information can be transmitted across such a channel. One last case of interest is the binary symmetric channel for which $p(0|0) = p(1|1) = p$ and $p(0|1) = p(1|0) = q$. It is left as an exercise (see problem P8.6) to show that in such a case the capacity can be found as $C = 1 + p \log p + q \log q$.

The capacity relationships discussed in this section and coding concepts of the previous section can be brought together in what is thought of as a fundamental theorem of information theory, which can be stated as

It is possible to encode a source to transmit information at a rate, R, that is close to, but less than, the channel capacity, C, of a channel into which the source is transmitting.

As referred to earlier, much of coding theory addresses the construction of specific coding schemes that get closer and closer to the channel capacity, for various types of channel noise characterizations and sources.

8.4 INFORMATION THEORY, DIAGNOSTICS, AND SEARCH

In a problem of diagnostics, we first observe a symptom and then take a set of measurements in order to localize a failure or a degraded set of conditions. Such types of problems are embedded in engineering (e.g., maintenance of a piece of equipment) as well as numerous other fields (e.g., why your car won't start, medical diagnosis). Information-theoretic concepts have been applied to diagnostics in the context of setting up sequential diagnostic steps such that the information obtained on each step is a maximum. In this case the so-called figure of merit (FOM) for efficient, though not necessarily optimum, diagnostic procedures is the maximum resolution of uncertainty per step. This makes intuitive sense in that this maximum resolution should lead quickly to identifying the source of the problem. For a variety of reasons that we will not address here, such an approach is not necessarily optimal but it can be expected to yield efficient procedures in many cases,. We also can conceive of a related approach that accepts as a FOM the maximization of information obtained (uncertainty resolved) per unit of test cost [6]. This likewise is intuitively appealing in that certain diagnostic procedures may be prohibitively costly even though they promise to bring considerable information each step of the way. Appealing variations in this theme can be found in which the FOMs relate to information measures in the numerator and cost or effort measures in the denominator.

By way of explaining the concept of sequential diagnostics or search utilizing information-theoretic concepts, we will pose and show a solution to a penny-weighing problem which can be stated as

> We have 12 coins all of which are the same, except for one which is either lighter or heavier than the others. We also have an equal arm balance, which allows us to compare weights on both sides of the balance. Devise a procedure such that, in three weighings, the dissimilar coin will be found as well as whether it is heavier or lighter than the other coins.

This can be viewed as a problem in diagnostics or search, both of which share a common thread of logic. We will use an information-theoretic approach to solve the problem, utilizing the notion of attempting to resolve as much uncertainty as possible for each of the three equal-arm-balance measurements.

We first observe that the uncertainty in such a problem is of measure log 24 since we have 12 coins and initially we do not know whether the false coin is heavier or lighter than the others. When a measurement is made, three possibilities are apparent: balance, the right side is heavier, the left side is heavier. We interpret this as a potential information gain (or uncertainty resolution) of at most log 3. If we made M such measurements we could obtain information of the most M log 3. For $M = 3$, 3 log 3 > log 24 so that we might expect that 3 (or more) measurements will be sufficient. Clearly for M less than 3 we cannot hope to resolve the uncertainty.

For the first measurement, any number from 1 to 6 coins can be placed on each side of the balance. It can be shown (problem P8.2) that if we initially measure 4 coins against 4 coins we obtain the greatest information (log 3). We note that the criterion for selection was indeed the requirement to maximize the resolution of uncertainty on the first as well as successive weighings.

As shown in Figure 8.3, for the first weighing of 4 versus 4 coins, we either have balance or no balance (e.g., right side heavier). If there is balance, we know that the false coin must be among those four not weighed, which will be designated as $s(1)$, $s(2)$, $s(3)$, and $s(4)$. By looking at the different possibilities at this point we can determine that the greatest potential information gain is 1.56 bits. This is seen by laying out the various plausible combinations for this second weighing and calculating the probabilities of

- the right side being heavier
- the left side being heavier
- balance

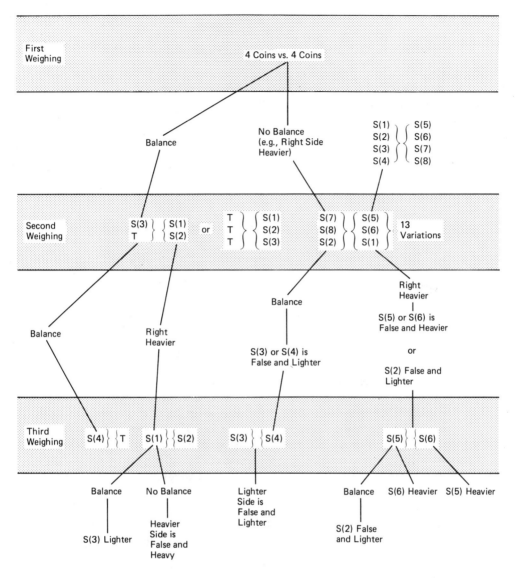

T = Coin Determined to be True

Figure 8.3 Coin search/weighing solution alternatives.

which represent the only three cases. We then calculate the entropy for the various combinations and look for those that have the greatest entropy, as listed in Table 8.3. The two alternatives yielding the highest entropy are shown in the figure and are represented as

1. measure $s(1)$, $s(2)$ against $s(3)$, T, or
2. measure $s(1)$, $s(2)$, $s(3)$ against T, T, T

TABLE 8.3 COIN COMBINATIONS AND ENTROPIES FOR PATH OF COIN-WEIGHING PROBLEM

Combinations of Coins*		Probability of Right Side being Heavier	Probability of Left Side being Heavier	Probability of Balance	Entropy (bits)
1	1	$\frac{1}{4}$	$\frac{1}{4}$	$\frac{1}{2}$	1.50
1	0	$\frac{1}{8}$	$\frac{1}{8}$	$\frac{3}{4}$	1.06
2	2	$\frac{1}{2}$	$\frac{1}{2}$	0	1.00
2	1	$\frac{3}{8}$	$\frac{3}{8}$	$\frac{1}{4}$	1.56
2	0	$\frac{1}{4}$	$\frac{1}{4}$	$\frac{1}{2}$	1.50
3	1	$\frac{1}{2}$	$\frac{1}{2}$	0	1.00
3	0	$\frac{3}{8}$	$\frac{3}{8}$	$\frac{1}{4}$	1.56
4	0	$\frac{1}{2}$	$\frac{1}{2}$	0	1.00

*Known true coins are added in each case for which there is an imbalance of coins on each side of the equal arm balance.

where T is a known true coin. Following the former (the latter is left as an exercise), if a balance is obtained the $s(4)$ is the false coin and whether it is heavier or lighter can be ascertained by measurement against a true coin, as shown. If there is not a balance (right side heavier), then clearly the next step is to measure $s(1)$ against $s(2)$. If a balance is obtained, then $s(1)$ and $s(2)$ are true coins and the false coin is $s(3)$, which is lighter than the others. If $s(1)$ and $s(2)$ do not balance then the heavier side contains the false coin, which is heavier than the others.

Returning to the 4 versus 4 case for which a balance is not achieved, Figure 8.3 shows the various measurement options that will lead to a solution. A diligent reader may wish to develop these alternatives by utilizing the principle that we wish to maximize the information obtained on each weighing. This of course is done by choosing measurement schemes that yield a maximum value of entropy or prospective uncertainty resolution, cases for which the probabilities of the various alternatives are as nearly equal as possible. Thus we can effectively utilize information theoretic principles to develop a specific solution to at least this class of search problems.

8.5 A SEARCH THEORY CONSTRUCT

We have indicated that there are relationships between information theory and search theory. In earlier portions of this chapter, in particular, it was shown that certain search problems could be successfully addressed by utilizing principles of information theory. In other words search routines could be devised whose figure of merit incorporated the notion of maximizing the information obtained for each step of the search process. However, it was pointed out that in this class of problems there is by no means a guarantee of optimization. Indeed, a very large number of sequential search problems are not amenable to optimal (and practical) solution, suffering from the possibility that a figure of merit approach can start one down a nonoptimal tree-like path from which it is impossible or at best difficult to recover. We will now cite a case that represents a variation on this theme, demonstrating that a search technique that establishes the highest overall detection probability is not necessarily the one that maximizes the expected information obtained.

There is a class of search problems that involves looking at various "cells," attempting to maximize the probability of detection of the object of the search while either constrained to a total level of search "effort" or minimizing the effort expended in the search process. The cells, for example, can be regions of the sky in which one is searching for a target (e.g., radar detection). In this introductory example [7], we consider 2 cells and postulate that the target is equally likely to be in either cell and also that the single look detection probability is 1/2. Further, we assume that only two looks are available. The detection probabilities for two looks in one cell versus one look in each of the two cells is evaluated as follows. For two looks, the probability of detection is 3/4 when the target is in the cell and 0 when it is not, yielding the overall detection probability $P_{20}(d)$ as

$$P_{20}(d) = \left(\frac{1}{2}\right)\left(\frac{3}{4}\right) + \left(\frac{1}{2}\right)(0) = \frac{3}{8}$$

If we look once in each cell, the detection probability is found as

$$P_{11}(d) = \left(\frac{1}{2}\right)\left(\frac{1}{2}\right) + \left(\frac{1}{2}\right)\left(\frac{1}{2}\right) = \frac{1}{2}$$

Thus for a given available level of effort (2 looks) the prodedure of looking once in each cell yields a higher overall probability of detection. Now the question is—if we follow a procedure that maximizes the resolution of uncertainty from an information theoretic viewpoint, do we get the same results? The answer is negative because we can obtain more information by looking twice in one cell, a case for which the detection probability, from the previous equation, is lower. This can be demonstrated as follows. For both cases we start out with an uncertainty (entropy) of one bit since the two probabilities are the same and equal to 1/2. In the two look case, after the two looks, if the target is detected (with probability 3/8) the probabilities are 0 and 1. If the target is not detected (with probability 5/8) the probabilities are 1/5 that it is in cell A and 4/5 for cell B. Thus the difference in information obtained, or the resolved uncertainty, can be found as

$$H\left(\frac{1}{2}, \frac{1}{2}\right) - \left[\frac{3}{8}H(1, 0) + \frac{5}{8}H\left(\frac{1}{5}, \frac{4}{5}\right)\right]$$

where $H(\ )$ is the entropy (Equation 8.2), which can be shown to be numerically equal to 0.38 bits. By a similar argument, for the one look in each cell situation the information obtained would be

$$H\left(\frac{1}{2}, \frac{1}{2}\right) - \left[\frac{1}{2}H(1, 0) + \frac{1}{2}H\left(\frac{1}{2}, \frac{1}{2}\right)\right]$$

which is 0.347 bits. Thus we obtain more information in the two-look case than in the one-look case whereas the latter yields the greater overall detection probability.

Therefore, if our criterion is to maximize the probability of detection of a target, we cannot necessarily look to the criterion of maximizing the information obtained to achieve that result. Thus, although there are linkages between information and search theory they must be used with care. In the remainder of this chapter we explore some additional facets of search theory quite apart from the subject of information theory.

8.6 OPTIMAL SEARCH CONCEPTS

There is considerable theory building in the area of optimal search, but further application work, as indicated earlier, is required in order to bring results into the domain of system synthesis and the search for optimum system configurations. For this reason, the matter is only touched upon here to provide a sense of direction about the concepts.

In this search problem, we define a space of many cells with a target presumed to be located in one of them. The ultimate objective of the search is to find the precise cell in which the target is located. We do not know in advance the target's location, of course, but we can characterize its location in terms of a probability density function or distribution. For example, if we think of the target as located on an x–y plane (equivalent to the search for a downed aircraft or lost ship in a search and rescue mission), we may have a priori information that allows us to describe the target's likely location by means of a probability function constructed "on top of" this plane. We then search for the target with a total available effort, E, portions of which we allocate to the cells in the plane. Each cell has an x–y dimension and we are interested in finding a way to allocate the available effort to the cells such that the probability of detection of the target is maximized. This presumes also that the probability of detection versus effort allocated function is known and increasing, that is, as more effort is applied to a given cell the likelihood of detecting the target, if it is in that cell, is monotonically increasing.

Earlier in this chapter we showed that using the criterion of maximizing the information obtained in a sequential search procedure did not lead to a maximum detection probability. In general, it may be said that at present, information theoretic concepts have not been widely applied to the type of search problem posed in the previous paragraph. Instead, the classical optimization technique of Lagrange multipliers has been the most-often-used method in this area, leading to the definition of so-called optimal search procedures.

A simple example will help to clarify the concept. Assume that over a period of time a variety of keys have been tossed into two tin cans. At some point a need for a particular key has come about and the searcher wishes to find this key. Assume further that the searcher believes that the desired key is twice as likely to be in one can (the red can) than the other can (the blue can). Thus the a priori probability function is simply

$$p(\text{red}) = \frac{2}{3} \qquad p(\text{blue}) = \frac{1}{3}$$

The effort available, E, for the search for the desired key can be thought of as having a given (maximum) amount of time available to conduct the search. The matter of allocation of effort is then considered in terms of dividing this available time among the red and blue cans so as to maximize the likelihood of finding the desired key. The more time spent searching through a given can, the greater the chances of finding the desired key if indeed it is located in the can being searched. There is no chance of "finding" the desired key in the other can, that is, there is a zero false alarm rate. The nature of the detection probability versus effort curve must be specified or assumed in order to proceed further. A typical curve would be exponential as

$$P(t) = 1 - \exp(-t) \qquad t \geq 0 \qquad (8.26)$$

where $P(t)$ is the probability of detection if searched for a time t, given that the search is carried out in the can with the desired key in it. We note that this form of detection function implicitly means that there is always some chance that we will fail to find the desired key in the proper can, even if we search for an indefinitely long time. However, as the time spent in search increases, the likelihood of finding it continues to increase, but at a slower rate—a rate of diminishing returns.

For the aforementioned situation it can be shown [8] that an optimal allocation for a total available time of, say, 4 hours, is

- allocate a total search time to the red can of $2 + (1/2)\ln 2$
- allocate a total search time to the blue can of $2 - (1/2)\ln 2$

We see that the total allocated time is four hours (the sum of the two times) and more time is spent in searching in the red can, which has an a priori probability of twice that of the blue can.

This specific result can be generalized for the exponential situation and any number of discrete cells. However, considering only two cells, let the detection function be different for each cell so that

$$P(j, t) = 1 - \exp[-a(j)t] \qquad \text{for } j = 1, 2 \text{ and } t \geq 0 \qquad (8.27)$$

This means that $a(1)$ and $a(2)$ are different in general, in other words, the detection functions are not the same for each of the two cells. If we let $a(1)p(1) > a(2)p(2)$ by ordering the cells, and

$$y(1) = \ln[a(1)p(1)] - \ln[a(2)p(2)] \qquad (8.28)$$

$$z(11) = \frac{y(1)}{a(1)} \qquad (8.29)$$

it can be shown [8] that the search time (effort) up to $z(11)$ should be allocated to cell 1 and the remaining time (effort) should be split between cells 1 and 2 such that the ratio in cells 1 to 2 is $a(2)/a(1)$.

For the latter example where $a(1) = a(2) = 1$, and $p(1) = p(\text{red}) = 2/3$ and $p(2) = p(\text{blue}) = 1/3$, we have

$$y(1) = \ln\left(\frac{2}{3}\right) - \ln\left(\frac{1}{3}\right) = \ln 2$$

and

$$z(11) = \ln\left(\frac{2}{1}\right) = \ln 2$$

Thus the remaining time is $(4 - \ln 2)$, which is split in the ratio $a(2)/a(1)$ or in this case equally among the two cells. Each cell thus receives $(4 - \ln 2)/2$ time units of the remainder. Thus cell one receives a total of

$$\ln 2 + \frac{4 - \ln 2}{2} = 2 + \left(\frac{1}{2}\right)\ln 2$$

and cell two receives

$$\frac{4 - \ln 2}{2} = 2 - \left(\frac{1}{2}\right)\ln 2$$

agreeing with the results cited previously. Other algorithms can be derived for other detection and target distribution functions (e.g., a circular normal probability density function for the latter) and the interested reader is encouraged to investigate these cases through the principal reference [8].

As indicated previously, search theory results of the type in this section have not generally been applied on a broad basis to issues of system synthesis in which one is searching for an optimum configuration from among a combinatorially large number of alternatives. It is believed that advances in this area will be fruitful and forthcoming with time. In real-time or near real-time areas, applications can likewise be plentiful. For example, radar search for targets in the sky to maximize detection probability is an arena in which this type of theory can be applied directly. The maximum available effort is clearly constrained by the time lines available in the postulated scenarios. Such scenarios will differ depending upon the application (e.g., air-traffic control vs. shipboard air defense vs. a "star wars" defense against ballistic missiles).

8.7 SEARCHING FOR AN OPTIMUM

The notions of search theory can be further generalized to that of searching for an optimum in a function in some n-dimensional space. Another way of looking at this issue is under the general category of optimization techniques. These techniques subsume an extremely broad class of problems and no attempt will be made in this text to deal systematically with this extensive subject. Some examples are cited here, however, to provide a brief perspective.

It is likely that the earliest optimization situations encountered by the engineering student are embedded in the field of calculus. Specifically, maxima and minima of functions of various types can be found utilizing straightforward differentiation procedures. For example, we can take a parabola whose apex points upward in the x–y plane, as

$$y = -x^2 + 6x - 5$$

and find its maximum point by differentiating with respect to x and setting the result equal to 0

$$\frac{dy}{dx} = -2x + 6 = 0$$

in which case $x = 3$. Thus the point at which this function is a maximum is found to be $(x, y) = (3, 4)$. That this is a maximum can be verified by observing the second derivative, which is negative and equal to -2. Thus in this simple fashion the optimization procedures of calculus provide us with a very powerful and broadly applicable set of tools.

Another important optimization technique, alluded to in earlier text, is that of the use of Lagrange multipliers. A simple example will illustrate this method. It was stated without proof in the section on information theory that the entropy (H) is a maximum when all probabilities are equal. The multiplier (lambda $= \lambda$) is introduced with the constraint that the sum of the probabilities is unity through the function

$$H + \lambda[p(1) + p(2) + \ldots, + p(n)] - 1$$

Differentiation with respect to the $p(i)$ yields the set of equations

$$- [\log p(1) + 1] + \lambda = 0$$
$$- [\log p(2) + 1] + \lambda = 0$$

$$\cdot \qquad \cdot \quad \cdot$$
$$\cdot \qquad \quad \cdot \quad \cdot$$
$$\cdot \qquad \quad \cdot \quad \cdot$$

$$- [\log p(n) + 1] + \lambda = 0$$

which implies that $\log p(i) = \lambda - 1$ for all $p(i)$ and therefore $p(1) = p(2) = \ldots . = p(n)$ [3]. This being the case then each $p(\ ) = 1/n$ and the maximum entropy is $\log n$, as stated earlier. Thus the method of Lagrange multipliers (used extensively in search theory) provides a direct, although sometimes complicated, procedure for obtaining an optimum.

Other direct optimization techniques are plentiful and can be found in the literature under such topics as linear, quadratic, integer, backtrack, and dynamic programming. In the context of search, we often deal with extremely complex problems that do not have clear methods of optimization. In such cases a technique known as heuristic programming or search is often employed. This technique in many cases is used sequentially, selecting a local optimum in a step-by-step fashion. If carried out in the proper sequence, such procedures can tend toward optimum selections but are most often not provably optimum. Heuristic techniques are embodied in a particular logic sequence within a computer program, thus defining a specialized algorithm that is the basis for the technique. The interested reader can explore this subject in greater detail in the literature that is currently available on heuristic methods.

8.8 CONVENTIONAL COMPUTER SEARCH

A conventional search situation that is very often imbedded in a computer application can be expressed in terms of a given record or file within which a selected data element is to be found. This can be a case for which one is searching a list to see if a particular name is on the list or one wishes to insert a new name in proper order in an alphabetic list or some variation on this general theme. The original list is ordered in some fashion, for example, numerically or alphabetically, and a search is executed upon such a list.

A much-used and very straightforward approach is to move through the list from the beginning and simply compare each element on the list in a sequential manner to the search criterion until a match is found. For example, to see if the name *Hancock* is on a list we move through the list, one name at a time, doing a *compare* against that name. When the name is found we may print out the name and any additional stored data (record) associated with that name. If we wish to insert a new name into a list, the process is approximately the same. A *value compare* is performed to see if the name to be added is greater than each name in the list. When the compare changes from a *yes* to *no,* the insertion location for the new name has been found. In this type of situation, all other factors being equal, the expected search time is the time to process through to the center of the list.

Although this method of search is straightforward and rather popular, it is not the most efficient process from the point of view of expected number of steps (number of compares). A simple example to demonstrate this is to compare, instead of on every element in sequence, on each tenth element. When the compare changes from *yes* to *no* one would do a sequential search backward from the *no* compare to find the element at which a *yes* compare is obtained, thus locating the proper placement of the new element. Thinking in terms of logic complexity, clearly the latter procedure requires more complicated coding, for which a price is paid. This is balanced against the more rapid search time plus the time for execution of the more complex logic. A precise calculation is needed to understand the nature of this trade-off, but from a pure search point of view the latter procedure is clearly more efficient.

We may continue with this general concept to find an even more efficient routine called the *binary* search. In this case we move to the center of the file and do a compare. If we get a *yes*, the new element is greater than the bottom half of the file in which case we move to the three-quarter point and continue the process. If we get a *no,* we move to the one-quarter point. In effect we are successively rejecting one half of the data elements and then dividing in two the other half in which we know the search-for position is located. In the binary search case, we determine the number of steps (compares) as the smallest integer that satisfies the following relationship:

$$2^s \geq n + 1 \qquad (8.30)$$

where s is the number of steps and n is the total number of elements in the list. To examine some simple numbers, if $n = 60$, then s would be 6. This means that we can uniquely locate a particular element on a 60-element list in a 6-step binary search. If we increase the list size to 65, for example, one additional search step

TABLE 8.4 NUMBERS OF STEPS FOR SEQUENTIAL AND BINARY SEARCH
PROCESSES

Number of Elements in List (n)	Number of Steps Required			
	Sequential Search		Binary Search	
	Maximum Number	Expected Number	Maximum Number	Expected Number
50	50	25	6	5
100	100	50	7	6
500	500	250	9	8
1,000	1,000	500	10	9
5,000	5,000	2,500	13	12
10,000	10,000	5,000	14	13
50,000	50,000	25,000	16	15
100,000	100,000	50,000	17	16
500,000	500,000	250,000	19	18
1,000,000	1,000,000	500,000	20	19

would be needed ($s = 7$), as per the latter relationship. Here again, the logic required to implement the binary search as well as the times needed to perform the various steps must be taken into account to detemine the more desirable approach.

A simple comparison may be made between the element-by-element sequential search and the binary search, as shown in Table 8.4. The expected number of steps for the sequential search goes up linearly with the number of elements in the list whereas for the binary search it is observed that the expected number of steps goes up rather slowly.

It should be noted that a variety of other efficient search procedures are possible in the overall context of *table look ups*. For example, utilization of pointers (see Chapter 9) together with the data elements can shorten search times considerably. The trade off here is against the additional storage space since there is always some price to be paid to obtain greater efficiency in search speed or in expected number of search steps. In addition, some rather sophisticated *key transformation* techniques have been devised to reduce the expected number of search steps. The reader is encouraged to explore these types of procedures if interest is strong in conventional table-look-up types of search problems.

8.9 SUMMARY

In this chapter we have examined a series of selected concepts from both information and search theory. It has been shown that these fields have some elements in common but also have constructs that are distinct and unrelated. We have also noted that the concepts of uncertainty and search bear relationships to the subject of systems engineering. In particular, correspondences can be established between information theoretic measures of uncertainty and search processes and such systems engineering areas as risk and decision analysis, diagnostic testing, probabilistic modeling, and the investigation of optimum design configurations when large numbers of alternatives are present. The connections between information and search theory and systems engineering are considered rudimentary and much developmental work is still to be carried out that promises significant advances for the future.

REFERENCES

1. Shannon, C. E., and W. Weaver, *The Mathematical Theory of Communication*. Urbana, IL: University of Illinois Press, 1949.

2. Brillouin, L., *Science and Information Theory*. New York: Academic Press, 1956.

3. Reza, F. M., *An Introduction to Information Theory*. New York: McGraw-Hill, 1961.

4. Raisbeck, G., *Information Theory*. Cambridge, MA: The MIT Press, 1963.

5. Attneave, F., *Applications of Information Theory To Psychology*. New York: Holt, Rinehart & Winston, 1959.

6. Brule, J. D., R. A. Johnson, and E. J. Kletsky, "Diagnosis of Equipment Failures," *IRE Transactions on Reliablility and Quality Control*, no. 17, November 1959; R. A. Johnson, "An Information Theory Approach to Diagnosis," IRE Reliability and Quality Control Symposium, January 1960.

7. Mela, D. F., "Information Theory and Search Theory As Special Cases of Decision Theory," Letters to the Editor, *Operations Research*, 9, no. 6, November-December 1961.

8. Stone, D. F. *Theory of Optimal Search*. New York: Academic Press, 1975.

*9. Hamming, R. W., *Coding and Information Theory*. Englewood Cliffs, NJ: Prentice-Hall, 1980.

*10. Lin, S., and D. J. Costello, Jr., *Error Control Coding: Fundamentals and Applications*. Englewood Cliffs, NJ: Prentice-Hall, 1983.

*11. Reingold, E. M., J. Nievargeit, and N. Deo, *Combinatorial Algorithms: Theory and Practice*. Englewood Cliffs, NJ: Prentice-Hall, 1977.

*12. Gray, M., "A Survey of Current Optimization Methods," *Report 3605,* Naval Ship Reserach and Development Center, Washington, D.C., January 1971.

*13. Roberts, F. S., *Applied Combinatorics*. Englewood Cliffs, NJ: Prentice-Hall, 1984.

*14. Koopman, B. O., "Search and Information Theory," U. S. Department of Commerce Clearinghouse, no. AD 687 534, U. S. Department of Commerce, Washington, D.C., 1967.

PROBLEMS

P8.1. Identify and briefly discuss five systems engineering tasks or activities in which uncertainty is present, with the potential that the information theoretic concept of entropy can be appropriately applied.

P8.2. For the coin weighing problem posed and discussed in this chapter show that the 4 versus 4 coin first measurement is proper and that the information thus obtained is of measure log 3.

P8.3. For the coin-weighing problem cited in P8.2, explain all weighing consequences and results for the second alternative on the second weighing, given balance on the first

*Reference is relevant to the text but is not specifically cited.

weighing. This alternative is defined in the text as measuring $s(1)$, $s(2)$, $s(3)$ against T, T, T, where T is a known true coin.

P8.4. For the coin-weighing problem cited in P8.2, verify Figure 8.3 with respect to the weighing sequences shown after the 4 versus 4 coin weighing for which a balance is not obtained. Identify the 13 alternatives referred to in Figure 8.3.

P8.5. Apply the Huffman and conventional binary encoding to the following set of eight messages:

Message	Source Probability
A	0.3
B	0.2
C	0.14
D	0.12
E	0.08
F	0.06
G	0.05
H	0.05

Calculate the efficiency of each of the two codes listed.

P8.6. Prove that the capacity of a binary symmetric channel is given by $C = 1 + p \log p + q \log q$ where p is the probability of no error and q is the probability of error.

P8.7. Prove Equations 8.17 through 8.20 from the basic definitions.

P8.8. Determine the optimum allocation of effort for exponential search among two cells when
  a. $a(1) = a(2) = 1$, $p(1) = 1/4$, $p(2) = 3/4$ and E (total effort) is 10 time units
  b. $a(1) = 1$, $a(2) = 2$, $p(1) = 1/4$, $p(2) = 3/4$ and $E = 10$.
Consult the text for the definitions of these terms.

P8.9. Find a set of conditions under which a generalized parabola ($y = ax^2 + bx + c$) is a maximum.

P8.10. Research the method of Lagrange multipliers and develop an example to demonstrate how it is used to find the maximum of a function.

P8.11. Develop logic flow diagrams that define table-look-up search procedures for
  a. one at a time sequential search
  b. one out of ten sequential search
  c. binary search

P8.12. A binary channel has a channel noise matrix characterized as follows: $p(0|0) = 1/2$, $p(1|0) = 1/2$, $p(0|1) = 1/3$, $p(1|1) = 2/3$.
  a. if the input symbols are transmitted with probabilities 1/4 and 3/4, find $H(X)$, $H(Y)$, $H(X|Y)$, $H(Y|X)$ and $I(X; Y)$
  b. Find the channel capacity and the corresponding input probabilities.

P8.13. Find the capacity of the binary channel represented by the noise matrix in the text, that is, $p(1|0) = 1/3$ and $p(0|1) = 1/4$. Verify the numerical values in the text for the entropies and transinformation when the input probabilities are both one half.

P8.14. Discuss the relationship between search theory concepts and the systems engineering problem of finding optimum design configurations from among a large number of alternatives.

Words differently arranged have a different meaning, and meanings differently arranged have a different effect

Pascal

PROGRAMMING LANGUAGES

9.1 INTRODUCTION

In this chapter we present two examples of specific programming languages. Many of the tasks and activities of systems engineering cannot be performed solely with existing, commercially available software packages and it is therefore necessary to be able to develop one's own programs. In addition, essentially all personal computers come equipped with built-in language compilers. Finally, other language packages are available commercially at prices less than $100. This makes the option of writing one's own programs for specialized applications extremely attractive. Besides, watching the computer properly execute a program that you, rather than someone else, have written turns out to be both challenging and exciting.

Our objectives in this chapter are relatively limited. Since there are numerous full texts available for the various programming languages, and other subjects are to be discussed in this text, only selected elements of two languages are presented. The first is the language BASIC, which stands for beginner's all-purpose symbolic instruction code. This language was developed in the mid-1960s at Dartmouth College and is attributed to John Kemeny and Thomas Kurtz. BASIC is a language that has exhibited broad acceptance and is often provided without additional charge when one purchases a personal computer. There are several different versions of BASIC depending upon the language author and computer offeror but such differences are relatively minor and can be easily mastered. The second language presented here is

Pascal, sometimes considered a spinoff of ALGOL and attributed to Niklaus Wirth [1] as the developer, working at the Technical University in Zurich, Switzerland, in the early 1970s. Pascal is introduced here since it is a modern language in relatively widespread usage, has a structured framework, and bears a strong resemblance to Ada [2] (a trademark of the United States government, Ada Joint Program Office, Department of Defense). Ada is potentially a strong language of the future due to its endorsement by the Department of Defense. Thus by gaining more facility with Pascal the reader can elect to dig more deeply into Pascal for its own special and rather powerful structured capabilities or move on ultimately to Ada which is likely to become increasingly important with time. With respect to both BASIC and Pascal, it is expected that the reader will gain only a rudimentary understanding here and is encouraged to move on to greater depths with other source materials. Examples are used as a means of demonstrating real programs since it is believed that this technique is the strongest one for gaining early appreciation for a computer language. These examples are chosen to illustrate aspects of the languages and therefore in several cases do not necessarily represent the most efficient code. After completing this chapter, the reader should be able to write a surprisingly large number of interesting and even complex programs in a diverse set of application areas.

Although there are many languages in use today, it has been shown that it is necessary to have only three logical constructs in order to develop any program [3]. These constructs are shown in Figure 9.1 and can be described in terms of a direct sequence, a decision, and a loop process. The direct sequence-process construct simply represents a series of steps or statements each of which will be implemented in the order in which they are depicted. The decision-process construct introduces two or more alternate paths for implementation. The *diamond* is the decision element

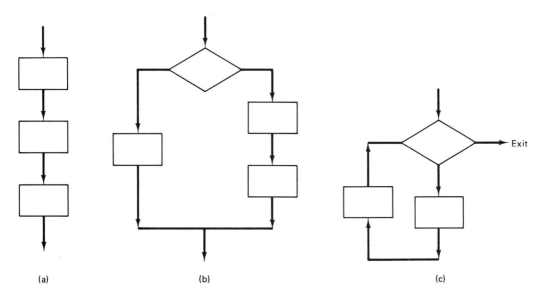

(a) (b) (c)

Figure 9.1 Logical constructs for programs: (a) direct sequence-process construct, (b) decision-process construct, (c) loop-process construct.

which, in Figure 9.1, shows two alternative paths. Typically, such an element is implemented in several languages by an IF. .THEN. .ELSE set of instructions. Finally, the loop-process construct iterates one or more sequential steps and exits the loop when a particular condition is satisfied. The diamond for the loop process in Figure 9.1 normally processes the loop several times and exits to the right when the condition is satisfied. Examples of a looping set of statements include DO. .WHILE, and REPEAT. .UNTIL. In the following brief descriptions of BASIC and Pascal we will find many examples of the three aforementioned logical constructs although the particular methods of implementation and specific instruction sets will display many differences.

9.2 BASIC

BASIC programs are written as a series of numbered statements as shown in Figure 9.2. This figure illustrates a simple program that demonstrates some key points and capabilities, also listed in the figure. The selection of particular numbers for the various statements is an option of the user, but statements are executed in sequence from the lower numbers to the higher ones. A BASIC program is often started with a statement numbered 10 but this is not essential. Numbers are omitted to allow for insertion of new statements at a later time so that corrections can be made without completely renumbering earlier statements.

```
10 X = 5
20 Y = 2
30 A = X + Y
40 B = X - Y
50 C = X * Y
60 D = X/Y
70 E = X^Y
80 F = SQR (E)
90 PRINT A" ";B" ";C" ";D" ";E" ";F
95 END

    RUN

7 3 10 2.5 25 5
```

Figure 9.2 Elementary operations.

9.2.1 Assignments, Operators, Functions, and PRINT

The first several statements in the Figure 9.2 program demonstrate variables and the notion of assignment. In statement 10 and 20, the variables X and Y are assigned the constants 5 and 2, respectively. In some versions of BASIC, the assignment statement includes the word *LET* as in *10 LET X=5*. Also, we may designate variables as *A1*, *B2*, *C3* and so on.

In the next five statements (30–70) further assignments are made, in these cases simply illustrating the operations of

- addition
- subtraction

- multiplication
- division
- exponentiation (sometimes shown as X**Y)

These five operations are only a few of the many that can be performed under BASIC. A more complete list of sample operations is provided in Table 9.1 together with a brief description of their meanings.

TABLE 9.1   SAMPLE OPERATIONS IN BASIC

Symbol of Operator	Meaning
+	Addition
−	Subtraction
*	Multiplication
/	Division
~ or **	Exponentiation
=	Equal to
< >	Not equal to
±	Plus or minus
>	Greater than
≯	Not greater than
<	Less than
≮	Not less than
≥	Greater than or equal to
≤	Less than or equal to
YES or Y	Yes
TRUE of T	True
NO or N	No
FALSE or F	False

Statement 80 involves a function, namely square root, written as SQR. After the value of variable E is computed in 70, its square root is taken and assigned to the variable F in 80. As with operators, BASIC has a variety of built-in functions, a list of which is shown in Table 9.2. Most versions of BASIC can be expected to have the listed functions within their structure. Other functions, in addition to SQR, will be discussed later in this chapter.

PRINT statement number 90 instructs the program to print out the computed and assigned values of the variables A through F. It is a very important statement since it is the only one in this program that provides an output on the screen. Without such a statement, even if all of the statements above it were properly executed, the user would have no way of confirming this or viewing the desired answers. The various quotation marks in statement 90 are ways of creating spaces in the output format. With one space between quotes we have one space between the printout of

TABLE 9.2 BUILT-IN FUNCTIONS IN BASIC

Function	Description
ABS (X)	Absolute value of X
EXP (X)	Natural exponential of X
LOG (X)	Natural logarithm of X
INT (X)	Integer value of X
SGN (X)	Algebraic sign of X
SQR (X)	Square root of X
RND (X)	Random number between 0 and 1
SIN (X)	Sine of X (radians)
COS (X)	Cosine of X (radians)
TAN (X)	Tangent of X (radians)
ATN (X)	Arc tangent of X (radians)

variables A and B and so on for two or more spaces. The semicolon separates the statement of variables. If statement 90 read

```
90 PRINT A;B;C;D;E;F
```

then the values of the six variables would be printed out next to one another with no space in between. This would clearly be a confusing and ambiguous output. Other mechanisms for providing output spacing are discussed later.

Finally, the program (Figure 9.2) is completed with an END statement that is also numbered. The statements 10 through 95 thus constitute the entire program.

After the user has reviewed the written program for typographical errors and possible logical errors, the program can be run by typing in the word RUN. This is a command that tells the computer to execute the program. Figure 9.2 also shows the output results from this program. Again, if the quotation marks in statement 90 were omitted, the output would be 73102.5255, hardly an intelligible one. The reader should review Figure 9.2 in detail to assure understanding before moving on. Problems at the end of the chapter are designed to reinforce overall grasp of the text.

9.2.2 NEW, String Variables, and Printing Messages

As a general rule, in moving from one program to another in the same session, it is necessary to instruct the computer that a new program is about to be written. This is done by the system command NEW. Thus for the next program, shown in Figure 9.3, we start out with the NEW command which is not strictly part of the new program code.

BASIC is not limited to the numeric variables shown in the first program. Names or other alphabetic labels may be used that are called string variables and are illustrated in the short program of Figure 9.3. In this program X$ is a string variable that has been designated as OPPIE. All string variables in BASIC must end with the

```
NEW

10 X$ = "OPPIE"
20 Y$ = "GROVES"
30 PRINT "HELLO ";X$
40 PRINT ; Y$" HAS A MESSAGE FOR YOU"
50 END

    RUN

HELLO OPPIE
GROVES HAS A MESSAGE FOR YOU
```

dollar sign ($). Thus in statements 10 and 20 we both define and assign values to the two string variables X$ and Y$.

We return in statement 30 to the PRINT statement. In this case, however, we print out only the (string) variable X$ but before it, in quotes, we call for the printing of the word HELLO. Any combination of letters or numbers in quotes, when included in a PRINT statement, will be provided as a printed output. We note that there is a blank space after HELLO, which is what we desire in order to have a space between HELLO and the value of X$, namely OPPIE. The same principle holds for statement 40 in which the alphanumeric phrase HAS A MESSAGE FOR YOU comes after the string variable and includes a blank space at the beginning, just after the leading quotation mark. Running this simple program yields the output shown after the program in Figure 9.3.

9.2.3 REM, READ. .DATA

The program in Figure 9.4 calculates the sum and average value of three scores, X, Y, and Z. A new statement, REM, allows the programmer to annotate the program so that anyone using the program can quickly identify what the intent of the program is. The remark is made directly after REM (REMARK in some versions of BASIC) and does not print as an output when the program is run. Explanatory remarks for portions of the program can be made anywhere in the program but they must be given identifying sequence numbers as with any of the program statements.

A new READ (statement 20) statement appears in this program whose purpose is to bring numeric data into the program. READ is coupled with DATA (statement

```
10 REM CALCULATES AN AVERAGE SCORE
20 READ X,Y,Z
30 S = X + Y + Z
40 A = S/3
50 PRINT "THE SCORES ARE ";X" ";Y" ";Z
60 PRINT "THE SUM IS ";S
70 PRINT "THE AVERAGE IS ";A
80 DATA 82,89,93
90 END

    RUN

THE SCORES ARE 82 89 93
THE SUM IS 264
THE AVERAGE IS 88
```

80) and both must appear in the program. The READ statement brings the number 82 from the DATA statement to the variable X, 89 to Y and 93 to Z. This represents an extremely efficient means of bringing small or moderate sized units of input or initial data into a program. The READ. .DATA combination must both contain the same number of elements in the same order to have the program operate properly. The results of running the program are shown in Figure 9.4.

9.2.4 INPUT, PRINT to Skip Line

The program in Figure 9.5 is identical to that in Figure 9.4 with just two exceptions. Line 20 in Figure 9.5 calls for inputting of the values of X, Y, and Z at the keyboard by means of the INPUT statement. Thus it takes the place of and performs the exact same function as the READ. .DATA statements (20 and 80) in Figure 9.4. When running this program, the user will be prompted by a question mark (?) to which the response should be the required input scores 83, 90, and 94. After a stroke of return at the keyboard, the rest of the program is executed, with results as shown in the figure.

```
10 REM CALCULATES AN AVERAGE SCORE
20 INPUT X,Y,Z
30 S = X + Y + Z
40 A = S/3
50 PRINT "THE SCORES ARE ";X" ";Y" ";Z
60 PRINT "THE SUM IS ";S
70 PRINT
80 PRINT "THE AVERAGE IS ";A
90 END

RUN

?83,90,94
THE SCORES ARE 83 90 94
THE SUM IS 267

THE AVERAGE IS 89
```

Figure 9.5 Alternative version of average score computation.

The question mark prompt that appears may be unclear to the user. It is possible to add text to the INPUT statement that clarifies what the program is requesting. This point is illustrated in the next program.

Finally, we note that the output in Figure 9.5 contains a space not present in Figure 9.4. This results directly from line 70 in Figure 9.5, which is a bare PRINT statement. This simple procedure skips lines and formats the output in a manner more pleasing to the user.

9.2.5 Other Commands

In reviewing the previous four illustrative programs, we note that there are commands to the computer and numbered statements that are specifically part of the program itself. In general, a command is not used as part of a program and a program statement cannot be used except as a numbered instruction within the program. The commands RUN and NEW have been discussed above; we now briefly describe some other useful BASIC commands.

After typing in a program or a portion of a program the user can see what has been entered, in a standard format, by using the LIST command. This is particularly helpful when several changes have been made and the program needs to be cleaned up. There are several ways in which such changes or corrections can be made. If a change is to be made on a line that one is currently still working on, corrections are made by simply backing up and retyping. If the line has been fully entered, after the RETURN stroke, the user can delete a line or more by using the DEL command. For example, to delete the printout of the sum in the program in Figure 9.4 one would enter the command DEL 60, followed by a keyboard return. Line 60 would then be removed from the program when next listed. Another way to delete a line is to type in the statement number followed by a keyboard return. This essentially instructs the computer that there is no statement associated with that number, thus deleting it as a program instruction.

Another set of useful commands is represented by SAVE, LOAD, LOCK, and UNLOCK. These commands come into play when a program is to be moved into a file for storage and later retrieval. Clearly, when writing a large program that takes several sessions at a terminal, the user does not wish to start typing in the program from the beginning each time. At the end of a session, by using the SAVE command, the user can save what has been developed on a diskette. At the next session the saved program is retrieved by means of a LOAD command. Additionally, the program may be LOCKed and UNLOCKed in the file. The reader is referred to other texts or manuals to gain facility with these commands in the context of writing programs in a development environment.

Finally, after a series of programs has been saved on a disk file, the user can find out what the disk contains by using the CATALOG command. This will call up the names of all programs that have been stored on a particular disk. This capability is clearly important; without it the user would have to keep a separate record of what programs are stored on the various disks.

9.2.6 INPUT message, IF. .THEN, GOTO

The program in Figure 9.6 allows the user to convert temperatures from Fahrenheit (F) to Celsius (C) or the reverse. The algorithm, which may not be familiar to the reader, is as follows:

a. add 40° to the given temperature, whether going from F to C or C to F
b. multiply the result by
  • five-ninths if converting from F to C
  • nine-fifths if converting from C to F
c. subtract 40° from the above result of b

This algorithm can be verified as correct and uses the fact that both the Fahrenheit and Celsius scales are the same temperature at minus 40°.

Statement 10 uses INPUT with a clarifying message to the user. This extremely important capability is a rudimentary *menu* and the user responds by typing in a 1 or a 2. This selection simultaneously assigns the typed number to the variable

A and is used later in the program to implement the above algorithm. Next, in line 15, the user is requested to enter the input temperature, with the question mark appearing as a consequence of line 20. In line 25 step a. of the algorithm occurs. In statement 30 a new capability is provided, that is, the IF. .THEN statement. As shown, if *A* = 1 then the user initially elected to convert from F to C. In such a case the program jumps to line 45 (THEN 45), which carries out the first part of step b. of the algorithm, namely, multiplication by five-ninths. Following to line 50 subtracts 40°, which is step c. of the algorithm. Another IF. .THEN at 55 jumps to line 75, which prints the desired conversion in °C.

Returning to line 30 we examine what happens if *A* is not equal to 1. In such a situation the user obviously selected *A* = 2 and desires to convert from C to F. Thus we go to line 35 (not 45) which implements the second part of step b. of the algorithm. Now the program must circumvent line 45, which in line 40 is executed by means of a GOTO statement. This instruction, considered a weakness in the overall quality of structured programming, nevertheless gets the job done. It forces a jump to the common statement in line 50, which is step c. of the algorithm. This step occurs in converting in either direction. The next statements if A = 2 would then be 60, 65, 70, and 80. Both cases (F to C and C to F) are illustrated by the run results in Figure 9.6.

```
 10 INPUT "DO YOU WISH TO CONVERT FAHRENHEIT INTO CELSIUS (TYPE 1) OR
CELSIUS INTO FAHRENHEIT (TYPE 2)? ";A
 15 PRINT "ENTER DEGREES"
 20 INPUT G
 25 H = G + 40
 30 IF A = 1 THEN 45
 35 K = 9 * H/5
 40 GOTO 50
 45 K = 5 * H/9
 50 L = K - 40
 55 IF A = 1 THEN 75
 60 IF A = 2 THEN 65
 65 PRINT "DEGREES FAHRENHEIT = ";L
 70 GOTO 80
 75 PRINT "DEGREES CELSIUS = ";L
 80 END


 RUN

 DO YOU WISH TO CONVERT FAHRENHEIT INTO CELSIUS (TYPE 1) OR CELSIUS
 INTO FAHRENHEIT (TYPE 2)? 1
 ENTER DEGREES
 ? 212
 DEGREES CELSIUS = 100

 RUN

 DO YOU WISH TO CONVERT FAHRENHEIT INTO CELSIUS (TYPE 1) OR CELSIUS
 INTO FAHRENHEIT (TYPE 2)? 2
 ENTER DEGREES
 ? -40
 DEGREES FAHRENHEIT = -40
```

Figure 9.6 Temperature conversions.

9.2.7 FOR. .TO,STEP. .NEXT, Comma, and Reassignment

We solve and print out a compound interest table in the program in Figure 9.7. In steps 10 and 15, the starting principal and interest rate are input by the user. The year parameter is initialized in statement 20 and the interest calculated in 25. The three column headings of principal, year, and interest are set up in step 30. We note here the use of a comma to separate these three headings. The comma establishes a standard output format that automatically tabulates at 16 spaces per field (although for a 40-character width, the third field would be eight spaces). The PRINT statement 35, which also uses commas, will place the computed values under the proper headings. The results are shown in the program output run also provided in Figure 9.7.

```
5   REM COMPOUND INTEREST
10  INPUT "PRINCIPAL IS ";P
15  INPUT "INTEREST RATE IS ";R
20  Y = 1
25  I = P * R
30  PRINT "PRINCIPAL","YEAR","INTEREST"
35  PRINT P,Y,I
40  FOR Y = 2 TO 5 STEP 1
45  P = P + I
50  I = P * R
55  PRINT P,Y,I
60  NEXT Y
70  END

    RUN

PRINCIPAL IS 1000
INTEREST RATE IS .1
PRINCIPAL        YEAR           INTEREST

1000            1              100
1100            2              110
1210            3              121
1331            4              133.1
1464.1          5              146.41
```

Figure 9.7 Compound interest calculation.

In step 40 the FOR. .TO,STEP. .NEXT statement is introduced. This allows for the automatic stepping of the year from $Y = 2$ to $Y = 5$ in steps of one year. The STEP statement may be used to step in any increment within the overall range of the parameter. Statements 40 through 60 represent a loop with 60 instructing the program to return to the next value of Y as defined in statement 40. By embedding the PRINT statement (55) in the loop, the principal, year, and interest values are output for each stepped value of year. This iteration process is inherently quite powerful since we can expand the tabulation for larger numbers of years without adding additional lines of code to the program.

One final note with respect to this program is statement 45 which in effect is a reassignment step. Clearly the statement $P = P + I$ does not make sense algebraically. In BASIC, as well as other languages, it has a well-defined and important meaning. It calculates a new value of P (the left side of the equal sign) as being the sum of the current value of P and the current value of I (the right side of the equal sign). This process continues as the looping occurs between statements 40 through

60. It is also possible and often more convenient to step an integer variable (such as years) directly by using the statement $Y = Y + 1$. Thus the new value of Y is set as the old value incremented by one. However, it is then necessary to create the looping process so that this statement is encountered repetitively. As an example, this can be achieved by using IF. .THEN and GOTO (see Figure 9.6) as well as other looping procedures.

9.2.8 Subscripted Variables and Stepping Summation

In the program of Figure 9.8 we illustrate a simple subscripted variable $L(I)$. This means that the variable L changes its value as a function of the integer subscript I. Normally the subscript I is stepped and values of $L(I)$ entered as data or computed. In the illustrative program we read in the data for five values of $L(I)$ (statements 10 and 15). The $L(I)$ are interpreted as failure rate or lambda values for five pieces of equipment characterized by the exponential failure distribution. These equipments are in series from a reliability point of view and in statement 20 we treat the number of series equipments as an input variable for illustrative purposes. Of course if M other than 5 is input then steps 10 and 15 would not be compatible and the program would not run correctly. In step 25 the user specifies the total time of operation of the equipment so that the overall program will compute the probabilities of successful operation without failure to time T.

```
10 READ L(1),L(2),L(3),L(4),L(5)
15 DATA .1,.2,.3,.4,.5
20 INPUT "MAX I IS ";M
25 INPUT "TIME OF OPERATION IS ";T
30 FOR I = 1 TO M STEP 1
35 R(I) = EXP(-L(I)*T)
40 PRINT R(I)
45 NEXT I
50 S(0) = 0
55 FOR I = 1 TO M STEP 1
60 S(I) = L(I) + S(I-1)
65 NEXT I
70 PRINT "OVERALL LAMBDA IS ";S(M)
75 RT = EXP(-S(M)*T)
80 PRINT "OVERALL RELIABILITY IS ";RT
90 END

RUN


MAX I IS 5
TIME OF OPERATION IS 1
.9048
.8187
.7408
.6703
.6065
OVERALL LAMBDA IS 1.5
OVERALL RELIABILITY IS .22313
```

Figure 9.8 Series reliability calculation.

In statement 30 we utilize the FOR. .TO,STEP. .NEXT option as before. In this case we run through the subscript from 1 to the maximum value which is 5. For each integer value of the subscript I, step 35 calculates the reliability $R(I)$ corresponding to each of the five equipments. Results are printed (40) and a new value of

I selected (45). When all values of I are stepped through the program continues on to statement 50. Here we are initializing the sum variable S, which will be the incremental sum of failure rates for the five equipments. This is demonstrated in the next loop running from 55 to 65. As we step the subscript I again (55), in statement 60 we calculate the sum variable S. For $I = 1$ we compute $S(I)$ as

$$S(1) = L(1) + S(1 - 1) = L(1) + S(0) = L(1)$$

Thus the sum to $I = 1$ is simply the lambda of the first equipment $L(1)$. Note that it was necessary to define $S(0)$ as zero as carried out in step 50. When I is stepped to 2, we have

$$S(2) = L(2) + S(2 - 1) = L(2) + S(1)$$

Therefore the sum to $I = 2$ is the earlier sum $S(1)$ plus the new lambda $L(2)$. It is clear then that this statement continues to find the new sum as the current sum plus the lambda for the next equipment. This may be called a *stepping summation*. When statement 70 is reached the overall lambda value has been determined as $S(5)$. Of course this could have been done directly by an assignment statement setting $S(5)$ equal to the sum of each of the lambdas but the point of this stepping summation would have been lost.

In statements 75 and 80 the overall reliability is computed and then printed as an output. The entire output of a sample run of this program is shown in Figure 9.8. The input time has been selected as $T = 1$, which may be interpreted as the reliability up to one year's operation. It is to be noted that the time units for the individual lambdas must be in the same dimension as that for the total time to operation T. This program allows new values of reliability to be automatically calculated and printed out for new values of total time. If the maximum number of equipments, M, is changed, the proper number of corresponding inputs must be provided. In general, if a subscripted variable has more than ten elements, a dimension statement must be provided in the program. Such a statement, for a subscript range up to twenty would read DIM L(20). This serves to reserve the proper amount of memory for the values of the subscripted variable.

9.2.9 Other BASIC Capabilities

The BASIC language has many features in addition to those already described. Here we touch upon the following further capabilities:

- nested loops
- subroutines
- restore
- random number generation
- user-defined functions

Nested loops are particularly important when stepping two variables in combination with one another. This can occur with two or more dimensional arrays or simply running through all combinations of variable values. Thus it is a significant capability in systems engineering applications where one is searching through various configurations in an attempt to find the best one. A nested loop is illustrated next

```
FOR I = 1 TO 4 STEP 1
    FOR J = 1 TO 3 STEP 1
    PRINT I  ",";  J
    NEXT J
NEXT I
```

We note an I loop stepping from 1 to 4 with a J loop contained completely within it. This containment or *nesting* is essential in order to have such loops operate properly. In this case I will first take on the value 1 and then J will step from 1 to 3. Thus the values $(I, J) = (1, 1), (1, 2)$, and $(1, 3)$ will be generated. After this occurs I will step to 2 and the values $(I, J) = (2, 1), (2, 2)$, and $(2, 3)$ will be developed, and so on. If I and J are subscripts for a double-subscripted variable, then all values of such a variable will be obtained.

We may find a direct application for nested loops when considering alternative design approaches for a variety of system functional areas. If the design approaches $D(I, J)$ for approach J to functional area I, we may systematically step through all design approaches in an attempt to find an optimum set (see Chapter 10).

BASIC also has the feature of containing embedded subroutines. A subroutine may be entered by the statement in the program GOSUB N where N is the number of the first statement in the subroutine. Thus the program will jump to statement number N and continue to execute successive statements until the last statement in the subroutine, which should be RETURN. When this is reached, the subroutine execution is complete and control is transferred back to the main program to the statement immediately following the GOSUB statement. In general, subroutines are used when a series of computations or set of logic is used several times, thus avoiding the necessity to repeat these steps more than once.

The RESTORE statement in a program is used in conjunction with READ. . DATA. When RESTORE is encountered in a program, a data pointer is reset to the top or beginning of the data elements. This allows the same data to be read more than once in association with the same or a different set of variables. After the RESTORE statement it is necessary to use another READ statement to make clear the assignment of data to a set of variables. An example is shown next.

```
10 DATA 1,2,3
20 READ U,V,W
30 RESTORE
40 READ X,Y,Z
```

In this portion of a program the numerical data values of 1,2,3 will first correspond to the variables U, V, and W and then, in addition, to the variables, X, Y, and Z.

Random number generation is a feature of all established languages. It is normally a built-in function just as sine or cosine would be built-in functions (see Table 9.2). It is called out here since random number generation can be particularly important in the systems engineering processes of modeling by means of simulation (see Chapter 12). The statement RND (X) in a program would develop a random number between 0 and 1 for the variable X. The selection would be made uniformly within the interval between 0 and 1 so that sampling is being performed from a uniform distribution. Methods of sampling from other distributions are presented in Chapter 12, Performance Modeling and Simulation.

User-defined functions may also be accommodated in BASIC. This is carried out by means of a statement such as

```
10 DEF FNG (X) = 2X + 1
```

DEF indicates that we are dealing with a new function as does FN. The single letter G names the new function and X is the independent variable. The illustrative function in this example is simply $2X + 1$ but may be any function using combinations of the built-in functions such as those in Table 9.2. In the case of string functions, which are acceptable in BASIC, the first three letters must be followed by a dollar sign as in FNG$. It is further noted that subscripted variables may not be used as arguments in a user-defined function.

9.3 PASCAL

The technique to be used here in exploring the features of Pascal will be to move from the specific to the general largely by means of examples. This will parallel the earlier examination of BASIC not only in principle but also in regard to the same illustrative programs provided in BASIC (the programs in Figures 9.2 through 9.8). Thus the differences as well as similarities between BASIC and Pascal will be clearly illustrated. Special features of Pascal will be discussed toward the end of the chapter. The illustrative programs are not always the most efficient but are developed to demonstrate a point or parallelism between the corresponding BASIC and Pascal programs.

9.3.1 CONST, VAR, Types, BEGIN. .END, WRITELN

Figure 9.9 shows a Pascal program that does the same computations as the BASIC program in Figure 9.2. The reader is urged in all cases to go back and compare the BASIC and Pascal programs that are designed to develop the same or similar results.

First we note that we name the program at the top of the program text (in this case we have PROGRAM OPERATORS). As we shall see, the program uses a transcendental function that we call into play by the declaration USES TRANSCEND. We then define the constants and variables in the program. The former is done using the reserved word CONST and the latter the reserved word VAR. Thus X and Y are constants and cannot change their values in the program whereas A through F are variables and, in general, can have different values for different parts of the program. We also have defined the types of variables by use of the word REAL in the program. This means that all variables are to be real numbers. In Pascal, this typing

```
PROGRAM OPERATORS;
  USES TRANSCEND;

CONST
  X = 5;
  Y = 2;

VAR
  A,B,C,D,E,F : REAL;

  BEGIN
    A := X + Y;
    B := X - Y;
    C := X * Y;
    D := X/Y;
    E := EXP(Y*LN(X));
    F := SQRT (E);
    WRITELN (A,B,C,D,E,F);
  END


RUNNING...

7.00000   3.00000   1.00000E1   2.50000   2.50000E1   5.00000
```

Figure 9.9 Elementary operations in Pascal.

is an essential part of every program. Further, a selected set of types that are available are listed here.

SIMPLE TYPES

- integer
- real
- boolean
- character
- string

STRUCTURED TYPES

- arrays
- records
- files
- sets
- pointers

Some of the more complex structured types will be explored in later portions of this chapter.

The formal structure of the program calls for a BEGIN and an END to the main program, as shown. Also, there are no numbered statements and indentation of lines is used so that various sections of code can be easily seen. Next we have assigned particular operations for *A* through *F*. Note the details of an assignment statement

- a space after the variables
- a colon after the space
- an equals sign after the colon
- a semicolon after each statement

Values for variables *A* through *F* are calculated using the same form as the aforementioned assignment statement and the appropriate operators and square root (SQRT) function. We note that Pascal does not have a formal exponentiation function so that variable *E* is calculated as "*X* to the *Y* power" by means of the expression EXP($Y*$LN(X)). Finally, the printed output for the program is provided by the WRITELN statement, which calls out the variables whose values are to be printed. Parentheses must be used (in distinction to the PRINT statement of BASIC) and the commas separate the variables.

The results of running the program are provided in Figure 9.9. The precise method of execution of a run can differ in various versions of Pascal and in general is not as straightforward as simply typing RUN. We will use the command RUNNING. . . here and urge the reader to consult the steps necessary to make a run for the particular Pascal package that is being used.

9.3.2 String Variables and WRITELN messages

In Figure 9.10 we illustrate the Pascal version of the BASIC program shown in Figure 9.3. The two variables *X* and *Y* are declared to be of type STRING. In the body of the program the string variable values are assigned by means of a single set of quotation marks as in 'OPPIE'. Finally, we provide printout by means of WRITELN( ), this time including an output message. The program run results are straightforward and shown in the figure.

```
PROGRAM MANHATTAN;

VAR
  X : STRING;
  Y : STRING;

  BEGIN
    X := 'OPPIE';
    Y := 'GROVES';
    WRITELN ('HELLO ',X);
    WRITELN (Y,' HAS A MESSAGE FOR YOU');
  END.



RUNNING...


HELLO OPPIE
GROVES HAS A MESSAGE FOR YOU
```

Figure 9.10 String variables in Pascal.

9.3.3 Annotations and READLN

The program in Figure 9.11 calculates a simple sum and average, paralleling the BASIC programs in Figures 9.4 and 9.5. Here we have added an annotation or comment feature similar to the REM or remark feature of BASIC. The annotation must be placed within asterisks and parentheses as is CALCULATES AN AVERAGE SCORE. This is strictly a note to the reviewer, user, or developer of the program to clarify the program's steps. Such annotations should be used liberally in a Pascal program and can be placed in any line of a program.

```
PROGRAM AVERAGE;

(*CALCULATES AN AVERAGE SCORE*)

  VAR
    X,Y,Z,S,A : REAL;

  BEGIN
    READLN (X,Y,Z);
    S := X + Y + Z;
    A := S/3;
    WRITELN ('THE SCORES ARE ',X,Y,Z);
    WRITELN ('THE SUM IS ',S);
    WRITELN;
    WRITELN ('THE AVERAGE IS ',A);
  END.



RUNNING...


82 89 93
THE SCORES ARE 8.20000E1  8.90000E1  9.30000E1
THE SUM IS 2.64000E2

THE AVERAGE IS 8.80000E1
```

Figure 9.11 Average score computation in Pascal.

The five variables are each defined to be of type REAL and the READLN (X,Y,Z) statement calls for the keyboard input of numeric values for X, Y, and Z. Thus READLN in Pascal more-or-less corresponds to INPUT in BASIC. We provide both messages and printed results in the WRITELN statements, as before. The WRITELN statement without an argument simply skips a line in formatting the output. There are also both READ and WRITE statements in Pascal, which we will not be using. In general terms, both READLN and WRITELN cause the current line to be terminated whereas READ and WRITE do not. Another way to describe this is to say that WRITELN puts an end-of-line (EOLN) marker when writing into a file whereas WRITE does not. In the same sense, READLN will read a whole line until the EOLN marker (put there by WRITELN) whereas READ, in most cases, will read a character at a time.

The output for the program provides the same results as did the corresponding BASIC program. In the Pascal program, we have shown the run results in scientific notation in which, for example, E1 is a multiplier and stands for "ten to the first power."

9.3.4 IF. .THEN. .ELSE

The temperature conversion program in Figure 9.12 (corresponding to the BASIC program in Figure 9.6) illustrates the IF. .THEN. .ELSE capability of Pascal. After the user selects the temperature conversion option and inputs the number of degrees, the IF. .THEN. .ELSE choice is employed. We note that in the BASIC program the IF. .THEN followed by GOTO was used. Pascal allows the use of GOTO statements but they do not lead, in general, to high-quality structured programs. In any case the ELSE option brings us to a point later in the program from which to proceed to a conversion from °C to °F. Also, it is to be noted that the IF. .THEN option sets up a nested BEGIN. .END section of code that in effect acts as a subroutine. Since it is used only once here it is not defined as a subroutine, the Pascal version of which

```
PROGRAM TEMPERATURE;

VAR
    G,H,K,L : REAL;
         A : INTEGER;

    BEGIN
      WRITELN('DO YOU WISH TO CONVERT');
      WRITELN('FAHRENHEIT INTO CELSIUS (TYPE 1)');
      WRITELN('OR CELSIUS INTO FAHRENHEIT (TYPE 2)?');
      READLN (A);
      WRITELN ('ENTER DEGREES');
      READLN (G);
      H := G + 40;
      IF A = 1
      THEN BEGIN
            K := 5 * H/9;
            L := K - 40;
            WRITELN ('DEGREES CELSIUS = ',L);
          END
      ELSE
          BEGIN
          K := 9 * H/5;
          L := K - 40;
          WRITELN ('DEGREES FAHRENHEIT = ',L);
          END
    END.



    RUNNING...


    1
    ENTER DEGREES
    212
    DEGREES CELSIUS = 1.00000E2

    RUNNING...


    2
    ENTER DEGREES
    -40
    DEGREES FAHRENHEIT = -4.00000E1
```

Figure 9.12 Temperature conversion in Pascal.

is called a PROCEDURE. More will be discussed about Pascal program structures and procedures later in this chapter.

9.3.5 Output Formats and REPEAT. .UNTIL

The compound interest Pascal program in Figure 9.13 is similar to the BASIC program in Figure 9.7. The first six steps of the program appear to be straightforward. The seventh line, calling for a WRITELN, is formatting three columns with the headings PRINCIPAL, YEAR, and INTEREST. The use of the colon followed by a

```
PROGRAM COMPINT;

(*CALCULATES COMPOUND INTEREST*)

    VAR
        P,R,I : REAL;
            Y : INTEGER;

    BEGIN
        WRITELN ('PRINCIPAL IS ');
        READLN (P);
        WRITELN ('INTEREST RATE IS ');
        READLN (R);
        Y := 1;
        I := P * R;
        WRITELN ('PRINCIPAL':9,'YEAR':5,'INTEREST':9);
        WRITELN (P:9,Y:5,I:9);
        REPEAT
            Y := Y + 1;
            P := P + I;
            I := P * R;
            WRITELN (P:9,Y:5,I:9);
        UNTIL Y = 5;
    END.



        RUNNING...


    PRINCIPAL IS
    100
    INTEREST RATE IS
    .1
    PRINCIPAL YEAR INTEREST
    1.00000E2   1 1.00000E1
    1.10000E2   2 1.10000E1
    1.21000E2   3 1.21000E1
    1.33100E2   4 1.33100E1
    1.46410E2   5 1.46410E1
```

Figure 9.13 Compound interest calculation in Pascal.

number is formatting the output printout so as to right justify the selected headings. The output format is shown below with the numbers representing the positions along the output line

The next WRITELN statement places the variable values in the same right-justified positions under the proper respective headings.

Next the REPEAT. .UNTIL loop is encountered. This loop first steps the value of year, Y, from 1 to 2. Then the current principal, P, and interest, I, are added to obtain the new value (for the second year) of principal. The interest for the new (second) year is then computed and the WRITELN statement for P, Y, and I are written on the screen. Since Y is less than 5 at this point the UNTIL statement returns the program control to REPEAT and the above process is repeated for the third year ($Y = 3$). When Y takes on the value 5 the calculations are performed for the fifth year as above. However, the UNTIL statement becomes satisfied, the REPEAT. .UNTIL loop is terminated and the program is ended. A run for the program is likewise shown in Figure 9.13.

9.3.6 ARRAYS and FOR . . DO

The Pascal program in Figure 9.14 performs the same function as the BASIC program in Figure 9.8, both of which compute the reliabilities of a series of equipments characterized by the exponential failure law. The program demonstrates a very simple use of the ARRAY feature, defining S, sum of lambdas; L, lambda of each

```
PROGRAM RELIABILITY;
USES TRANSCEND;

  CONST
    M = 5;

  VAR
        I : INTEGER;
    S,L,R : ARRAY[0..M] OF REAL;
    RT,T  : REAL;

  BEGIN
    READLN (L[1],L[2],L[3],L[4],L[5]);
    WRITELN ('MAX I IS ',M);
    WRITELN ('TIME OF OPERATION IS ');
    READLN (T);
    FOR I := 1 TO M DO
     BEGIN
       R[I] := EXP (-L[I]*T);
       WRITELN (R[I]);
     END
    S[0] := 0;
    FOR I := 1 TO M DO
     BEGIN
       S[I] := L[I] + S[I-1];
     END
    WRITELN ('OVERALL LAMBDA IS ',S[M]);
    RT := EXP (-S[M]*T);
    WRITELN ('OVERALL RELIABILITY IS ',RT);

  END.
```

Figure 9.14 Series reliability computation (continued on next page).

```
.1 .2 .3 .4 .5
MAX I IS 5
TIME OF OPERATION IS
1
9.04837E-1
8.18731E-1
7.40818E-1
6.70320E-1
6.06531E-1
OVERALL LAMBDA IS 1.50000
OVERALL RELIABILITY IS 2.23130E-1
```

Figure 8.14 (*cont.*)

equipment; and R, reliability of each equipment, as of type ARRAY [0. .M] OF REAL. This means that they are subscripted variables, they each have M elements (the subscript can move to a maximum of M) and that they are real numbers.

We read in the values of lambda for each equipment (without a screen message) as before. We also read in the time of operation T. All of this is performed in response to a prompt at the keyboard. We then enter a FOR. .DO loop that steps through integer values of the subscript I up to and including the constant value M. The loop has a BEGIN. .END subroutine, which is a convention in Pascal and defines as well the end of the FOR. .DO loop so that control can return to the beginning of the loop when the END statement is reached. After this loop is completed the sum variable S is initialized to zero and a second FOR. .DO loop is entered. This loop calculates the sum of lambdas variable S until the sum for all equipments is obtained. That sum is written as an output (OVERALL LAMBDA IS) and the overall reliability RT is both calculated and printed out on the screen. This Pascal program bears a strong resemblance to the BASIC program in order to demonstrate these Pascal features in distinction to developing optimized code.

9.3.7 WHILE. .DO

The WHILE. .DO instructions continue to carry out a sequence of statements while a particular condition holds true, whereas the previous REPEAT. .UNTIL waited for a particular condition to come about. An often-used situation in which WHILE. .DO is used is that of reading data in from a file. A typical structure for a piece of code in this regard might be

```
WHILE NOT EOF DO
    BEGIN
        statement
          .      .
          .      .
        statement
        READLN statement
    END
```

In this code EOF stands for *end of file* so that data from a file is continuously read in by the READLN statement until the end of file is reached. At that point the BEGIN. .END is exited and the program moves beyond the control of the WHILE. .DO statement.

9.3.8 Program Structures and Procedures

So far we have examined Pascal in relation to a specific set of illustrative programs that have also been written in BASIC. This has provided insight into some of the details of Pascal as well as the similarities and differences between Pascal and BASIC. We have seen, in particular, the fact that Pascal is much more highly structured. Pascal is also designed to be quite modular in its essential form and we briefly examine here both the general structure and inherent modularity of Pascal.

A key feature of a general Pascal program is the PROCEDURE. A procedure is itself a complete program which is contained within the overall Pascal program. A simple structure for a program containing a procedure is shown in Figure 9.15(a). As before, the program starts with its name, annotations or comments, and the declarations of constants and variables. Next a procedure is listed and named. The procedure also continues with declarations of constants and variables local to the procedure and one or more BEGIN. .END sequences of instructions. Finally we come to the *main body* of the program, which usually reads in data and calls out the procedure by name. The procedure is thus invoked in its entirety and, since it is contained in the main body of the program it looks very much like a subroutine. Such a procedure may be invoked more than once in the program's main body, thus enhancing the modularity of the overall program.

If two procedures are to be used in sequence, the structure of the overall program would require adding the second procedure in full just after the first procedure and then invoking the second procedure in the program's main body by calling out the name of the procedure in the proper location. It is also possible to *nest* procedures, as illustrated in Figure 9.15(b). This shows procedures two and three nested within procedure one, which means that two and three must be called out within the main body of procedure one. Procedure one is called out, as before, within the main body of the overall program.

9.3.9 Functions

The Pascal language has built-in functions, listed in Table 9.3, in much the same manner and variety as does BASIC (see Table 9.2). Such functions may be called out in a direct statement such as the square root (SQRT ( )) statement in Figure 9.9. In some versions of Pascal an additional statement must be made just after the PROGRAM statement in order to utilize a particular set of functions. For example, in Apple Pascal, in order to utilize transcendental functions in the main body of the program the statement USES TRANSCEND (see Figure 9.9) would have to be included after the initial PROGRAM statement. This is also true for other special capabilities such as graphics, for example, the statement USES TURTLEGRAPHICS for Apple Pascal.

```
PROGRAM SAMPLE;
(*COMMENT*)
CONST
VAR
    PROCEDURE ONE;
       CONST
       VAR
          BEGIN
          ---
          END

    BEGIN(*MAIN PROGRAM*)
       READLN( );
       PROCEDURE ONE;
       ---
    END.
```

(a)

```
PROGRAM SAMPLE;
(*COMMENT*)
CONST
VAR
    PROCEDURE ONE;
    CONST
    VAR
       PROCEDURE TWO;
       CONST
       VAR
          BEGIN
          ---
          END(*TWO*)

       PROCEDURE THREE;
       CONST
       VAR
          BEGIN
          ---
          END(*THREE*)

    BEGIN
    ---
       PROCEDURE TWO;
       ---
       PROCEDURE THREE;
    END(*ONE*)

BEGIN(*MAIN*)
---
    PROCEDURE ONE;
---
END.
```

(b)

Figure 9.15 Pascal program constructs. (a) an embodied procedure, (b) two procedures nested in a third procedure.

A user-defined function in Pascal can also be constructed, but with some further degree of formalism. A *header* is required that both names the function and the type of its output. If we were calculating, for example, the sample variance of a set of numbers, we might set up the header as

```
FUNCTION VARIANCE : REAL;
```

The remainder of the user-defined function would then require the formal definition

TABLE 9.3 BUILT-IN PASCAL FUNCTIONS

Function	Calculates
ABS (X)	The absolute value of X
ARCTAN (X)	The arc tangent of X
CHR (X)	The character represented by X
COS (X)	The cosine of X
EOF	If end-of-file (returns true for end-of-file and false otherwise)
EOLN	If end-of-line (returns true for end-of-line and false otherwise)
EXP (X)	Natural exponentiation
LN (X)	The natural logarithm of X
ODD (X)	If X is odd or even (returns true for odd and false otherwise)
ORD (X)	The decimal integer that represents the character X
PRED (X)	The predecessor of X
ROUND (X)	The value of X rounded to the nearest integer
SIN (X)	The sine of X (radians)
SQR (X)	The square of X
SQRT (X)	The square root of X
SUCC (X)	The successor to X
TRUNC (X)	The truncated value of X (drops the decimal part of X)

of variables (VAR) and the BEGIN. .END structure of a subroutine or procedure that carried out the variance calculations. This user-defined function would then be inserted in its appropriate position in the overall program.

9.3.10 Files

A file in Pascal is usually thought of as residing externally (as on a diskette) which can then be accessed in order to read in a relatively large volume of data into a program. The processes that are clearly important in working with Pascal programs that require access to external files are at least

1. inputting data from the keyboard into an external (diskette) file
2. retrieving data from such a file
3. sending output to a printer

These processes will not be examined in detail here. Pascal has specific statements that facilitate these three processes and therefore allow the user to store, retrieve, and print out relatively large amounts of data that are used in a program. For example, to

- input data from the keyboard one may use the REWRITE statement
- retrieve data from an external file one may use the RESET statement
- send output to a printer one may also use the REWRITE statement

In some of the aforementioned cases new types of variables are defined, for example, that open the keyboard as an input device and set up the printer as an output device. The variable type INTERACTIVE may be used in these kinds of situations. In addition, the WHILE NOT EOF is often used to read an entire file into a program. The reader is referred to the large number of references available to gain facility with the specifics of setting up and working with external files.

9.3.11 Records

A record, which has a very specific meaning and method of implementation in Pascal, can be observed by referring back to a figure in Chapter 3 dealing with a simple personnel listing. In Figure 3.6 we identified a hypothetical employee listing along with the following data regarding these persons:

- name
- title
- department
- program assignment
- percent time assigned
- office location
- program supervisor

This typical list constitutes a record in which we identify a person and certain characteristics about the person that we wish to track and manipulate.

Next we show how we would set up a record for a more limited data set for an employee listing such as the one shown.

```
TYPE
     NSTRING = PACKED ARRAY [1..20] OF CHAR;
     EMPLOYEE = RECORD
          NAME : NSTRING
          TITLE : NSTRING
          PTIME : REAL

VAR
     PERSON : EMPLOYEE
```

The record is first declared by identifying types as we have seen before with Pascal variables. The record itself is shown by the name EMPLOYEE and it is stated to be a record. Under the record we list the data elements or fields that we wish to maintain. In this case we have selected only three of the elements in the data list: name, title, and percent of time assigned (the last of which we call by the name PTIME). The first two of these can be described by a string of characters and we can declare them to be of type NSTRING, where we define NSTRING to be a packed array of characters that, in this case, may contain up to 20 characters. PTIME, which we expect to be a number from 1 to 100 can be declared to be of type integer or of type real. We will use the latter since it is more general and in the program body we may wish to sum percentages of time for a list of people and then convert to the overall decimal equivalent.

All of these steps are preparatory to actually defining the principal variable, which we will take to be PERSON. PERSON is thus a variable of type EMPLOYEE, which is itself defined to be a RECORD that has three fields or data elements within it: NAME, TITLE, and PTIME. This may appear to be complex but it is structured and logical and allows us to manipulate each of the fields in a rather direct way that uses the operators and constructs that we have already seen as available to us in Pascal.

The next important aspect of records is to be able to reference an individual field which in the previous example would be NAME, TITLE, or PTIME. The method of doing so is illustrated here by assigning particular values to each of the fields in the example

```
PERSON.NAME := 'JOHN M. JACKSON    ';
PERSON.TITLE := 'SENIOR ENGINEER    ';
PERSON.PTIME := 50;
```

This is an assignment of a value to the individual fields of the variable PERSON. We note the way in which this is done—the variable PERSON is written followed by a period which in turn is followed by the name of the subordinate field. If we wish to read or write a particular data element we would then typically use either of the following two expressions as

```
READLN (PERSON.NAME);
WRITELN (PERSON.PTIME);
```

The fact that both PERSON.NAME and PERSON.TITLE are defined to be of type array means that we normally would read them in a character at a time using an index or subscript. Typically this can be done by an instruction such as

```
FOR I := 1 TO 20 DO
  READLN (PERSON.NAME(I));
```

This concept has been examined previously with respect to some of the sample programs in Pascal.

Under the latter procedure it is noted that anytime one wishes to call out a field both the variable name and the field are contained in the statement. In Pascal it is possible to simplify this process by using the WITH statement. This might be done as follows:

```
WITH PERSON DO
  BEGIN
    FOR I := 1 TO 20 DO
      READLN (NAME(I));
    FOR I := 1 TO 20 DO
      READLN (TITLE(I));
    READLN (PTIME);
  END (*WITH*)
```

The WITH statement calls out the variable at the beginning and allows for the manipulation of the fields by their own individual names without having to repeat the name of the variable, which in this case is PERSON. This is a considerable time saver in producing Pascal programs.

Several other capabilities are present in Pascal with respect to records, which will be cited here but not explained in detail. References 1, 7, 8, and 9 contain additional information regarding these kinds of capabilities which include

- copying an entire record—all the fields of one record can be copied into the corresponding fields of another record by using a simple assignment statement, provided both records are of the same type
- setting up arrays of records—using an array whose elements are themselves records
- using the set data type—manipulations of sets of data elements such as taking unions and intersections of sets
- searching—finding a particular data item within a record

9.3.12 Pointers

Pointers in Pascal, as well as other languages, create dynamic data structures. These types of structures are capable of growth within the context of other fixed elements in a program. For this reason they are especially useful when data is constantly changing, as in a real-time or near real-time application. Essentially, as new data are called into a program, these data can be easily inserted and operated upon without having to modify the program itself. A simple example is that of inserting a new person's name and an associated record into a list in alphabetic order.

For pointers Pascal uses the concept of a *node*, which represents a data element in a dynamic data structure. A node is of type record and can be declared in a program as follows:

```
TYPE
  NODE = RECORD
      WORD : PACKED ARRAY (1..4) OF CHAR;
      LOCATION : INTEGER;
  END;
  NPOINTER = ↑NODE;

VAR
  X, Y, Z : NPOINTER
```

As seen in this declaration, NODE is of type record and it has two subordinate fields, WORD and LOCATION. WORD will contain a name of up to four characters and will be the essential data element in a particular application area, that is, a list of names of people on a project. The second field, LOCATION, will contain an integer whose purpose may be to carry data or point to another node. The pointer type NPOINTER is associated with the record of type NODE by the declaration

NPOINTER = ↑NODE. Finally, pointer variables X, Y, and Z are declared to be of type NPOINTER. Here again we see a rather complex declaration for pointers built upon the features of a record.

In the body of a program it is necessary to be able to set up a node. This is done by a statement such as NEW(X) or NEW(Y). Such statements allocate storage to new records that are pointed to by pointers X and Y. This can be represented as shown in Figure 9.16 where X is the pointer variable, pointing to the node which has two cells, one of which will contain a data element for WORD and the other for LOCATION.

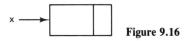

Figure 9.16

Having established a node and a pointer, it is of interest to place data in the cells of the node. The insertion of data can be achieved by statements such as

```
X↑.WORD := 'GREG';
X↑.LOCATION := 23;
```

These are typical assignment statements as we have seen before with respect to records, with the only difference being that the pointer, ↑, designation is used to signify that the variable X is of a pointer type. After these assignments, the node has the inserted data elements and looks like Figure 9.17. Another type of assignment is simply to copy the so-called value of one pointer into another pointer. We can do this by introducing variable Y and making the statement

```
Y := X;
```

which results in Figure 9.18. This is interpreted to be copying the value of pointer X into pointer Y. We note that X had already been set up as a node with data entered into it. The effect of Figure 9.18 is to point Y to the node already pointed to by X. If we were to set up another node Z and place data into it as

```
NEW (Z);
Z↑.WORD := 'THOM';
Z↑.LOCATION := 47;
```

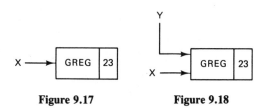

Figure 9.17   Figure 9.18

we would then have the results shown in Figure 9.19 and could proceed to exchange the nodes pointed to by X and Z by the statements

$$X := Z;$$
$$Z := Y;$$

This would lead to the diagram in Figure 9.20. The feature shown in Figure 9.20 provide sufficient capability to create what is sometimes called a *linked list*. We will now examine this process one step at a time.

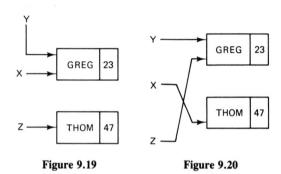

Figure 9.19   **Figure 9.20**

In order to set up a linked list it is necessary to have at least one field of a record defined as a pointer. This means that we will have to modify our initial declaration of types as follows:

```
TYPE
  NPOINTER = ↑NODE
  NODE = RECORD
    WORD : PACKED ARRAY(1..4) OF CHAR;
    LOCATION : NPOINTER;
  END;

VAR
  X, Y, Z : NPOINTER
```

We note the only real change that has been made is that the second field, namely LOCATION, has been defined to be of type NPOINTER. This is the essence of establishing a link between the various nodes and the data elements carried within the nodes, that is, a linked list.

With the pointer variables previously defined we will start with setting up a new node and placing data into it as before (Figure 9.21). We now set up another new node and place data into it, as shown in Figure 9.22. A linkage may be established as shown in the statement and diagram in Figure 9.23. This links the location field of X to the node pointed to by Y and is the essential feature of the linkage of

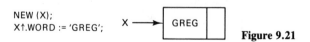

Figure 9.21

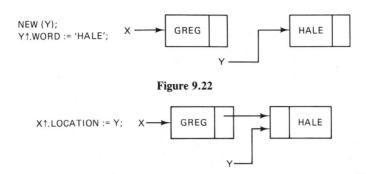

Figure 9.22

X↑.LOCATION := Y;

Figure 9.23

nodes. The process may then be continued with the statement and diagram shown in Figure 9.24.

Having defined a means of setting up a linked list we will now examine displaying data elements and traversing a list. The display of data elements is carried out by a statement we already know, namely WRITELN or WRITE. Reference to the field in which data is contained is performed in a straightforward manner as

WRITELN (X↑. WORD, Y↑. WORD, Z↑WORD)

which will lead to the output GREGHALETHOM. In short, we can provide a screen display by appropriate reference to the data element using the conventional WRITELN statement.

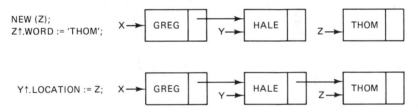

Figure 9.24

The significance of being able to traverse a list is that we desire to display and print out the WORD field of all of the elements of a list. A simple WHILE statement allows us to do this with a short set of program instructions. Prior to doing this, however, we will add two features to the previously illustrated linked list with variables X, Y, and Z; and data elements GREG, HALE, and THOM. For convention's sake we add another pointer variable, which we will call HEAD. HEAD will point to the first node (as does X) by adding the statement

HEAD := X;

In addition we will set up the value NIL in the LOCATION cell of the last node by means of the statement

Z↑. LOCATION := NIL;

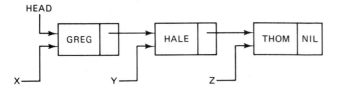

Figure 9.25

We therefore start with a diagram (Figure 9.25) for the linked list. We note that the additions we have made (adding HEAD and NIL) have not altered the positions of X, Y, and Z or the data in the WORD fields of the nodes. We can now traverse the linked list by means of the following short procedure:

```
PROCEDURE TRAVERSE

BEGIN
    WHILE HEAD < > NIL DO
        BEGIN
            WRITELN (HEAD↑.WORD);
            HEAD := HEAD↑.LOCATION;
        END; (*WHILE*)
END (*TRAVERSE*)
```

The WHILE statement establishes the inner BEGIN. .END sequence as long as HEAD is not equal to NIL. We WRITELN the WORD field of the node to which HEAD is pointed initially, namely, GREG. The next statement, HEAD := HEAD ↑ .LOCATION moves the pointer HEAD to the next node. In other words, HEAD is reassigned to where HEAD ↑ .LOCATION has been pointed. Thus we have the new diagram, Figure 9.26. At this point we print out HEAD ↑ .WORD, which has the value HALE and move HEAD to the next node. At the third (and in this example, last) node we print out THOM and find that the assignment of HEAD coincides with NIL and the WHILE process is terminated. For a long linked list, the aforementioned process continues automatically by means of the WHILE loop. Additional fields can be added to each record so that a large number of data elements would be included implicitly in the list.

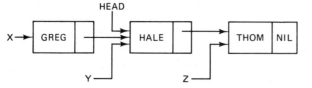

Figure 9.26

Two additional features of interest for pointers are deleting and adding nodes to a linked list. We will briefly discuss the role of the latter in setting up an ordered list.

We return to the previous diagram and add two pointers, which we will call FIRST and SECOND, as shown in Figure 9.27. We assume that we wish to delete

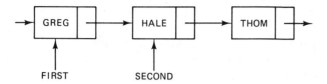

Figure 9.27

the node containing HALE from the list, which can be achieved in part by means of the following initial statement:

```
FIRST↑.LOCATION := SECOND↑.LOCATION;
```

This has the effect of taking the linkage from the LOCATION field pointed to by FIRST and connecting it to the node pointed to by the LOCATION field of the node pointed to by SECOND, resulting in Figure 9.28. The final step in the deletion process is to eliminate the middle node, which is done by the statement:

```
DISPOSE (SECOND);
```

This statement eliminates the node pointed to by SECOND and the deletion process is complete.

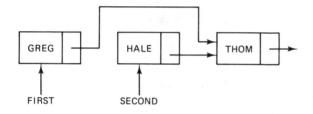

Figure 9.28

In the case of adding a node, we assume that we wish to add a node between the nodes containing GREG and HALE and we will leave the added pointer variables FIRST and SECOND. The required statements for the insertion of a node are:

```
NEW (FIRST↑.LOCATION);
FIRST↑.LOCATION↑.LOCATION := SECOND;
```

The first statement sets up the new node pointed to by the LOCATION field of the node pointed to by FIRST, as shown by Figure 9.29(a). Finally, the LOCATION field

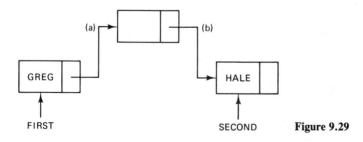

Figure 9.29

of the new node is connected to the (final) node pointed to by SECOND, shown by the link in Figure 9.29(b).

The process of inserting nodes is particularly relevant to the matter of creating an ordered list such as a list of names in alphabetic order. If we have such an alphabetic list, as a new name is to be inserted, we scan the list from the beginning until we find the appropriate location for the new name, which would fall between two existing nodes. The precise method for performing this scan is straightforward and can be found in the many texts on Pascal. Clearly the method for inserting a new node that is described in this section would then be employed such that the alphabetic name is contained in the new node. This process continues indefinitely as new names are to be added in alphabetic order. An analogous technique is available if a name is to be deleted, utilizing the aforementioned method for deleting a node.

9.4 SUMMARY

This chapter has reviewed the BASIC and Pascal languages as specific capabilities that can be brought to bear in the context of computer-aided systems engineering. The pedagogical method of working through small but concrete sample problems has been utilized. Further, the same set of calculations has been addressed in both BASIC and Pascal so that similarities and differences can easily be seen.

Computer packages for a variety of programming languages are readily available at modest prices. Thus, the systems engineer who gains facility with one or more of these languages will have a powerful tool with which to address numerous problems and issues not easily handled by other commercially available software packages. Specific types of programming language packages are cited in Appendix A.

This chapter also completes the explicit discussion of Part II of this text dealing with tools and techniques. Chapter 10, which follows, begins Part III, Systems Engineering and Applications, that addresses more specifically the tasks and activities of systems engineering.

REFERENCES

1. Jensen, K., and N. Wirth, *Pascal, User Manual and Report*. New York: Springer-Verlag, 1974.
2. *Reference Manual for the Ada Programming Language*. ANSI/MIL-STD-1815A-1983, U. S. Department of Defense, February 17, 1983.
3. Weinberg, V., *Structured Analysis*. Englewood Cliffs, NJ: Prentice-Hall, 1980.
*4. *Introduction To Programming In Basic*. Rate training Manual and Officer-Enlisted Correspondence Course, Naval Education and Training Program Development Center, 1983.

*5. Presley, B., *A Guide to Programming the IBM Personal Computer*. Lawrenceville, NJ: Lawrenceville Press, 1982.

*6. Presley, B., *A Guide to Programming in Applesoft*. Lawrenceville, NJ: Lawrenceville Press, 1982.

*7. Anderson, R. W., *From Basic to Pascal*. Blue Ridge Summit, PA: Tab Books, Inc., 1982.

*8. Koffman, E. B., *Problem Solving and Structured Programming in Pascal*. Reading, MA: Addison-Wesley, 1981.

*9. Gottfried, B. S., *Programming with Pascal*. Schaum's Outline Series in Computers, New York: McGraw-Hill, 1985.

*10. Lewis, T. G., *Pascal Programming for the Apple*. Reston, VA: Reston, 1981.

*11. Welsh, J., and J. Elder, *Introduction to Pascal*. Englewood Cliffs, NJ: Prentice-Hall, 1979.

PROBLEMS

For all programs called for in these problems, incorporate maximum use of remarks or annotations, titling of inputs and outputs, and include a set of expected program run results.

P9.1. Demonstrate in one or more programs of your own design and selection the following features of BASIC:
   a. the addition, subtraction, multiplication, division, and exponentiation operations
   b. three functions, including random number generation
   c. PRINTing of numerical results, titles, and messages
   d. numeric and string variables
   e. READ. .DATA
   f. INPUT of data and messages
   g. IF. .THEN
   h. GOTO
   i. FOR. .TO,STEP. .NEXT
   j. semicolon and colon use
   k. stepping and reassignment
   l. subscripted variables
   m. stepping summation
   n. dimension statement
   o. nested loops
   p. a subroutine
   q. RESTORE
   r. user-defined functions

P9.2. Write a BASIC program to convert feet to meters and pounds to kilograms.

P9.3. Write a BASIC program to simulate the tossing of 2 dice ten times.

P9.4. Write a BASIC program to calculate the sample mean and variance of ten input numbers. Show how both READ. .DATA and INPUT would be used to provide the ten inputs.

*Reference is relevant to the text but is not specifically cited.

P9.5. Write a BASIC program to generate all product elements representing a 4 by 5 morphological box, using row elements $R(I)$ and column elements $C(J)$. Print out the results in a two-dimensional array.

P9.6. Write a BASIC program that will implement a weighting and rating evaluation matrix for three systems against six criteria. Define the criteria by name and develop your own rating procedure as well as appropriate column labels for the printed outputs.

P9.7. Write a BASIC program to calculate the probabilities of 0, 1, 2, 3, and 4 errors of a four-bit digital message where the individual bit error probability is 0.01.

P9.8. Demonstrate in one or more programs the following features of Pascal:
    a. addition, subtraction, multiplication, division, and exponentiation operations
    b. three functions, including random number generation
    c. printing of numerical results, titles, and messages using WRITELN
    d. numeric and string variables
    e. inputting of data using READLN
    f. IF..THEN..ELSE
    g. REPEAT..UNTIL
    h. FOR..DO
    i. arrays
    j. WHILE..DO
    k. formatting output results
    l. use of a PROCEDURE
    m. use of user-defined functions
    n. records
    o. pointers

P9.9–9.14. Write Pascal programs corresponding to the problems cited in P9.2 through P9.7.

Part III   Systems Engineering and Applications

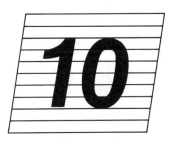

Better one safe way than a hundred on which you cannot reckon

Aesop

TOP-LEVEL SYSTEMS ENGINEERING

10.1 OVERVIEW

Top-level systems engineering is a process in which the key elements of systems design are defined and assessed so as to arrive at fundamental approaches to the overall design of the system. It is a sorting-out process that defines the most promising directions for top-level design configurations, and it is usually performed at a functional level. That is, the design choices are developed and selected for each of the major functional categories into which the overall system can be decomposed. This is sometimes also called the top-level architecture definition for a large-scale system.

We can discuss top-level systems engineering in term of five major factors, as shown in Figure 10.1. The process begins with a statement of requirements by the user of the system. In this case we shred out the top-level requirements, leaving behind the detailed requirements for the future process of detailed system design. We also show the top-level specifications in the figure since they are closely coupled to the requirements. The requirements and specifications are defined for the major functions of the system, as illustrated in earlier chapters as well as in Section 10.2. The key design alternatives for the system flow from the requirements/specifications through the inventiveness of the systems engineering design team. They are called alternatives at this point since the purpose of top-level systems engineering is to define, evaluate, and ultimately select a set of alternatives that best satisfies the user's requirements.

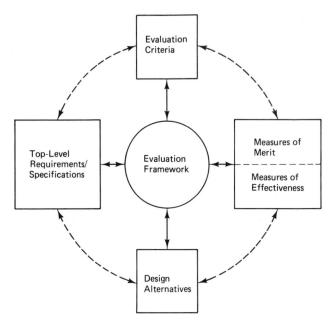

Figure 10.1 Top-level systems-engineering factors.

Measures of merit (MOMs) and measures of effectiveness (MOEs) are shown as important factors in the process. They are listed in the same box since we make no distinction between the two concepts in this context. Either term has been used in the literature to mean more or less the same thing. They represent high-level indicators of how well the system performs its functions. In addition, we take them here to be measurable in the sense that they can be computed or calculated in rigorous and nonsubjective terms. This may be achieved through either some process of analysis (e.g., modeling and simulation) or the use of test or empirical data embedded in an analytic process, or both. We further distinguish between these MOMs and MOEs and the use of a set of evaluation criteria. The latter are also clearly used to assess the merits of the design alternatives, but they are construed as possibly containing subjective elements that cannot be readily and rigorously quantified. Thus we see them as separable, largely due to their potential qualitative nature. Finally, these factors are brought together in some type of evaluation framework that is supported by an evaluation procedure.

Again, the purpose of the entire process is to sort through and narrow down the available design alternatives so as to ultimately select the most promising alternative. Given such a selection, the systems engineering process continues to lower levels of design, but the major approaches have been well defined. Examples of top-level design choices might be selecting time-division multiplex versus frequency-division multiplexing for a communications system, or a centralized versus decentralized approach to a large-scale multinode computer system.

Our approach in this chapter is to examine each of the five factors in some detail, showing examples of each in turn. Illustrations of various ways in which an evaluation framework can be structured are provided, using real-world cases in order to demonstrate the process as well as the potential results. The detailed analyses that lead to the calculations of the MOMs and MOEs are reserved, as a general rule, for considerations in later chapters.

10.2 REQUIREMENTS ANALYSIS

Requirements analysis plays a central role in all large-scale systems engineering programs. As alluded to in Chapter 2, the statement of requirements flows from the overall user needs and represents the first definitive articulation of what it is that the user wants. It may be the last such overall formal statement made by the user in cases for which the user turns over to a contractor the job of writing the top-level specification for the system, which is often the situation. In any case, the requirements are to be satisfied by the system design and implementation and therefore it is necessary to assure that this is true for each and every requirement.

A top-level outline of the major areas that might be covered in a requirements document are listed here [1]:

- scope
- applicable documents
- systems definition
- quality-assurance provisions
- preparation for delivery for hardware and software
- notes and definitions

The most important description in the list is the systems definition. In essentially all cases, the system is described hierarchically by going through the elements in a tree-like numbering scheme. In Table 10.1 we show a typical systems-definition portion of a requirements document, going only to a three-number coding; the actual document will normally go to six- and seven-level subdescriptions.

We note several interesting features in the requirements definition illustrated in Table 10.1. First, the key technical aspects of the system are defined in one major section, which goes through the various functional subsystems. Second, certain support requirements are spelled out under the system definition such as logistics, personnel, and training. Third, under the subtitle of precedence there is some consideration given to life-cycle cost as well as requirements weighting factors. In the former case, it is indicated that life-cycle cost shall be used as a key factor in performing trade-off analyses to establish the order of precedence among the various system alternatives that are developed. As we shall investigate later, this is what leads to so-called cost-effectiveness assessments of the system. In the latter area, it is recognized explicitly that various requirements have different importance in the overall system-development process. An attempt is made, in the requirements document, to provide guidance as to the weighting that is to be given, as follows:

- most important requirements—uses the word *shall* to designate mandatory requirements

- next most important requirements—uses the words *shall, where practicable*
- next most important requirements—uses the words *preferred* or *should*
- least important requirements—uses the word *may*

Thus we have important distinctions being made by the user in an attempt to convey to the developer of the system the relative importance of the various requirements. This is relevant to the various computer-aided preference evaluation and weighting schemes that we have considered.

The previous discussion and illustration of a particular technical requirements document is typical for a large-scale system, whether it is in the civil or defense arena. We are dealing with a hierarchy of requirements, and this is the key statement telling the developer what the user wishes to field. It is therefore incumbent upon the developer to analyze these requirements in considerable detail to assure that they have all been satisfied by the preferred system design. This necessitates computer-aided techniques, which also provide the functions of completeness, consistency, and traceability.

TABLE 10.1   SYSTEM DEFINITION PORTION OF A
REQUIREMENTS DOCUMENT

3 SYSTEM DEFINITION
  3.1 SYSTEM SEGMENT DEFINITION
      3.1.1   Communications Segment Description
      3.1.2   Communications Segment Mission
      3.1.3   Threat
      3.1.4   System Segment Diagrams
      3.1.5   Interface Definition
      3.1.6   Furnished Equipment
      3.1.7   Operational and Organizational Concepts
  3.2 CHARACTERISTICS
      3.2.1   Performance Characteristics
  3.3 DESIGN AND CONSTRUCTION
      3.3.1   Materials, Processes, and Parts
      3.3.2   Electromagnetic Radiation
      3.3.3   Product Marking
      3.3.4   Workmanship
      3.3.5   Interchangeability
      3.3.6   Safety
      3.3.7   Human Performance and Engineering
      3.3.8   Computer Resources
      3.3.9   General Design Requirements
      3.3.10 Grounding
  3.4 DOCUMENTATION
      3.4.1   Documentation Management
      3.4.2   Specifications
      3.4.3   Drawings
      3.4.4   Interface Control Documentation
      3.4.5   Test Plans and Procedures
      3.4.6   Technical Manuals

Table 10.1 *(cont.)*

3.5 LOGISTICS
    3.5.1 Logistics Support
    3.5.2 Maintenance
    3.5.3 Supply
    3.5.4 Facilities and Equipment

3.6 PERSONNEL AND TRAINING
    3.6.1 Personnel
    3.6.2 Training

3.7 FUNCTIONAL AREA CHARACTERISTICS
    3.7.1 Satellite/Terrestrial Communications
    3.7.2 Modulation/Demodulation
    3.7.3 Multiplexing/Demultiplexing
    3.7.4 Switching and Routing
    3.7.5 Encryption/Decryption
    3.7.6 Recording, Storage, and Playback
    3.7.7 Contol and Monitor
    3.7.8 Formatting and Conversion
    3.7.9 User Termination
    3.7.10 Dual Node Commanding

3.8 PRECEDENCE
    3.8.1 Conflicts
    3.8.2 Requirement Weighting Factors
    3.8.3 Life-Cycle Cost
    3.8.4 Supplementary Specification and
           Standards

10.2.1 Requirements versus Measures of Merit, Measures of Effectiveness, and Evaluation Criteria

The systems approach discussion in Chapter 2 suggests an evaluation process wherein the system alternative attributes are compared against a set of requirements and a set of evaluation criteria. This comparison process needs further explanation in order to define the evaluation more precisely.

The requirements document, as shown in Table 10.1, provides a base line statement by the user as to what is needed. If complete, this document will establish what is important to the user, including the desired attributes of the system and the (minimum) numerical levels of these attributes. When the system is being designed, sufficient information is generated to allow the developer to predict (and later test) the performance levels of the system in relation to the required attributes. This is done at the overall system level and, as well, at subsystems and lower levels.

The attributes of the system are also sometimes called *measures of merit* (MOMs) or *measures of effectiveness* (MOEs). They are defined in the terms and with the same dimensionality of the requirements document. For example, if we are dealing with a city's metro system, we may specify that a typical user during rush hour should not wait more than some period of time, on the average, for the next train, or that the mean time between arrivals of a train at a station should not be more than some number. In such a situation, the mean waiting time and time between arrivals become part of the requirements document for the system. If they are important enough, they are defined as system-level requirements. By the same token, the developer of the system must now demonstrate that the system design is such that these requirements can be met. This means that for the developer, both the waiting time and arrival time become attributes of the system whose performance levels must be satisfied. As important attributes, they then become defined as MOMs or MOEs for the system. They are measured in the same units (dimensions) and under the conditions specified in the requirements document. In the early stages of design, these MOMs or MOEs are predicted by the use of models, simulations, and other such analytical techniques. As the system progresses into hardware and software, more and more of the prediction is based upon test data and greater confidence is built in terms of the ability of the system to satisfy the requirements. The point is that the requirement is (or should be) stated in precisely the same terms as the MOM or the MOE, except that the requirement comes from the user and the MOM or MOE comes from the developer as a measure, over time, of the extent to which the requirement is being met.

The next question is: What is the role and context of the set of evaluation criteria in relation to both the requirements and the MOMs and MOEs? Although there is generally no firm agreement on the precise answer to this question, we offer the following explanation. The evaluation criteria become the highest level of final system evaluation, taking all factors into account, including cost. This means that after all the MOM, MOE, and requirements comparisons are made, with the latter as the controlling agenda, systems engineers will embody in the set of evaluation criteria all of the factors that they consider important, and perform a top-level system's comparison on the basis of these factors, then called evaluation criteria. These criteria may include a variety of subjective considerations, such as growth potential, risk, environmental compatibility, and aesthetic quality. In short, the assessment against the evaluation criteria is the final check on the desirability of the system and may include factors or attributes that are not in the requirements document but are normalized values that are considered important by the developer. The exception to such a process is, of course, when the user establishes the set of evaluation criteria for the system and makes this known to the user. However, even in such a case, these criteria would normally be the highest-level factors that go into the judgment as to the pros and cons of the system. More complete and illustrative examples of MOMs, MOEs, and evaluation criteria are discussed later in this chapter.

10.2.2 Requirements versus Specifications

There is also a close relationship between requirements and specifications. The latter, in particular, is derived from the former, and represents a hierarchy of statements as to what is to be contained in the system. For defense systems, specifications follow the framework defined in Mil. St. 490 and (to some extent) in 483 and often the requirements and specifications documents follow each other, even down to the numbering scheme. This is helpful in that it assists in establishing correlation and traceability between requirements and specifications. Indeed, in some programs the top-level system specification is functional in nature and is used in place of a separate requirements document.

As an example, the following list is the set of top-level chapters of a functional system specification for a communications system [2]:

- scope
- documents
- requirements
- quality assurance provisions
- notes
- definitions and terminology

This bears a close relationship to the outline of the major portions of the requirements document discussed earlier in this chapter. The second-level definitions differ from system to system, in general, as is illustrated by the second level listing shown in Table 10.2 for a functional specification [2], which can be compared with the second-level (two-digit) listing provided in Table 10.1.

TABLE 10.2 SECTIONS OF A FUNCTIONAL SPECIFICATION (Section 3.0—Requirements)

3.1 Attribute Tables	3.6 Logistics
3.2 System Services and Functions	3.7 Design Priorities
3.3 Design and Development Requirements	3.8 Preplanned Product Improvement
3.4 Performance Requirements	3.9 Training Requirements
3.5 Human Interfaces	

10.2.3 Requirements Traceability

Requirements traceability is used as a generic term for providing the capability to correlate the requirements with other important aspects of the systems engineering process. If the requirements document is the expression of need on the part of the user, then the rest of the process, in effect, is proof that the need is ultimately being satisfied. Requirements traceability simply helps to make that proof more explicit.

The issue may be illustrated by reference to Figure 10.2. On the left of the figure is the requirements document and we show a particular requirements statement circled. Examples of such a requirement may be

- the system shall have a pointing accuracy of plus or minus 5 milliradians (mrad)
- the average waiting time at a station during rush hour shall be no greater than 10 min. with a standard deviation not to exceed 5 min.

All requirements are then to be explicitly correlated with an appropriate statement in the specifications that describe the system, shown illustratively in Figure 10.1. The specifications, in general, are multilevel so that there is a need to tie each level back to the requirements document. By so doing, the items in the various levels of specification are also tied to one another. The same is true for other portions of

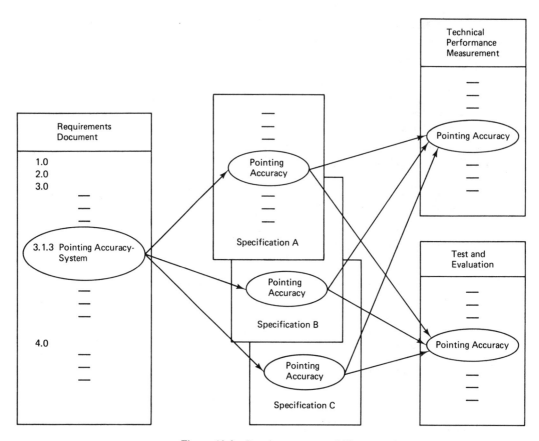

Figure 10.2 Requirements traceability example.

the systems engineering process such as technical performance measurement and the test and evaluation program. Indeed, everything that is done in the program that supports the satisfaction of a requirement should, in principle, be tied back to the specific requirement in question.

In Chapter 3, Figure 3.8 showed an illustration of a DBMS set of fields that attempts to provide requirements traceability. In many cases, the correlation can be established in precisely that manner. In more complex situations, it may be necessary to use the language of the automated DBMS as an overlay in order to provide the precise capability desired. In such a case, that would constitute writing a new program in the particular language provided by the DBMS. Various systems engineering organizations have written requirements traceability packages using a DBMS as a basic structure with special applications overlays [3]. In even more complex situations we may purchase a so-called *requirements language package* for which the manufacturer has written a general-purpose program, similar to the one mentioned above, to handle back-and-forth traceability. More will be said in Section 10.2.5 about the matter of requirements languages, which have become a rather well known set of commercially available software packages.

10.2.4 Requirements Allocation

Implicit in the matter of traceability of requirements is the task of requirements allocation. In general, allocation of requirements is performed by the developer of the system as the synthesis process is carried out. It is an allocation from top to bottom that applies a requirement at a higher level of system indenture to successive lower levels.

To illustrate, if a top-level requirement for a functional subsystem is to point a solar telescope to the center of the sun with an error no greater than 10 arc-seconds, then this *error budget* is normally allocated to the major subsystems that contribute to the overall error. If there are three such subsystems, then the allocation might simply be on the basis of a root-sum-square computation, implicitly assuming independence and additivity of error source random variables (see Chapter 5). This of course leads to the specification of the characteristics of lower levels of equipments so that a target is established for the designers of all equipments at all appropriate levels. In effect, the requirements allocation at the same time becomes an allocation of specifications.

The process is illustrated in Figure 10.3. Here we assume that we are attempting to point a solar pointing instrument such as a telescope at the center of the sun. For simplicity, we consider three major sources of error

1. error in pointing resulting from errors in the overall spacecraft control system
2. errors in the telescope control loop, which is carried as an instrument aboard the overall spacecraft
3. errors resulting from possible optical-path distortion within the telescope instrument itself

Although each of these three may have subordinate error sources, we consider only

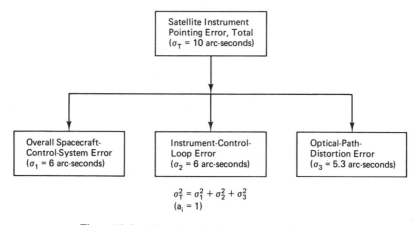

Figure 10.3 Allocation of pointing-error requirements.

these major sources and assume them to be generated by independent and additive random variables. Under these conditions, the overall error may be expressed as

$$\sigma_T^2 = a_1^2 \sigma_1^2 + a_2^2 \sigma_2^2 + a_3^2 \sigma_3^2 \qquad (10.1)$$

where σ_i refers to the three error sources listed previously and the a_i are coefficients to be estimated. The first issue is that of associating the required value of 10 arc-seconds with the total error variance, σ_T^2 (see Equation 10.1). Although there is no firm convention for this, we will associate the required value with the one sigma value of total error, namely, 10 arc-seconds $= \sigma_T$. If the total error were normally distributed, the $\pm \sigma$ value contains 68% of the distribution (see Chapter 5). The same procedure can be applied to two sigma values which would contain 95% of the distribution. If we do not analyze or otherwise model the interaction between the three major error sources, typically the coefficients would be taken as unity. This leaves us with one equation with the three individual error sources as unknowns. Here the allocation process is usually based upon subjective judgments as to the complexity of the prospective design, history with other programs, state-of-the-art assessments, and the like. In any case, for a complex system with numerous error allocations to be made, so-called engineering judgment might lead to an allocation of one-sigma values as

$$\sigma_1 = 6 \text{ arc-seconds}$$

$$\sigma_2 = 6 \text{ arc-seconds}$$

$$\sigma_3 = 5.3 \text{ arc-seconds}$$

This satisfies the overall constraint and provides the designer of each of the relevant subsystems with a required error target. This process is continued into lower levels of detailed design, as appropriate to the situation and the vagaries of the systems engineering approach.

Another example of requirements allocation is that of reliability allocation. A simple case, shown in Figure 10.4, lists a required value of MTBF of 1000 hr. The associated failure rate, for exponential failure, is the reciprocal of the MTBF or $\lambda = 0.001$. From Figure 10.4 we require an allocation for this example to three "black boxes" that make up the overall unit. Since the failure rates are additive (see Chapter 5), we have one equation with three unknowns, as we had in the aforementioned pointing example. An allocation is made, using the same arguments as before, but this time it is made in the context of propensity toward failure.

We see in the aforementioned examples that requirements allocation is necessary in order to develop lower levels of requirements through the hierarchy of systems, subsystems, and components. However, we usually wind up with, in a formal sense, more unknowns than we have equations. Unless other constraints are developed or the problem analyzed in more detail, the allocation results ultimately become a matter of engineering judgment. In many situations the allocations that are made are considered preliminary and as the design progresses and more data are obtained the subordinate allocated values are adjusted such that the overall required value is still satisfied.

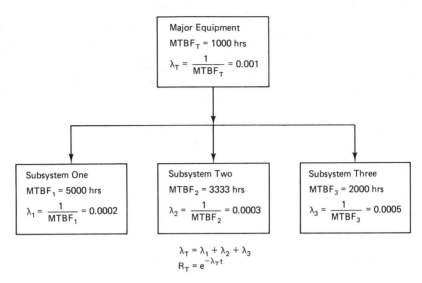

Figure 10.4 Reliability-allocation example.

10.2.5 Requirements Tools and Languages

A variety of requirements tools and even languages have been constructed to bridge the gap between the statement of requirements, the specifications for the system, and the system design and analysis. Many of the tools and languages are rather formal and are available as software systems, although such systems have not been generally available for the p-c. Many have been developed with the ostensible motivation to bridge the requirements–system design gap in regard to software development. Others are more generalized, have not been structured in software, and take the form of a diagrammatic discipline. In essentially all cases, however, they focus on turning requirements into a coherent system design in the most efficient and complete manner.

At the low end of the spectrum of such tools and languages is the requirements traceability matrix discussed earlier in Section 10.2.2. At the high end there are sentence parsers, processors, design constructs, and report generators that attempt to automate to the maximum extent the translation from requirement to design element. One simple way to look at this class of tools and languages is to examine the two dimensions of size versus capability, as illustrated in Figure 10.5. By size we mean the number of data elements that have to be part of the system itself as well as part of the data base set up by the system. Thus if the tool has to accommodate 1500 pages of a requirements document, it will have to contain a relatively large data base. By capability we refer to the ability of the tool to carry out a variety of specialized functions such as sentence parsing, checking for consistency and completeness, and report generation. At the low end of the size-capability coordinates we have the simple DBMS, even that available for the p-c such as dBASE III Plus. It is a relatively small DBMS and has a very limited capability to perform specialized func-

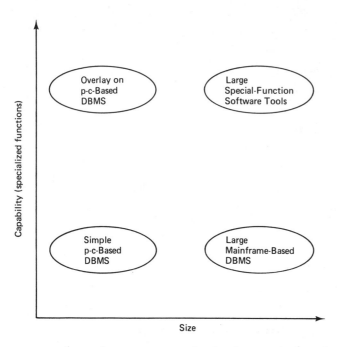

Figure 10.5 Size versus capability dimensions for requirements tools and languages.

The figure shows a plot with "Capability (specialized functions)" on the vertical axis and "Size" on the horizontal axis, containing four labeled ovals:
- Overlay on p-c-Based DBMS (upper left)
- Large Special-Function Software Tools (upper right)
- Simple p-c-Based DBMS (lower left)
- Large Mainframe-Based DBMS (lower right)

tions. As we move up in size but not in (requirements analysis) capability, we are led to the more sophisticated DBMS for the larger computers, such as db2, IMS, and IDMS. More capable but smaller size tools are those that have been built by constructing an overlay upon a commercially available p-c based DBMS. Finally, to get both size and capability we find large special function tools such as Problem Statement Language/Problem Statement Analyzer (PSL/PSA) [4]; and others such as Software Requirements Engineering Methodology (SREM) and Input/Output Requirements Language (IORL) [5].

A keen observer of the scene of requirements tools, A. M. Davis, has looked at two other dimensions, namely readability and formality [6]. For high readability and low formality he cites the English language, the program design language (PDL), and simple text formatters. High formality and low readability leads to various axiomatic and operational techniques. He places PSL and IORL somewhere in the middle. Various other analysts and developers of requirements tools have examined other dimensions and these are abundantly explored in the literature [5, 7]. Since many of these tools have been oriented to the software development arena and, in addition, are not available for the p-c, we reserve further discussion of them to Chapter 14. The latter may be used also to examine the literature more fully with respect to a relatively broad variety of formal requirements tools and languages.

10.3 DESIGN ALTERNATIVES

In addition to requirements and specification analysis, the definition of design alternatives is the second major factor in top-level systems engineering (Figure 10.1). A convenient way to examine this factor is at a functional level. The requirements and specifications for a system are usually developed by system function so that a functional decomposition of the system, at least at a top level, is immediately apparent. The systems-design engineers therefore look for alternative approaches to the implementation of each of the system functions. The process is illustrated in Figure 10.6 in which various approaches are listed for selected functions of a communications system. The same general diagram applies as well to other types of systems.

System Functions (Communications)	Alternative Design Approaches [(DIJ) = Design J for Ith Function]
1. Satellite/Terrestrial Communications	D11, D12
2. Modulation/Demodulation	D21, D22, D23
3. Multiplexing/Demultiplexing	D31, D32
4. Switching and Routing	D41, D42, D43
5. Encryption/Decryption	D51, D52
6. Recording and Playback	D61, D62
7. Control and Monitoring	D71, D72, D73
8. Formatting/Signal Conversion	D81, D82

Figure 10.6 Design alternatives for system functions.

For the given example, in principle, there are a total of (2)(3)(2)(3)(2)(2)(3)(2) = 864 combinations that constitute possible alternative systems. Most likely, some of these will not be feasible since not all functional approaches are compatible with one another. Even if this fact reduces the total number of possibilities by a factor of ten, we are left with 86 feasible alternatives. This is also a relatively large number, which is why the matter of top-level systems engineering is considered a significant subject in its own right. As referred to earlier in Chapter 2, the combinatorial issues of systems engineering make the design and evaluation process a very complex one and in a very real sense a great unsolved problem.

10.3.1 Air Transportation Example

By way of illustrating the matter of defining alternatives for a large-scale system, we set forth some alternative concepts for the possible growth of our air transportation system. Four such concepts are illustrated in Table 10.3. Alternative A is essentially a direct extension of our current system. Alternative B emphasizes a short-haul supplement that adds V/STOLports in high density short-haul areas to unload traffic from the conventional system. The third alternative moves in the direction of accommodating long-haul traffic by the specific addition of remote transfer airports. These airports are used to transfer and consolidate long-haul traffic and need not be associated with a high-density metropolitan area. Alternative D places large exchangeports in optimized geographic positions and adds local terminals in high-density areas that are designed to move traffic efficiently to the exchangeport.

The various system functions are represented in the subcategory headings of Table 10.3 but are more explicit in terms of what we know to be the so-called major subsystems of our national air transportation system, namely, airports, air vehicles, airspace, and ground access. The operational notion explains in brief form the essential idea behind each alternative. This can be placed in a typical matrix layout such as we have considered in various other chapters of this book. Two major points are

TABLE 10.3 ALTERNATIVE CONCEPTS FOR GROWTH OF AIR-TRANSPORTATION SYSTEM

Concept A Extension of Current Operation

Operational Notions

Mix of public/private long-haul (LH) and short-haul (SH); continuation of city pair airport network; small satellite airports (SA) mainly for private light aircraft; extension of current airport and operational concepts to year 2000 in attempt to satisfy demand for service to passengers (pax), cargo, and private transport.

Airports

City-pairs (CP) with separate runways for short-takeoff-or-land (STOL) and private aircraft as well as parallel runways for conventional aircraft (CTOL); small private transports airports (SA); limited separate STOL, and rotary wing (RW) aircraft in suburban areas outside large metropolitan cities

Air Vehicles

CTOL, including supersonic transport (SST); STOL; RW

Airspace

Full air-traffic control (ATC) or optimized airport system, e.g., dual-microwave ILS for dual-runway IFR operations; capability for noninterfering CTOL, STOL and RW operations; implemented advanced automation system to support ATC operations and resultant increase in capacity and reduction in delay

Ground Access

Highways and some extensions of rapid transit but entry into system is primarily at airports

Typical Trips

Concept B High Density Short-Haul Supplement

Operational Notions

Same as Concept A but supplementing city-pairs (CP) with V/STOLports and operations in high-density areas to unload short-haul (SH) traffic from CTOLports and provide some feeder service; emphasizes growth in system through improved services designed to optimize short-haul traffic service

Airports

Same as Concept A except for extensive use of V/STOLports in both central business district (CBD) and suburbs

Air Vehicles

CTOL/SST; V/STOL; RW

Airspace

Same as Concept A except that full ATC service is provided at V/STOLports

Ground Access

Same as Concept A except that V/STOLports tend to be closer to true pax/cargo origins and destinations

TABLE 10.3 (*cont.*)

Typical Trips

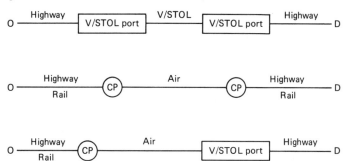

Concept C Remote Transfer Airport Supplement

Operational Notions

Supplement city-pair airports with remote transfer airports (TP) for pax and cargo; V/STOL feeder service to TP for long haul (LH) traffic consolidation; LH city pair service where sufficient origin-destination (OD) demand; TP access almost exclusively by air

Airports

Same as Concept A but supplemented with pax/cargo transfer airports (TP) remotely located; terminal design optimized for transfer operations

Air Vehicles

CTOL/SST; V/STOL and RW feeder service; new-generation wide-body LH pax aircraft

Airspace

Same as Concept A; full ATC service at TPs

Ground Access

Same as Concept A for CP access; rudimentary but not primary ground access for TPs

Typical Trips

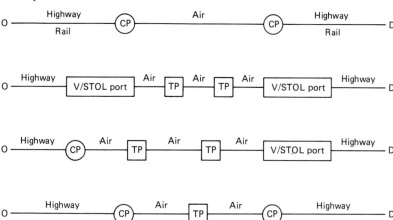

TABLE 10.3 (*cont.*)

Concept D Local Terminal and Exchangeport Supplement

Operational Notions

Same as Concept C except that
- exchangeport (XP) location optimized based upon accessibility and environmental conditions
- local terminals (LT) near true O-D set up as public-system entry-exit points; one LT may be at XP itself
- access to XP by mixed public mode (both air and ground) as dictated by individual situation

Airports
Same as Concept C except as noted above

Air Vehicles
Same as Concept C

Airspace
Same as Concept C

Ground Access
Same as Concept A for access to LT; LT consists of public terminal for local access near O-D; access by auto, rapid transit, high-speed ground-rail transit, and optimized transfer to all types of airports

Typical Trips

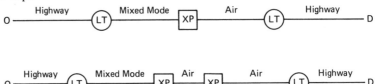

to be kept in mind with regard to top-level systems engineering in regard to the definition of the design alternatives.

1. the various alternative designs are framed in terms of the functional breakdown of the overall system, which often can be put into a one-to-one correspondence with the major subsystems of the system
2. the matrix format (see HEMAT from Chapter 2) is a convenient way to develop the alternatives, which itself assists in the process of evaluating the alternatives and ultimately applying computer-aided tools.

10.4 MEASURES OF MERIT AND MEASURES OF EFFECTIVENESS

There is no basic distinction made in this text between measures of merit (MOMs) and measures of effectiveness (MOEs), although there is literature that attempts to do so. For our purpose they are synonymous and define, in quantitative, measurable terms, the high-level attributes of the system. The notion therefore is that it is necessary for purposes of comparing alternatives to be able to calculate the various values of these measures. Next, we show a couple of examples of these measures for two distinctly different types of systems.

10.4.1 Strategic Defense Initiative Measures

In March of 1983 President Reagan set forth the beginnings of what has been called the strategic defense initiative (SDI), sometimes referred to as "Star Wars." Without consideration of the advantages or disadvantages of such an approach to the matter of strategic defense, we focus here only on the brief articulation of MOMs or MOEs for such a system.

The overall system is designed to provide a defense against enemy nuclear missiles [8]. It is to perform such a mission by intercepting such missiles by means of a variety of space, air, and ground-based assets that operate in three layers in relation to the phases of flight of the potential enemy missiles: boost, midcourse, and terminal (in some cases a fourth post-boost phase is considered). Table 10.4 lists some of the measures that appear to be of interest in ultimately assessing the technical capability of a defense system to cope with the various postulated threats. Selected cost-related issues are also reflected in the table.

TABLE 10.4 POSSIBLE MEASURES OF MERIT OR EFFECTIVENESS FOR THE STRATEGIC DEFENSE INITIATIVE

- Leakage rate, by phase
- Probability of detection
- Probability of engagement
- Probability of kill
- Percent targets detected
- Percent targets engaged
- Percent targets killed
- Probability of survival
- Absentee ratio
- Cost exchange ratio
- Cost of countermeasures
- Cost, surveillance system
- Cost, logistics system
- Cost, weapon system
- Cost, battle management/C^3 system

Two key measures for this type of system include *probability of kill* and *leakage rate*. The former is defined locally in terms of the specific engagement of a defensive weapon against an enemy missile and the likelihood that such an engagement will result in destruction of the missile. This measure is also used in virtually all analyses of weapon systems. It is also clear that an extensive set of computations are required in order to make such an assessment in the SDI context and various modeling and simulation activities are required in order to do so. The leakage rate is considered to be an important high-level measure since it addresses the matter of how many enemy missiles might come through the layers of the defensive shield corresponding to each of the three phases of flight, that is, boost, midcourse, and terminal. For example, if the probability of penetrating each of the layers is 0.1, then the probability of penetrating all of the layers is 0.001, assuming a simple binomial model representation. This means that the defense would allow one in one thousand missiles to penetrate. Here again the matter of actually calculating leakage rates is seen as an extremely complex one requiring massive modeling and analysis of the various alternative system designs against the postulated threats, including enemy countermeasures to the defense.

10.4.2 Intercity Transportation Example

Another illustration of MOMs or MOEs can be drawn from the field of transportation in relation to a typical trip [9]. Although the issue of providing passenger or freight service between two points can be viewed as a simple situation, a relatively large number of measures can be construed as significant, as illustrated in Table 10.5. This table shows several conventional service and cost measures but also defines some relatively new types of basic, composite, and normalized measures, as follows:

- passenger mobility index (PMI)—defined as the product of service frequency and capacity, divided by the trip time
- freight mobility index (FMI)—same as for the PMI, but applicable to freight service
- normalized PMI (NPMI)—defined as the PMI normalized by (divided by) the cost per passenger-mile
- NPMI per unit cost—same as for the NPMI, but normalized by the cost per ton-mile
- NPMI per energy usage—energy usage defined as the number of BTUs per passenger or ton-mile
- NPMI per energy/cost expenditure—cost expenditures are of the several types listed in Table 10.5
- NPMI per pollutant/noise contribution—pollutant by type is measured in tons per year; noise is measured by population affected within a particular noise exposure factor
- viability index—defined as the net present value (NPV) divided by the present value of R&D and investment cost,

Several of the measures listed in Table 10.5 are considered further in this chapter with regard to evaluation frameworks. The transportation and "star wars" examples described previously show what is meant by an MOM or MOE and that the issue of computation of the specific values of these attributes can be an exceedingly complex

TABLE 10.5 POSSIBLE MEASURES OF MERIT OR EFFECTIVENESS FOR INTERCITY TRANSPORTATION

Basic Measures

Cost
- Total R&D cost
- Total investment cost
- Total operations & maintenance cost

Service/Demand
- Passenger demand
- Passenger fares
- Freight demand
- Freight fares
- Door-to-door trip time
- Dock-to-dock trip time
- Capacity
- Load factor
- Service frequency

Energy
- Energy usage/passenger-mile
- Energy usage/ton-mile

Environmental Impact
- Pollution (by type)/year
- Population (affected by noise) within NEF 30

Composite Measures

Cost
- Total cost
- Revenues
- Net profit (loss)
- Cost per passenger-mile
- Cost per ton-mile
- Return on investment (ROI)
- Net present value (NPV)
- Payback period
- Viability index

Composite Measures (cont.)

Service/Demand
- Passenger mobility index (PMI)
- Freight service index (FSI)

Cost and Service/Demand
- Normalized PMI (NPMI)
- Normalized FSI (NFSI)
- NPMI per R&D cost
- NPMI per total cost
- NFSI per R&D cost
- NFSI per total cost

Cost, Service/Demand, and Energy
- NPMI per energy usage
- NFSI per energy usage
- NPMI per energy usage per R&D cost
- NPMI per energy usage per total cost
- NFSI per energy usage per R&D cost
- NFSI per energy usage per total cost

Cost, Service/Demand, and Environment
- NPMI per pollutant emission
- NFSI per pollutant emission
- NPMI per noise population
- NFSI per noise population

one. Some of the methods for making such computations are considered further in Chapter 12, Performance Modeling and Simulation.

10.5 EVALUATION CRITERIA

Evaluation criteria are taken here to represent a broader class of system attributes than MOMs or MOEs and include both quantitative and qualitative factors. These may be given as weighted in order of importance by the user or they may not. Simple preferences may be defined by the user or the issue of importance may be left to the system developer to deal with. In any case, the evaluation criteria are factors to be used by the developer to

- rationalize the selection of alternative design configurations for purposes of developing the best possible design of the system, and
- demonstrate to the user that the above has been done and that at a systems level the top-level requirements have been satisfied

10.5.1 Aviation-System Evaluation Criteria

Earlier in this chapter in Table 10.3 we identified a series of alternative concepts for the possible growth of our national air-transportation system. In Table 10.6 we list a set of criteria against which the alternative concepts can be evaluated. These criteria are developed in a two-level hierarchy and include cost considerations. A weighting and rating scheme has in fact been applied to these alternative concepts in relation to the given criteria in order to develop directions for the planning of our national air-transportation system. As previously discussed, such a rating and weighting scheme is particularly simple to implement in p-c software in the form of spreadsheets and the variety of preference and alternative evaluation aids that are available.

10.5.2 Communications System Evaluation Criteria

Further illustration of a set of evaluation criteria is shown in Table 10.7 in reference to a large-scale communications system, as cited in Reference [1]. These 18 criteria contain a mix of qualitative and quantitative factors but all are considered important with respect to the top-level assessment of a communications system. The quantitative criteria may be synonymously defined as MOMs or MOEs and a series of subordinate parameters would have to be developed in order to evaluate them. Such parameters might include bit error rate (BER), RDT&E costs, number of voice and data channels at various transmission rates, intermodulation distortion, and dynamic range. As with the air transportation example, the 18 evaluation criteria can be set in the context of a matrix evaluation, which can be carried out with available software packages.

TABLE 10.6 EVALUATION CRITERIA FOR A NATIONAL
AIR-TRANSPORTATION SYSTEM

A. International/Economic
   1. Supports International Trade and Commerce
   2. Supports Defense Needs and Interests
   3. Encourages Economic Growth of Existing Metropolitan Centers
   4. Stimulates New Economic Growth Centers
   5. Encourages Financial Stability of All Sectors
   6. Utilizes Existing Investments in Resources

B. Social
   1. Enhances Population Distribution
   2. Improves Personal Mobility
   3. Avoids Social Disruption

C. Environmental
   1. Minimizes Noise Impact
   2. Minimizes Pollution Effects
   3. Improves Aesthetic Value

D. Quality of Service
   1. Average Door-to-Door Speed
   2. Frequency of Service
   3. Reliability of Service
   4. Safety and Security (Passengers/Cargo)
   5. Comfort and Convenience

E. System Capacity
   1. Community Service and Accessibility
   2. Capacity-to-Demand Ratio and Growth Capability

F. Investment Costs
   1. Airports
   2. Airways
   3. Vehicles
   4. Ground Access

G. Operating Costs
   1. Airports
   2. Airways
   3. Vehicles
   4. Ground Access

H. Human Factors
   1. Acceptability to Labor Sector
   2. Compatibility With Human Skills and Orientation

TABLE 10.7 ILLUSTRATIVE EVALUATION CRITERIA
FOR A COMMUNICATIONS SYSTEM

1. Capacity
2. Connectivity
3. Expandability/growth potential
4. Interface compatibility
5. Interoperability
6. Life-cycle cost
7. Mission readiness
8. Quality of service
9. Reliability-maintainability-availability
10. Response time
11. Schedule risk
12. Security
13. Software complexity/size
14. Speed of service
15. Supportability
16. Survivability
17. Technical risk
18. Use of existing resources

10.6 EVALUATION FRAMEWORKS

The central core of Figure 10.1 shows an evaluation framework into which the other top-level systems engineering factors feed. Numerous frameworks can be developed with different levels of complexity and insight into the merits of alternative systems. We cite here some examples of approaches, all of which can be implemented through some application of computer-aided techniques.

10.6.1 Intercity Freight Transportation

In Table 10.5 we showed a set of possible MOMs or MOEs for intercity transportation. We now consider these same measures in relation to a set of specific design alternatives. In Figure 10.7 we show the resulting calculations for these design alternatives (systems) in relation to 18 measures (from Table 10.5). In Figure 10.8 we normalize 12 of these measures about the values of the 40-foot motor carrier alternative, taken to be the base value of unity. In the latter figure we illustrate both in tabular and graphical form an overall evaluation framework [9]. For the selected set of 12 measures, we show illustrative numerical values for each of the five freight systems

1. a 40-foot motor carrier
2. TRAILS, a conceptual high-speed freight system defined as state-of-the-art in high technology guideway operation
3. TOFC, a trailer or flat car dedicated 70-car train

ROW	MEASURE	40' Motor Carrier[1] (LTL), O-D Terminal	Trails[1]	TOFC[1] (Dedicated 35-car train)	NB Igloo Air Freight[3] (Container)	WBLH[3]
		SYSTEM				
1.	FSI	2.5×10^7	3.2×10^7	1.7×10^7	3.7×10^7	3.7×10^7
2.	Cost/ton-mile	$0.06	$0.08	$0.05	$0.22	$0.13
3.	NFSI	4.2×10^8	4.0×10^8	3.2×10^8	1.7×10^8	2.8×10^8
4.	EI: (BTU/ton-mile)[2]	2000	2100	700	12,900	9000
5.	NFSI/EI	2.1×10^5	1.9×10^5	4.6×10^5	1.3×10^4	3.1×10^4
6.	Total R&D Cost	10^5	2.5×10^8	10^5	10^5	10^5
7.	NFSI/R&D Cost	4.2×10^3	1.6	3.2×10^3	1.7×10^3	2.8×10^3
8.	Total System Cost	11×10^9	15×10^9	9×10^9	41×10^9	24×10^9
9.	NFSI/Total Cost	3.8×10^{-2}	2.7×10^{-2}	3.6×10^{-2}	4.1×10^{-3}	1.2×10^{-2}
10.	NFSI/EI/R&D Cost	2.1	7.6×10^{-4}	4.6	0.13	0.31
11.	NFSI/EI/Total Cost	1.9×10^{-5}	1.3×10^{-5}	5.1×10^{-5}	3.2×10^{-7}	1.3×10^{-6}
12.	Tons Per Year[4] (CO) Pollutant	2000		950		
13.	NFSI/Pollutant Rate	2.1×10^5		3.4×10^5		
14.	Noise-Level-Area (sq. mile)	900	910	890		
15.	NFSI/Noise-Level-Area	4.7×10^5	4.4×10^5	3.6×10^5		
16.	ROI (%)	1.1	0.5	1.0	1.0	1.0
17.	NPV ($)	-1.60×10^9	-3.26×10^9	-2.50×10^9	-5.90×10^9	-3.37×10^9
18.	Viability Index	-0.73	-0.71	-0.73	-0.75	-0.74

[1] TSC's Advanced Freight Systems Study is principal source of operations data for surface systems.

[2] Transportation Energy Conservation (TEC) Data Book is source for energy data.

[3] Trends and Choices is principal source of operations data for air systems.

[4] Compilation of Air Pollutant Emission Factors (CAPE) is source for pollutant data when combined with TEC.

Figure 10.7 Comparison framework for intercity freight transportation (600-mile trip, 10Mton system capacity). [9]

4. NB Igloo, a narrow-body containerized air-freight system
5. WBLH, a wide-body lower-hold containerized air freight system.

The graphical profile in Figure 10.8 allows one to compare these five freight systems a measure at a time. No one system appears to be uniformly better than the others although the TOFC is highly competitive. The extreme low-value peaks for the TRAILS system shows up in relation to measures that are normalized by R&D cost. This reflects the fact that this high-technology alternative suffers by comparison since it requires a very large research and development investment relative to the other alternatives. The graphical format and normalization with respect to one system tend to reveal differences that can be easily observed at a glance whereas the

Measure	SYSTEM				
	40' Motor Carrier	Trails	TOFC	NB Igloo	WBLH
FSI	1	1.3	0.7	1.5	1.5
NFSI	1	1.0	0.8	0.4	0.7
NFSI/EI	1	0.9	2.2	0.1	0.2
NFSI/R&D Cost	1	4×10^{-4}	0.8	0.4	0.7
NFSI/Total Cost	1	0.7	0.9	0.1	0.3
NFSI/EI/R&D Cost	1	3.6×10^{-4}	2.2	0.06	0.15
NFSI/EI/Total Cost	1	0.7	2.7	0.02	0.07
NFSI/Pollutant Rate	1	—	1.6	—	—
NFSI/Noise-Level-Area	1	0.9	0.8	—	—
ROI	1	0.5	0.9	0.9	0.9
NPV*	1	2.0	1.6	3.7	2.1
Viability Index**	1	1	1	1	1

* All computed values of NPV were negative. Thus smallest values of these ratios are "best."

** All computed values of viability index were negative and within 3% of each other.

(a)

Figure 10.8 Normalized and graphical intercity freight-transportation comparisons: (a) ratios of composite measures using 40-ft motor carrier as the base, (b) relative values normalized to 40-ft motor carrier. [9] (Continued on next page.)

tabulation in the same figure requires more careful study. The importance of graphical outputs should not be underestimated in terms of its role in conveying information quickly to both the analyst and the decision maker.

The previous concrete example uses certain features in the overall comparison framework that can be summarized as

- a series of related normalized measures for purposes of comparison
- normalizing values for these measures about one of the system alternatives
- tabular outputs (compatible with computer-aided spreadsheets)
- graphical output profiles (that can be generated with special-purpose p-c based software)
- no attempt to include qualitative factors, resulting in a correspondence between MOMs or MOEs and the set of evaluation criteria
- no attempt to establish commensurability or weights between measures so that a composite score cannot be obtained by combining scores across the set of measures

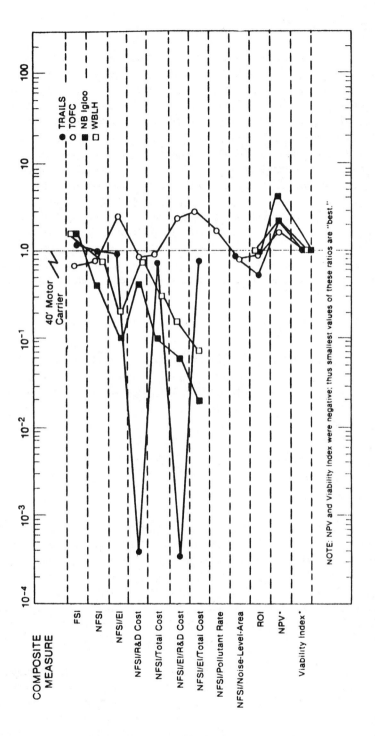

Figure 10.8 (*cont.*)

10.6.2 Commensurability

The issue of commensurability is a well-recognized one in decision theory and the matter of evaluation of alternative systems, courses of action, and the like. It is illustrated in the comparison framework described previously for alternative freight-transportation systems. The various measures or criteria for evaluation are dimensionally disparate and therefore cannot be combined in the evaluation process, which is the essential meaning of incommensurability. In such a situation, each criterion must be examined separately and the evaluator must assess the merits of the alternatives in some type of mental integration process. It is possible to make the measures or criteria commensurable by relating them to one another through a weighting or preference procedure. This has been shown in several previous portions of this text and represents another evaluation framework that is available and that is also accessible through computer-aided techniques. In particular, the weighting and rating schemes discussed in Chapters 3 and 6 demonstrate the use of spreadsheets as well as alternatives and preference-evaluator software for this purpose. The generic class of decision support software systems are also designed to handle various evaluation frameworks.

10.6.3 HEMATs Revisited

The notion of heuristic matrix (HEMAT) evaluation was introduced in Chapter 2 and is relevant to the matter of overall evaluation frameworks. We recall the matrices in Figure 2.2 that mapped the following system characteristics against one another:

- requirements
- specifications
- functional capabilities
- system/subsystem alternatives
- preferred system/subsystems

In the context of this chapter we have examined principally the mapping of measures or evaluation criteria against the system/subsystem alternatives in order to arrive at a preferred system. This is another mapping that can be made more explicit in a HEMAT format. The subsystems have been explored with respect to their correspondence with the functions that they constitute. In other words, in the context of top-level systems engineering we have considered the merits of functional design alternatives. The functional categories into which the overall system can be decomposed often can be put into a one-to-one correspondence with the subsystems of the overall system.

The HEMAT can be configured so as to provide a top-level examination of the cost-effectiveness of a system. In order to do that, it is generally necessary to be able to weight the effectiveness (or merit) measures to allow them to be combined. That is precisely what was done in the discussion provided in Chapter 6 with respect to the alternatives and preference evaluator software tools. Here we briefly examine a

HEMAT framework that looks for the "knees" in the cost-effectiveness curve. The concept is illustrated in Figure 10.9, which shows a typical cost-effectiveness curve. For lower values of cost we may expect a steep improvement in effectiveness. For higher values of cost we usually expect diminishing returns and a much reduced effectiveness slope. Somewhere in between we look for the knee in the curve such that we have a good compromise in effectiveness improvement per unit of cost increase and still satisfy the overall effectiveness requirements for the system.

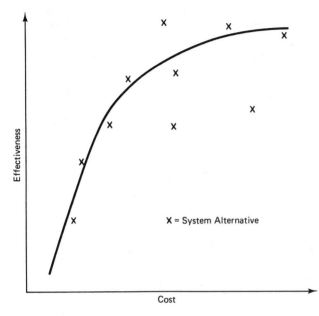

Figure 10.9 Cost-effectiveness curve.

We shall now illustrate a HEMAT application to a cost-effectiveness evaluation framework based upon the format in Figure 10.6. In that figure we examined alternative design approaches in relation to a set of system functions. To simplify the example, we will take only three functions and two alternative design approaches for each of these functions. We will assume that we are able to measure the effectiveness and cost for these alternatives as follows, noting that DIJ is the Jth approach to function I:

Design Alternative	Effectiveness Units	Cost Units
D11	5	30
D12	6	40
D21	4	20
D22	7	40
D31	6	20
D32	5	30

For both functions one and two the approaches yield improvements in effectiveness for cost increases. This is not the case for the third function for which D31 is uniformly better than D32, namely, it exhibits higher effectiveness at a lower cost. Therefore, for this one function only we can select D31 as better than D32 no matter what approaches they might be combined with to form an overall system. In other words, for this one case we can "suboptimize" and know in advance that the selection lies in the domain of the overall optimum on a system-wide basis. This result is illustrated in Figure 10.10. Here we show the three functions and the eight systems that result from taking all combinations of design approaches. As we observed, systems V through VIII can be ruled out (containing D32) since other systems have higher effectiveness at lower cost. Systems I through IV show monotonic increases in both effectiveness and cost. In moving from system II to III, however, we see an increase of twenty "units" of effectiveness for only ten incremental cost units. One might therefore select system III if no further conditions are placed upon the choice.

We summarize by noting the following points relative to this evaluation framework:

- it shows a specific example of how the heuristic matrix may be used as an evaluation framework
- it assumes commensurability of MOEs
- it does not include any qualitative evaluation criteria and assessment
- it displays the principle of "looking for the knee" in the cost-effectiveness curve
- it demonstrates the combinatorial nature of design by allowing all design approaches in combination for the various functional areas of the system
- it shows how the HEMAT maps directly into a spreadsheet format so that computer aids may be directly employed in the evaluation framework

The last point to be made in regard to this example is the manner in which the various systems alternatives were constructed from the design-approach alternatives. Two design approaches were cited for each of the three functional areas. Since all combinations were acceptable, we had a total of $2^3 = 8$ system alternatives. The basic cost and effectiveness data were associated with a total of six design alternatives. If we add two more design alternatives (for example D41 and D42) for a fourth functional area, we would increase the number of system alternatives to $2^4 = 16$. Thus the number of functions and design alternatives may increase arithmetically whereas the number of system alternatives will thereby increase exponentially. This point has been made several times before but in this connection we now observe that we can instruct a computer to develop the system alternatives from the basic data regarding the functional design alternatives, generate the HEMAT tabulation in Figure 10.10 and, in addition, plot the results as points on a cost-effectiveness curve. In this situation the computer instructions would likely be a specific computer program (see problems at end of chapter).

Function/Subsystem	System I			System II			System III			System IV			System V			System VI			System VII			System VIII		
	Design	Effect.	Cost	Design	Effect.	Cost	Design	Effect.	Cost	Design	Effect.	Cost	Design	Effect.	Cost	Design	Effect.	Cost	Design	Effect.	Cost	Design	Effect.	Cost
One	D11	5	30	D12	6	40	D11	5	30	D12	6	40	D11	5	30	D12	6	40	D11	5	30	D12	6	40
Two	D21	4	20	D21	4	20	D22	7	40	D22	7	40	D21	4	20	D21	4	20	D22	7	40	D22	7	40
Three	D31	6	20	D31	6	20	D31	6	20	D31	6	20	D32	5	30	D32	5	30	D32	5	30	D32	5	30
		15	70		16	80		18	90		19	100		14	80		15	90		17	100		18	110

Figure 10.10 Cost-effectiveness HEMAT spreadsheet.

257

10.7 CASE APPLICATION SUMMARY

CASE may be directly applied to the various elements of top-level systems engineering. We briefly cite here some of the principal application areas.

The initial elements of top-level systems engineering focused on requirements and specifications. The formats for these types of documentation were defined and are amenable to handling by simple word processors. However, if keyed into a word-processing software package, the ability to manipulate data elements becomes extremely limited. Manipulation of these elements, as necessary in terms of establishing a requirements traceability capability, can be achieved through the use of a DBMS. One solution, therefore, is to go to an integrated package that can handle both word processing and data-element manipulation by means of an embedded DMBS. Another possible solution is to use the so-called multitask applications manager that can overlay other software packages. In both cases we limit our considerations to the p-c; moving to a larger machine and more expensive and capable software changes the selections. This is the situation with respect to requirements tools and languages that requires both more capable computers as well as software packages.

HEMAT is easily constructed in a spreadsheet format. As an evaluation framework for top-level systems engineering, various types of HEMATs can be constructed (see Chapter 2), all of which can be put into a spreadsheet. However, if we wish to move back and forth between HEMAT applications we would like to have some database management capability as well. In this situation we have two immediate solutions similar to those already described. However, in this case we would wish to have both spreadsheet and DBMS capabilities. This can be captured with an integrated package or the multitask applications manager.

The calculation of MOMs or MOEs, as indicated, may require significant modeling and analysis for a large-scale system. To the extent that this is the situation, we may consider two choices: (1) developing special purpose programs by writing computer programs (see Chapter 9), or (2) using some of the mathematical tools that are available and discussed in Chapter 7 (and later in Chapter 12). If the analysis effort is not too great, another possibility is to use the limited programming capability provided by the more sophisticated spreadsheets. The programs obtained in such a context are sometimes called macros.

Finally, a variety of special software tools (Chapter 6) are available to support many of the aspects of top-level systems engineering. These latter tools must be examined in some detail to see how they apply to the specific analyses suggested here; generally they are viewed, with some exceptions, as broadly supportive of the tasks and activities of the systems engineering process as a whole.

REFERENCES

1. *Technical Requirements Document (TRD) for the System Segment Specification for the Communications Segment of the Consolidated Space Operations Center.* TRD-SSS-CSOC-00004, U.S. Air Force, Space Division, February 18, 1983.

2. *Functional System Specification (Draft),* Inter-Service/Agency Automated Message Processing Exchange Program, Air Force Automated Systems Project Office, Gunter Air Force Station, AL, November 1983.

3. Dorfman, M., and R. F. Flynn, "ARTS—An Automated Requirements Traceability System," *J. Systems and Software,* vol. 4, pp. 63–74, 1984.

4. Teichroew, D., and A. Hershey III, "PSL/PSA, A Computer-Aided Technique for Structured Documentation and Analysis of Information Processing Systems," *IEEE Trans. Software Engineering,* SE-3, no. 1, January 1977.

5. See various articles in full issue devoted to "Requirements Engineering Environments: Software Tools for Modeling User Needs," *COMPUTER Magazine,* April 1985.

6. Davis, A. M., "The Design of a Family of Application-Oriented Requirements Languages," *COMPUTER Magazine,* May 1982.

7. Lauber, R. J., "Development Support Systems," *COMPUTER Magazine,* May 1982.

8. "Defense Against Ballistic Missiles: An Assessment of Technologies and Policy Implications," and "The Strategic Defense Initiative: Defensive Technologies Study," Department of Defense, April 1984. See also "SDI, A Technical Progress Report," Submitted to the Secretary of Defense, by the director, Strategic Defense Initiative Organization, June 1985.

9. Eisner, H., and P. J. Steen, "A Framework for the Comparison of Intercity Passenger and Freight Transportation Systems," 1980 Proceedings of the International Conference on Cybernetics and Society, Cambridge, MA, October 1980, Institute of Electrical and Electronics Engineers, Inc., (IEEE), New York, NY.

PROBLEMS

P10.1. Define and discuss an additional ten measures of merit or effectiveness for the strategic defense initiative (Star Wars).

P10.2. Define and discuss ten measures of merit or effectiveness for a large-scale communications system such as that discussed in Tables 10.1 and 10.7.

P10.3. Define ten evaluation criteria for
  a. an automobile
  b. a stereo system
  c. a home telephone system
  d. a personal computer

P10.4. Define and carry out a HEMAT weighting and rating evaluation for three real alternative personal computer systems using at least ten evaluation criteria. Obtain your data from a local computer retail store or other reference materials.

P10.5. Write a two-page specification for a requirements traceability matrix that will relate a requirements document, a specification, and a test and evaluation procedure.

P10.6. Define and discuss six criteria that might be used in order to carry out an allocation of requirements.

P10.7. You are told by the chief engineer that a total permissible error (2σ value) is 8 units and to allocate error values equally to three error sources that you determine to be additive, independent, uniformly distributed, and with a maximum variable excursion of 12 units. What can you tell the chief engineer about the implication of the instructions to you in terms of constraints upon the mean and variance of these error distributions?

P10.8. Investigate three of the requirements tools and languages from the referenced literature and develop a set of evaluation criteria for them. Evaluate these alternative tools/languages against the criteria using your own scoring system.

P10.9. Carry out a cost-effectiveness assessment similar to that developed in the text for four functions and the following alternative design approach data:

Design Approach	Effectiveness Units	Cost Units
D11	50	100
D12	70	120
D21	60	140
D22	80	160
D31	40	90
D32	60	110
D41	75	100
D42	70	105

Lay out the results in a spreadsheet format and plot these results with effectiveness as the ordinate and cost as the abscissa. Interpret the results.

P10.10. Carry out a cost-effectiveness assessment similar to that shown in the text for four functions and the following design approach relationships:

Functions		Designs	
	One	Two	Three
One			
Effectiveness	X	$3X/2$	
Cost	C	$2C$	
Two			
Effectiveness	Y	$4Y/3$	
Cost	D	$5D/4$	
Three			
Effectiveness	Z	$2Z$	$3Z/2$
Cost	E	$6E/5$	$7E/5$
Four			
Effectiveness	V	$3V/2$	
Cost	F	$4F/3$	

where: $Y = 2X$ $D = 3C/2$
        $Z = 3Y/4$ $E = 4D/3$
        $V = Z/2$ $F = 5E/4$

Lay out the results in a spreadsheet format and plot in X (effectiveness) and C (cost) units. Interpret the results.

P10.11. Construct logic-flow diagrams that can be used to develop the costs and effectiveness for each of the system alternatives that are obtained by considering all design approaches within a function against all system functions.

P10.12. Write a BASIC or Pascal program that will implement the logic flow diagrams in P10.11. If you have access to a computer, run the program with the data in the text to verify the results in Figure 10.10.

P10.13. Add to the diagrams in P10.11 the necessary logic to test for and reject all design approaches within a function that are uniformly poorer than other design approaches. Assure that the system alternatives containing these poorer design approaches are also rejected.

P10.14. In Chapter 2 (Figure 2.2), a series of ten heuristic matrices (HEMATs) are shown that map the following against each other:
- requirements
- specifications
- system/subsystem alternatives
- preferred system/subsystems

For each of these ten matrices, define a typical evaluation for the matrix cells. Discuss the rationale for your selection.

Time is money

Bulwer-Lytton

SCHEDULING AND COSTING

11.1 INTRODUCTION

Scheduling and costing are two extremely important aspects of large-scale systems engineering. In relation to the top-level considerations of the last chapter, the schedule can embody the overall time sequence of a system life cycle and the system cost, as reflected in the curve of Figure 10.9, is a key parameter in the top-level as well as more detailed cost-effectiveness evaluations. As referred to in Chapter 2, both are considered in a definitive military standard [1] dealing with systems engineering management. Under technical program planning and control, scheduling is implicit in the technical reviews which include

- system requirements reviews (SSRs)
- system design review (SDR)
- preliminary design review (PDR)
- critical design review (CDR)

Each of these reviews needs to be scheduled and related to all other activities for a particular project. In addition, under the technical performance measurement (TPM) requirement, TPMs are to be specifically related to cost and schedule. With respect to costing, life-cycle cost analysis is considered to be part of the systems engineering process (see Figure 2.3), and under the category of optimization a system/cost effectiveness analysis is to be carried out [1]. Cost analyses are likewise implicit in several other aspects of the three major categories of systems engineering managment, namely,

1. technical program planning and control
2. system engineering process
3. specialty engineering

Another touchstone for schedule and cost requirements for a large-scale program is the DoD's Cost/Schedule Control Systems Criteria (C/SCSC) document [2]. It has been necessary to follow the provisions of this document since 1967 on all major DoD programs. For nonmajor programs (above $2 million and below C/SCSC thresholds) a less-intensive version known as Cost/Schedule Status Reports (C/SSR) has been used. Implementations of these cost and schedule "systems" are active and indeed are the subjects of studies, evaluations, seminars, and considerable industry interest.

We will not labor here over the specific schedule and cost aspects of either of the two DoD requirements described here. Instead, we will focus on some fundamental features of scheduling and costing as well as how these can be facilitated by the use of the computer, with emphasis on the p-c, which is the specific thrust of this text.

11.2 SCHEDULING

A schedule is a description of a set of activities and events along a time line. The latter may be defined as days, and either, or both, months and years from a specific start date or may be reflected against particular days, months, and years on the calendar. Two scheduling procedures are in widespread use today—the Gantt chart and the PERT chart, both of which will be considered below. Most other scheduling procedures represent variations, in one way or another, on these two themes.

11.2.1 Gantt Charts

A Gantt chart is a horizontal bar chart that graphically shows the time periods over which various activities and events are programmed to occur (see Figure 6.2). The activity is shown by the bar; specific events can be depicted by triangles anywhere along the bar. These events can thus begin an activity (such as *start system testing*), end an activity (such as *system testing completed*) or define an interim milestone (such as *testing of control subsystem completed*). They are normally both prospective and retrospective, that is, they show future plans as well as what has occurred in the past.

Gantt charting is a rather old and well-established scheduling technique. It is straightforward in its design and conception and leaves a lot of room for variations on the basic idea, such as coding through symbols (e.g., cross-hatching), footnoting, and correlating (e.g., coding activities with the Dewey-decimal numbering system that relates to a project's work breakdown structure).

In terms of p-c software that can assist in scheduling by means of Gantt charts, the options available to the systems engineer are numerous and inexpensive. This category of software was referred to in Section 6.2, which deals with project management tools. Specific packages can be found in Appendix A.

11.2.2 PERT Charts

PERT charts deal explicitly with activity interdependencies [3] in a manner not readily available in the Gantt chart. This important feature has allowed the PERT chart to supplant the Gantt chart in many large-scale systems engineering applications. The significance of this PERT capability may be illustrated by reference to Figure 11.1. This figure shows both Gantt and PERT charts for the same selected set of activities and events. The organizational structure for computer software for a typical system may be thought of as containing (see Chapter 14)

- CSCIs (computer software configuration items), which consist of
- TLCSCs (top-level computer software components—CSCs), which consist of
- LLCSCs (lower-level CSCs), which consist of
- units

We assume in Figure 11.1 that various units are being worked on in parallel, that they will have to be integrated to form LLCSCs, and that the illustrated subnetwork is part of a larger PERT network. For each unit, a coding and testing activity is part of the network.

In the Gantt chart, these activities are shown as horizontal bars with the following time units:

- coding Unit 1.1—3 time units
- testing Unit 1.1—3 time units
- coding Unit 1.2—3 time units
- testing Unit 1.2—2 time units
- coding Unit 1.3—2 time units
- testing Unit 1.3—2 time units

Integration of the three units into LLCSC 1 is scheduled to occur at time = 6.

In the PERT chart, three time estimates are provided for each activity whose time distribution is assumed to be of the Beta form [3]. These estimates are interpreted as

- t_o = optimistic activity time
- t_l = most likely activity time
- t_p = pessimistic activity time

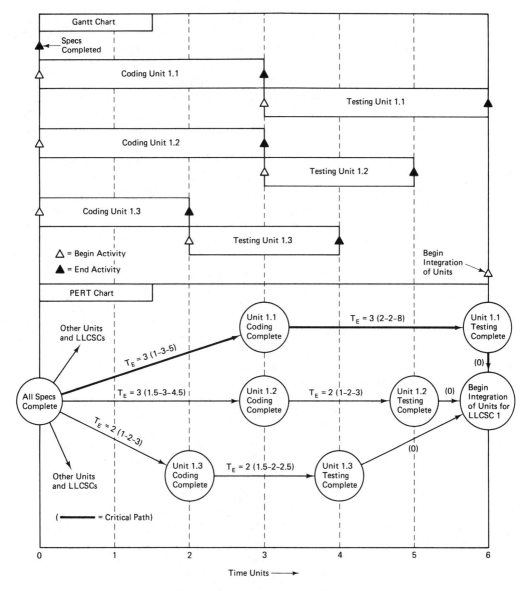

Figure 11.1 A Gantt and PERT chart comparison.

and the expected activity time (T_e) and activity time variance (σ^2) are given by

$$T_e = \frac{t_o + 4t_l + t_p}{6} \tag{11.1}$$

$$\sigma^2 = \left(\frac{t_p - t_o}{6}\right)^2 \tag{11.2}$$

Equation (11.1) is used to determine the expected time for each activity, as shown in the figure. The two activities for unit 1.1 are in sequence and take a total of 6 time units. This is longer than the time sequences for units 1.2 (3 + 2 = 5) and 1.3 (2 + 2 = 4) so that the coding and testing for unit 1.1 establish this subnetwork's critical path. No further time is required after testing is completed in order to initiate integration of the 3 units into LLCSC 1.

The critical path notion is a very important one for PERT networks. All parallel paths require less total time, which means that there is some slack in these paths. Specifically, the slack in the paths for units 1.2 and 1.3 can be seen from the figure to be 1 and 2 time units, respectively. This, for example, means that the coding for unit 1.2 could start one time unit later or the testing for that unit could start one time unit later (assuming coding was completed on time), but not both. Thus the interdependency displayed by the arrow-type diagram for the PERT network places in evidence the critical path as well as the slack available in parallel paths. Whereas these notions are easy to see on a simple illustrative network, they can be more obscure in a complex network with many interdependencies. Formal forward and backward passes through the network are well-defined techniques for finding the critical path(s) as well as the slack that might be available on parallel paths. Reference sources can be reviewed to investigate the formalisms, which also contain information regarding the role and utility of the Beta distribution and variance characterization.

As with Gantt charts, PERT charting capabilities have been incorporated into many commercially available software packages for the p-c. In general, the PERT feature will cost somewhat more since it is more complex in its inherent logic and computational capability. Again, both Chapter 6 and Appendix A refer to such software packages under the specific category of project-management tools.

11.2.3 Decision and Stochastic Networks

The original PERT chart dealt with activity uncertainty by assuming that its time was characterized by a Beta distribution. Each activity was therefore part of a project network with time periods known only in a probability sense. A generalization of the PERT procedure [4] introduced the notion that entire chains of events and activities might be replaced by others, depending upon either success or failure of preceding events and activities. In such a case, the PERT network generalizes to a decision or stochastic network [4,5]. Further information on these types of networks can be found in Chapters 4, 12, and 13. Dealing with this type of uncertainty it is thus possible through usage of decision and stochastic networks, to consider schedule alternatives and risks in a more explicit and complete manner.

Software for the p-c, as applied to decision and stochastic networks, is available commercially in only a rudimentary sense. The alternatives evaluators and decision support systems referred to in Chapter 6 and Appendix A do not really have the features that are represented in the software for, say, Q-GERT [5,6]. It may be expected, however, that a Q-GERT or similar type of capability will soon be available at the p-c level as the latter's price–performance characteristics improve. In this particular arena, system engineers should examine the nature of their requirement very carefully to see if existing and inexpensive software represented by project management; and either alternatives evaluators or decision support systems, or both, can do the job. In the area of pure schedule analysis, it may be that currently available p-c software can be adapted. As more extensive alternative, risk, and feedback elements are added, the current capability of the p-c and p-c software may be exceeded.

11.3 COSTING

Costing requirements to be addressed by the systems engineer can be extensive and varied. A large-scale program will typically be concerned with life-cycle costs (LCCs) and their minimization over the system's life expectancy (e.g., 20 years). The systems engineer will normally have to design and implement an overall life-cycle cost program that will contain such elements as

- an LCC plan
- system LCC trade studies
- identification of LCC drivers
- an LCC impact assessment
- an LCC tracking system

Thus, a broad systems engineering context for the issue of costing is the construction of a life-cycle cost model (LCCM) for a large-scale system. In principle, an LCCM should contain all the elements required as part of the systems engineering process. Such a model has two essential dimensions: a cost element breakdown or structure, as in Table 11.1, and a time dimension, with the former mapped into the latter. An example of these two top-level dimensions was cited in Figure 3.3, showing the cost elements as rows of a spreadsheet format and time periods (e.g., years) as columns of this format. Thus it is natural to think of the spreadsheet, or the integrated package, as the software that can and should be used in constructing an LCCM. This is certainly the case and spreadsheets are being extensively used to develop such LCCMs. However, one is not likely to find an LCCM in a typical p-c software retail store, but this will probably change in the near future.

At the current time, therefore, the systems engineer with an LCCM requirement and a desire to implement at the p-c level might consider three alternatives

1. obtain an integrated software package (spreadsheet, DBMS, graphics, etc.) or a multitask manager (see Chapter 6) and develop an LCCM "on top of" them
2. obtain the packages cited in number 1 and borrow or purchase a p-c-based LCCM from a (generally noncommercial) source that has already done the development
3. borrow or purchase a mini or mainframe-based LCCM and adapt it by some type of simplification to a p-c environment. In this case the LCCM may have been developed in some standard language such as Fortran in which case a Fortran compiler for the p-c would be the mechanism for transfer to a p-c environment

TABLE 11.1 SAMPLE COST-ELEMENT BREAKDOWN FOR A
LIFE-CYCLE COST MODEL

1 Research, Development, Test, & Evaluation
   1.1 Research and Development
      Preliminary Studies
      Design Engineering
      Hardware
      Software
      Other Personnel Costs
   1.2 Test and Evaluation
      Test Planning
      Test Hardware
      Test Software
      Test Operations
      Test Evaluation
      Other Personnel Costs
2 Procurement
   2.1 Installations
      New Construction
      Modification and Renovation
   2.2 Equipment (Hardware and Software)
      Primary Mission
      Mission Support
      Other Specialized
   2.3 Stocks
      Initial Stock—Primary Mission
      Initial Stock—Support Mission
      Spares—Primary and Support
   2.4 Initial Training
      Training and Support Personnel
      Training Materials and Equipment
      Training Facilities
   2.5 Other Procurement (e.g., Transportation) Costs
3 Operations & Maintenance
   3.1 Equipment Replacement (Hardware and Software)
      Primary Mission
      Mission Support
      Other Specialized
   3.2 Maintenance
      Primary Mission
      Mission Support
      Other Specialized
   3.3 Training
      Training and Support Personnel
      Training Materials and Equipment
      Training Facilities
   3.4 Salaries (Operators)
      System Operators
      Other Operational Support
   3.5 Material
      Expendables
      Other Support Material
   3.6 Other Operations & Maintenance Costs (e.g., Transportation)

The top-level manipulation of an LCCM in the three cost element-time dimensions cited is relatively straightforward. With all of the cells of a spreadsheet or matrix filled with cost estimates, it is a simple matter to obtain such slices as

- the RDT&E costs by year or for the entire program
- the O&M costs for a particular year or for a budgetary time period
- the unit costs of the system, introducing the additional data of the number of systems purchased (see Chapter 3)

However, the LCCM requires that one be able to estimate, as accurately as possible, the costs in the various cells of the cost element-time spreadsheet. Various methods for obtaining such estimates therefore become the heart of the matter in constructing an LCCM and emphasis is placed on such techniques in remaining sections of this chapter. Various examples are drawn from a variety of LCCMs that have been addressed in the open literature and whose development has been sponsored, at least in part, by the federal government.

11.3.1 Quantity–Unit Cost Products

Perhaps the simplest formulation is the situation in which a cost element is the product of a unit cost and the number of items that have such a unit cost, with the proviso (for the time being) that the two are independent. Such situations may be represented by

- the annual cost of system operators is the estimated number of such operators multiplied by the average annual cost per operator
- the cost of producing a set of computer programs is the estimated number of such programs multiplied by the average number of lines of code per program multiplied by the average cost of producing a line of code for that type (e.g., complexity) of program
- the cost of testing a set of assemblies is the estimated number of assemblies used in the system multiplied by the average cost per assembly

The generalized relationship represented by the above, of course, is:

$$\text{Cost} = \sum (\text{Number of items}) \times (\text{Unit cost per item}) \qquad (11.3)$$

where the summation implicitly recognizes the diversity in types of items that are normally taken into account.

Another variation on this theme is the case in which the quantity and unit price are not independent. The general assumption in this situation is that as quantity increases, unit price decreases as a consequence of overall economies of larger scale production and learning. In this case a trade-off may exist in which the systems-acquisition agent may be attempting to optimize between the unit costs of procurement, the number of units (systems) procured and the (life-cycle) cost of supporting the number of units in an operational setting. It may also be the case that a requirement exists for a minimum number of units (systems) that are to be procured. This is often true for large-scale systems acquired by the federal government, especially the DoD. These types of economies of scale are likely to be thought of for entire system buys, where one is procuring some number of, for example, airport surveillance radars, rapid transit system cars, or sea-based communications systems. In many cases, as one moves into the component level, the dependence virtually disappears as the components (ICs, etc.) are provided across a broad marketplace for the given system as well as numerous other systems.

A general relationship that can be used for cost estimating when the quantity of production affects cost is the so-called log-linear curve [7], given as

$$Y = aX^b \qquad (11.4)$$

where Y is the cost of the ith unit (or the cumulative average cost of n units), X is the cumulative unit number, a is the cost of the first unit and b is the slope of the curve. The log-linear phrase is used because when logarithms are taken of the Equation 11.4, we obtain the linear form

$$\log Y = \log a + b \log X$$

with b as the slope and ($\log a$) as the ($\log Y$) intercept term. Typically, this type of log-linear relationship means that as the total number of units produced increases by some factor (i.e., doubles, triples, etc.), the cost per unit is reduced by some constant percentage. The cited reference [7] provides tabulations of the log-linear equation as well as a *Project Rand* cost–quantity calculator. These tabulations can be easily duplicated by writing a relatively simple set of computer programs that can be run on a p-c.

11.3.2 Bottoms-Up Unit Cost Estimation

From the discussion in Section 11.3.1, it is observed that estimating the number of units (e.g., number of operators, number of programs/lines of code) is not primarily a cost-estimating issue. Rather, it is a systems engineering matter whereby these numbers flow from the base line and parametric design of the system. The construction of unit costs, however, is firmly a cost-estimating issue and in many ways can be viewed as the central core of developing an LCCM. We will deal here with two basic methods for constructing unit costs—a bottoms-up construction and a top-down parametric construction. The latter is sometimes also called a cost-estimating relationship (CER).

For the bottoms-up procedure we first recognize that a system in one context may be only one of many subsystems in another. For example, a search radar at an airport may be a full and complete system but in a larger context it is only a subsystem in the overall air-traffic-control system and an even smaller part of our national aviation system. At whatever level of system or subsystem one is operating, a bottoms-up procedure adds products of numbers of items and unit costs (Equation 11.3) until an overall unit cost is obtained. For example, if the unit cost is that of a search radar, this approach calls for

1. breaking the radar down into all its component parts
2. estimating the numbers of each class of part (e.g., ICs, circuit boards)
3. estimating the unit cost for each class of part
4. taking the sums of the products of numbers 2 and 3 (Equation 11.3)
5. adding additional nonunit component costs (e.g., RDT&E), amortized over the number of radars to be sold

Such an approach is extremely common and can be quite reliable. This is particularly true in a nondevelopmental systems-integration situation where all the system elements are commercial off-the-shelf (COTS) elements. If most of the system elements (subsystems, components, parts, etc.) can be found in a commercial catalog, it is relatively straightforward to build up the costs in a bottoms-up fashion.

11.3.3 Parametric Cost-Estimating Relationships

The top-down parametric construction [cost-estimating relationship (CER)], on the other hand, is much more complex and requires considerably more formalism. Examples of CER variables are provided in Table 11.2 in which it is observed that the essence of a CER lies in the selection of key parameters against which costs can be properly correlated. Such parameters are called *cost drivers* in the sense that their values strongly affect the overall cost of the system. CER construction is top-down in the sense that there is no piece-part buildup of costs; instead there is an attempt at inferring costs by extrapolating from the cost data gleaned from other similar programs and systems. Figure 11.2 illustrates a CER graph for a tropospheric scatter-communications system. In this example, cost is plotted against distance and number of channels as key cost drivers.

Linear cost-estimating relationships—two variables. A linear CER assumes that the relationship between cost and another key (cost driver) variable can be approximated by the linear slope-y intercept equation

$$Y = mX + b \qquad (11.5)$$

The normal procedure for calculating the values of m and b is to use a linear regression method that minimizes the mean square error. The values of m and b can be found by solving previously defined relationships (see Equation 7.1)

$$m = \frac{n\Sigma xy - \Sigma x \Sigma y}{n\Sigma x^2 - (\Sigma x)^2} \qquad (11.6)$$

$$b = \frac{\Sigma y \Sigma x^2 - \Sigma x \Sigma xy}{n\Sigma x^2 - (\Sigma x)^2} \qquad (11.7)$$

TABLE 11.2 ILLUSTRATIVE COST-ESTIMATING
RELATIONSHIP VARIABLES

COST OF . . .	AS A FUNCTION OF . . . (Assumed Cost Drivers)
Radar System	Output Power
	Frequency
	Bandwidth (pulse width)
	Weight
Missile Booster	Weight
	Type of Propellant
Satellite Terminal	Output Power
	Antenna Size
	Frequency
	Receiver Sensitivity
Airframe	Weight
	Type of Material
Aircraft Engine	Thrust
	Bypass Ratio
Software (in Person-Months)	Delivered Source Instructions
	Lines of Source Code
Satellite Payload Cost	Payload Weight
	Design Complexity
Annual Repair of Equipment	Mean Time to Failure
	Type of Equipment
Satellite Subsystems	Weight of Satellite
	Type of Satellite
Radio Equipment	Number of Channels
	Frequency
	Power
Dish Antenna	Size of Antenna
	Frequency

To illustrate the computation we will assume that historical data for the costs, Y, and the key cost driver, X, is as tabulated below:

Key Cost Driver (X)	Cost (Y)
4	30
3	60
7	90
9	70
10	100
10	120
11	140
14	140
15	170
17	140

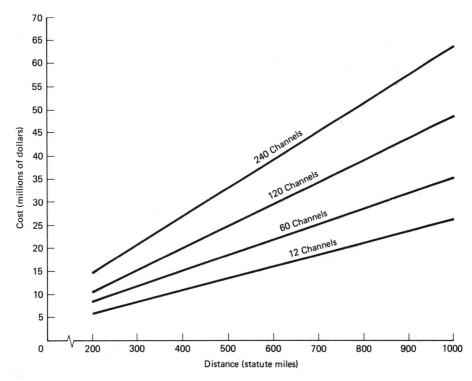

Figure 11.2 Tropospheric scatter communications (4400–5000 MHz).

Application of Equations 11.6 and 11.7 yields the following values:

$$\sum x = 100, \qquad \sum y = 1060, \qquad n = 10$$

$$\sum xy = 12,190, \qquad \sum x^2 = 1186$$

$$m = 8.5, \qquad b = 20.5$$

and

$$y = 8.5x + 20.5$$

Thus we have the formal CER plotted in Figure 11.3, together with the original ten historical data points. The linear CER computation is thus seen as quite direct, leading to immediate results that can hopefully be used to extrapolate to new values of the key cost-driver variable.

Nonlinear cost-estimating relationships—two variables. It may also be the case that the historical data belies the assumption of a linear relationship. This is often discovered by "eyeballing" this data on an x–y plot although more formal techniques are available. If a linear CER is considered inappropriate, various other

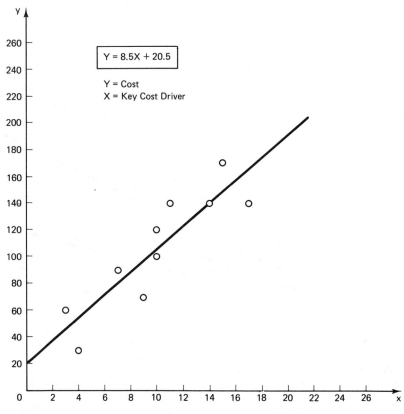

Figure 11.3 Linear CER example.

nonlinear forms may be used, such as

$$Y = a(0) X^{a(1)} \qquad \text{geometric or log-linear}$$

$$Y = a(0) a(1)^x \qquad \text{exponential}$$

$$Y = a(0) + a(1)X + a(2)X^2 \qquad \text{parabolic}$$

$$Y = \frac{1}{a(0) + a(1)X} \qquad \text{hyperbolic}$$

where $a(0)$, $a(1)$, and $a(2)$ are constants to be determined by a least-square fit procedure. All of these four equations and others have been utilized to produce CERs and the interested reader may consult the various references in this specialized area. Potentially applicable software packages that address this type of curve fitting are referenced in Chapter 7 and Appendix A.

Multiple variables. It may also be the situation that cost is viewed as driven by more than one principal cost driver (see Table 11.2), namely, $Y = f[X(1), X(2), \ldots]$. Multiple-regression techniques can then be used to calculate the values of the unknown constants in order to produce the desired CERs. These are sorted into linear and nonlinear techniques, paralleling the previous discussion. Software packages are likewise available for the p-c that will allow the user to deal with these cases (see Chapter 7, Appendix A, and related software catalogs).

Software development cost-estimating relationship example. One of the most complex issues of cost estimation is contained within the overall subject of software development, an activity in the overall domain of systems engineering. This field will be explored further in the software development and analysis portion of this text (Chapter 14). Here we cite the results of one investigator [8] in specific relation to CERs.

A variety of factors presumed to influence software development costs (potential cost drivers) were examined in relation to the following set of software application areas:

- command and control
- scientific
- business
- utility

Those factors with an overall high- and medium-impact rating are listed in Table 11.3, based upon ratings in terms of degree of impact for the 4 aforementioned ap-

---

TABLE 11.3 SIGNIFICANT FACTORS AFFECTING SOFTWARE COSTS [8]

High Significant-Impact Factors

• Communication	• Language requirements
• Time and memory constraints	• Maintainability requirements
• Developer's first time on specified computer	• Operational requirements changes
• Concurrent development of hardware	• Operational requirements/design interfaces
• Design stability	• Quality requirements
• Definition of instructions	• Sizing error
• Degree of innovation	• Software development schedule
• Programmer testing	• Software support availability
• Programming environment	• Target CPU designation
• Programming facilities	• Work breakdown structure
• Modern programming techniques	

Medium Significant-Impact Factors

• CPU time constraints	• Design complexity
• Program memory size constraints	• Reliability requirements
• Data management techniques	• Testing requirements
• Developer using another activity's computer	• Transportability of requirements
• Development and target computer different	• User-considered requirements
• Development personnel mix	• Multiple software utilization at sites
• Development site	• Specified response time
• Number of development sites	

---

plication areas [8]. Notwithstanding this bewildering set of possible variables, many CERs in the area of software development have taken the form

$$C = a(0)\ S^{a(1)} \tag{11.8}$$

where C is the development cost, S is the size of the program in number of source instructions or words of object code or some similar variable, and $a(0)$ and $a(1)$ are constants to be determined. The bottom-line conclusion of the cited investigator is that Equation 11.8 is the most practical, replacing cost by person–months (PM) of effort for various application areas and phases of development. Illustrative results are shown in Figure 11.4 for both source instructions and object code. These results are provided here to illustrate the methods being used and the form of some of the CERs in the software arena. Additional information specifically related to this software area is provided in Chapter 14.

Application Area	Concept-Formulation Phase	Analysis and Design Phase	
All Software			
Object	PM = $4.790I^{0.991}$	PM = $4.790I^{0.991}$	
Source	PM = $5.258I^{1.047}$	PM = $5.258I^{1.047}$	(I $\geq$ 10,000)
Command and Control			
Object	PM = $4.573I^{1.228}$	PM = $4.573I^{1.228}$	
Source	PM = $4.089I^{1.263}$	PM = $4.089I^{1.263}$	(I $\geq$ 10,000)
Scientific			
Object	PM = $4.495I^{1.068}$	PM = $4.495I^{1.068}$	
Source	PM = $7.054I^{1.019}$	PM = $7.054I^{1.019}$	(I $\geq$ 10,000)
Business			
Object	PM = $2.895I^{0.784}$	PM = $2.895I^{0.784}$	
Source	PM = $4.495I^{0.781}$	PM = $4.495I^{0.781}$	(I $\geq$ 10,000)
Utility			
Object	PM = $12.039I^{0.719}$	PM = $12.039I^{0.719}$	
Source	PM = $10.078I^{0.811}$	PM = $10.078I^{0.811}$	(I $\geq$ 10,000)

Figure 11.4 Estimating relationship examples for development efforts.

The limits of cost-estimating relationships. Some words of caution are necessary with regard to developing CERs. The fundamental basis for this development is that the expected costs for new similar types of systems can be extrapolated from the costs of existing systems. The key word is *similar,* keeping in mind that new systems can be dissimilar in numerous ways such as

- different contractors
- new contracting terms and conditions
- new embedded technologies
- different manufacturing facilities

In addition, adjustments may be necessary to account for inflation, new cost drivers that may have emerged, sparse data points, and skewed results as a consequence of a couple of outlying points. Thus the cost analyst must exercise considerable judgment in the selection and use of raw data as well as the fundamental form of the CER equation.

A humorous extrapolation of cost relationships is provided by N. Augustine [9] who estimated aircraft unit costs to be close to the entire defense budget by about the year 2060, and of the same order as the gross national product by about 2125! Aside from lamenting the steep cost increases of tactical aircraft, this is a tongue-in-cheek comment on the matter of CERs and cost extrapolations.

11.3.4 Discounting and Cost Escalation

The concept of discounting recognizes that the value of a unit of currency changes with inflation, which in turn is usually considered to be a function of time. Spending a dollar today is therefore different from spending a dollar, say, ten years from now since the latter will likely be worth considerably less than today's dollar. Future expenditures are thus discounted to reflect the reduced future worth. This can be done formally by calculating a present value (PV), today's value of an anticipated future expenditure, as

$$PV = \frac{\text{Future Expenditure}}{(1 + r)^n} \tag{11.9}$$

where r is the annual rate of inflation (assumed to be constant) and n is the number of years into the future at which time the expenditure is expected. The denominator term is sometimes called the discount factor. Table 11.4 lists the PV of a dollar spent in each of the next 20 years for discount rates of 5% and 10%. In a sense a converse to Table 11.4 is the matter of cost escalation. This relates to the same inflationary factor, as a minimum, and the potential increases in costs beyond that. If both factors are incorporated in an annual rate of cost growth, r, the *then year* dollars required to make a purchase in the future are related to this year's dollars through the familiar *interest* formula, namely

$$D = D_o(1 + r)^n \tag{11.10}$$

where D_o is the current dollar amount and D is the *then year* dollars. Moving from Equation 11.9 to 11.10, the discount factor moves from the denominator into the numerator. Therefore the reciprocals of the numbers in Table 11.4 will yield appropriate values for Equation 11.10. Cost escalation is an important factor in budgeting for a system over an extended time period. Dollars to be budgeted in future years must be sufficient to cover possible cost escalation. Many large-scale systems-program managers have gotten into difficulty by failing to recognize and account for this simple fact.

11.3.5 Uncertainty in Costing

All of the previous discussion addresses cost analyses for which one obtains a series of "point" estimates in a largely deterministic procedure. Thus the estimates can be viewed as expected values of cost unless other ground rules are used. It is often of

**TABLE 11.4 PRESENT VALUES OF A
DOLLAR SPENT IN THE NEXT 20
YEARS**

Years From Now Expended	Net Present Value ($1)	
	5%	10%
1	0.952	0.909
2	0.907	0.826
3	0.864	0.751
4	0.823	0.683
5	0.784	0.621
6	0.746	0.564
7	0.711	0.513
8	0.677	0.467
9	0.645	0.424
10	0.614	0.386
11	0.585	0.350
12	0.557	0.319
13	0.530	0.289
14	0.505	0.263
15	0.481	0.239
16	0.458	0.218
17	0.436	0.198
18	0.416	0.179
19	0.396	0.164
20	0.377	0.149

interest to a program manager to explore upper and lower bounds on costs by asking the following type of question: What are the chances of program costs exceeding C_1, C_2, C_3, etc.? Given a budget, if the probability distribution of costs can be estimated, the program manager will have a way of estimating the cost risk, that is, the likelihood of exceeding budget—by total program, by year, by cost category, and so on. This brings cost analysis into the domain of probability theory (see Chapter 5) and identifies cost parameters as random variables. A relatively simple way of examining cost uncertainty is to rewrite Equation 11.3 as

$$C_T = n_1 C_1 + n_2 C_2 + \cdots + n_m C_m \qquad (11.11)$$

reinterpreting n_i as the number of items of type i and C_i as the unit cost of such an item. Total cost (or a particular subtotal such as RDT&E) is then a weighted sum of random variables, where the weights are assumed to be estimated constants. An approach from this point involves the following steps:

1. estimate the upper and lower bounds for each of the unit costs (C_i)
2. relate these bounds to estimates of the unit cost variances (σ_i^2)

3. assuming independence of the unit costs, compute the variance of the total cost (see Chapter 5) as

$$\sigma_T^2 = n_1^2 \sigma_1^2 + n_2^2 \sigma_2^2 + \cdots + n_m^2 \sigma_m^2 \qquad (11.12)$$

4. interpret total cost uncertainty in terms of C_T, σ_T^2 and other assumptions and considerations

For step 1, this approach may be likened to the PERT activity time-estimation technique with the lower bound representing an optimistic cost and the upper bound a pessimistic cost. Whatever this interpretation, the key is step 2, which translates the bounds into a variance. In order to do this, some assumptions are required. The PERT time assumption involved a Beta distribution in which case the variance was obtained through Equation 11.2. The upper and lower cost bounds could be interpreted as two or three sigma values of a normal distribution, or of some other assumed distribution. Although there are many alternatives for this step, it needs to be taken and justified in terms of the nature of the unit costs and how they are likely to be distributed. Step 3, then, becomes a simple computation with Equation 11.12. Finally, step 4 can be treated in several ways, the most popular of which is to use the central limit theorem if there are many cost elements, which is likely to be the case for a large cost model and program. This theorem suggests that the distribution of the total cost would tend toward the normal distribution, whatever the distributions of the separate cost elements. The variance estimate (σ_T^2) together with the expected value estimate of cost (C_{TE}), then, allow the user to make well-defined statements concerning the a priori likelihood of staying within budget or any other postulated cost limits.

These four steps, although by no means standardized, represent a reasonable approach to dealing with cost uncertainty for a large-scale LCCM. They are, however, not the only approach. Other purely statistical techniques can and have been employed (e.g., distribution-free methods) as have simulation approaches using Monte Carlo procedures (see Chapter 12). The systems engineer and cost analyst must select an approach appropriate to the specific application at hand, including the requirements for proper program management and the priorities of the program manager.

11.3.6 Life-Cycle Cost Model Examples

Numerous LCCMs have been constructed, especially under the overall sponsorship of the federal government. Unfortunately, it is not a simple matter to obtain a complete off-the-shelf LCCM that operates at the p-c level and is at the same time applicable to a variety of systems. The characteristics of some well-known LCCMs are briefly described in the next section.

FLEX. FLEX is the name of a Department of the Navy LCCM. It is a model with a considerable amount of history (developed in 1974) and it has been improved over time. It is considered to be a flexible model in terms of at least the following types of characteristics:

- the model's breakdown of costs (as per Table 11.1) can be modified to correspond to the WBS of a typical program
- various unit cost estimating techniques (e.g., bottoms-up and parametric CERs) can be easily inserted
- a common data base is used, which helps with issues of consistency and uniformity of data
- multiple runs are facilitated through the model's program so that sensitivity and trade-off analyses can be executed easily

The model has been programmed in Fortran and thus needs a Fortran compiler to run. It is a recommended LCCM by the Navy in the sense that it is cited in Mil. Hdbk. 259 [10] and can be accessed through two recommended source documents [11,12]. In addition, the model

- is deterministic but can be linked to other programs that operate on a Monte-Carlo (see Chapter 12) basis or that contain data, and report-writing capabilities
- currently operates on mainframes and minis, but is adaptable to a p-c environment
- is applicable to all phases of the system-acquisition process
- incorporates discount and inflation factors
- can be run interactively to aid in the processes of sensitivity and trade-off analysis

Thus, the FLEX model is one that is highly portable and adaptable, through a user-defined simplification process, to a p-c environment.

Defense Communications Agency Model. The defense communications agency (DCA) model [13] is actually a cost and planning factor manual for use by personnel involved in cost and economic analysis of communications systems. The main body of the manual contains six major sections incorporating the following types of information:

- cost-estimating procedures appropriate for large transmission and switching systems
- tables of cost data, pricing data, and CERs for various types of communications equipment
- methods for estimating support-cost elements as part of a system's acquisition cost
- guidance for estimating annual O&M costs
- procedures for estimated leased communications costs
- supplemental cost and economic analysis considerations

Table 11.5 lists various aspects of communications systems for which specific data have been developed and presented. This compilation is seen as an extremely useful and comprehensive source for systems engineers working on military and related communications systems.

Cost-element breakdowns, such as that shown here in Table 11.1, are generally not standardized from model to model. Perhaps the only three elements that can be viewed as similar for all models are the three top-level categories of RDT&E, procurement, and O&M costs. Moving to cost elements below this top level normally leads to categories that become particularized to the type of system under consideration. As an example, the DCA model contains life-cycle costs for illustrative types of communications systems such as the line-of-sight (LOS) microwave system for which a full 12-year cost spread is listed in Table 11.6. Note the absence of RDT&E costs and the cost elements under the categories of *investment, military construction,* and *annual operating* (O&M) costs.

A relatively large number of CERs are embedded in the DCA model, specifically related to communications systems. Table 11.7 provides a representative listing of such CERs. It is noted that many are nonlinear and limits on the variables for applicability of the CERs are not shown since this listing is for illustrative purposes only. The relationships shown in the table also demonstrate the large body of raw data that must be compiled and analyzed in order to develop appropriate CERs.

RCA PRICE. THE RCA cost model (developed and operated by the RCA Corporation) is a kind of industry touchstone in that it is extremely well known and considered to be accurate and substantial. The model actually contains three major submodels dealing with

- hardware
- software
- life-cycle cost

The parameters for these models are numerous and a large amount of design information is required in order to utilize them. This is entirely appropriate since model results generally improve as the design data becomes better defined. Clearly, using this model has little to do with operating at a personal computer level. However, it is at least mentioned here since it is a major source of methodology and data with respect to cost analyses and modeling. The interested reader with the resources to consider such a model can make direct contact with the RCA Corporation. With the commitment made by RCA, it is likely that this model will be accessible and kept current for its existing and future subscribers.

Military handbook 259. This handbook [10] deals with life-cycle costs in Navy acquisitions and points toward the Navy FLEX model described previously as the recommended computerized LCCM. However, a complex LCCM has not only an imbedded set of CERs, it also has what might be referred to as a series of life-cycle cost submodels. These submodels address various cost elements (Table 11.1)

TABLE 11.5 DEFENSE COMMUNICATIONS AGENCY COST
DATA FOR COMMUNICATIONS SYSTEMS [13]

1 Cost-Estimating Procedures
   1.1 LOS Microwave Systems
   1.2 Tropospheric Scatter Systems
   1.3 High-Frequency Radio Systems
   1.4 Satellite Communications Systems
   1.5 Cable Systems
   1.6 Fiber Optic Systems
   1.7 Advanced Concepts
2 Communications Prime Mission Equipment
   2.1 Transmission Systems Equipment
   2.2 Multiplex Equipment
   2.3 Switched Systems Equipment
   2.4 Control System Equipment
   2.5 Auxiliary Equipment
3 Communications Systems Support Costs
   3.1 Integration and Assembly
   3.2 Contractor Training
   3.3 Test, Peculiar, and Common Support Equipment
   3.4 System Test and Evaluation
   3.5 System/Project Management
   3.6 Data-Technical Support Documentation
   3.7 Operational Site Activation
   3.8 Initial Spares and Repair Parts
4 Annual Operating Costs
   4.1 Military Personnel Rates
   4.2 Operations and Maintenance
   4.3 Recurring Investment
   4.4 Operating Support
5 Leased Communications Costs and Subscriber Rates
   5.1 Planning for Leased Services
   5.2 Communications Services Industrial Fund Subscriber Rates
   5.3 International Commercial Service
   5.4 Domestic Commercial Service
6 General Cost Considerations
   6.1 Automatic Data Processing (ADP) Cost Estimating
   6.2 Residual Value
   6.3 Manpower/Equipment Rates
   6.4 Equipment Installation Schedule Factors
   6.5 International Monetary Rates of Exchange
   6.6 Construction Price Indexes
   6.7 Cost–Quantity Relationships
   6.8 Economic Escalation
   6.9 Equipment Lease Factors
   6.10 Fiscal-Year Time Phasing of Cost Estimate
   6.11 Discounting
   6.12 Report Costing and Freedom of Information Requests
   6.13 Analysis of Commercial Activities
   6.14 Capital Equipment Costs
   6.15 Transportable Communications Uniits

TABLE 11.6 LINE-OF-SIGHT MICROWAVE COMMUNICATIONS SYSTEM
LIFE-CYCLE COSTS [13]

Cost Element	FY 1 ($)	FY 2 ($)	FY 3 to 12 ($)	Total Cost ($)
RDT&E	0	0	0	0
Investment				
Procurement				
Microwave equipment	111.8	166.0	—	277.8
Multiplex	150.2	219.0	—	369.2
Tech control & P&T	192.4	247.6	—	440.0
Orderwire	21.4	21.4	—	42.8
Alarm system	7.4	6.5	—	13.9
Electric power	93.9	93.9	—	187.8
Integration & assembly	28.8	37.8	—	66.6
Training	17.5	30.0	—	47.5
Test equipment	86.5	113.3	—	199.8
System test & evaluation	28.8	37.8	—	66.6
System engineering	231.4	176.8	—	408.2
Project management	75.5	57.7	—	133.2
Data	119.5	156.4	—	275.9
Contractor tech supt.	53.4	39.8	—	93.2
Assembly, installation & checkout of site	230.6	302.0	—	532.6
Initial spares & repair parts	389.2	509.2	—	898.4
Transportation	102.0	133.5	—	235.5
Military Construction				
Site activation	232.0	172.3	—	404.3
Annual Operating				
Military personnel	—	—	308.0	3,080.0
Operations & maintenance	—	—	232.8	2,328.0
Recurring investment	—	—	107.2	1,072.0
Operating support	—	—	154.9	1,549.0
Total (including 10-year system cost)	2,172.3	2,521.0	8,029.0	12,722.3

within the overall LCCM. A listing of typical submodels, cited in the handbook, is
provided here

- economic analysis model
- accounting model
- logistic-support cost-simulation model
- maintenance model
- reliability model
- inventory-management model
- maintenance-personnel model

TABLE 11.7 ILLUSTRATIVE DEFENSE COMMUNICATIONS AGENCY COST-ESTIMATING
RELATIONSHIPS FOR COMMUNICATIONS SYSTEMS [13]

Category of Equipment	Cost-Estimating Relationship ($)	Variable Definitions
Analog LOS Radio Equipment	$6578 \exp[0.0001(3.94N + 208F)]$	N = No. of voice-grade channels F = frequency in GHz
Parabolic Dish Antenna	$65.9D^2 + 2317$	D = antenna diameter in feet
Coaxial Cable	$15.75L$	L = linear feet
Tower-Mounted Passive Reflector	$20.5A^{0.835} + 419$	A = area in square feet
Communications Satellite T, T, & C*	$145 + 16.3W$ ($K)	W = weight in pounds
Communications Satellite Communications Subsystem	$86.6 + 35.85W^{0.87}$ ($K)	W = weight in pounds
Super-High Frequency (SHF) Earth Terminal	$0.0835D + 0.157P + 0.679$ ($M)	D = antenna diameter in feet P = transmitter power kW
Buffered CRT Terminal	$2,560M^{0.199}$	M = memory in Kbytes
Facsimile Terminal Transceiver	$203T^{0.931}S^{0.874}$	T = transmission time in minutes per page S = resolution in lines per inch
Communications Satellite Electrical Power Supply	$166P^{0.501}$ ($K)	P = beginning of life power in watts
Communications Satellite Attitude Control Subsystem	$434 + 97.8W^{0.9}$ ($K)	W = weight in pounds

*Telemetry, tracking, and command functions

- warranty model
- cost factors for key system parameters

11.3.7 Coupling Schedule and Cost

A very common consideration is that of coupling schedule and cost, especially with respect to personnel assignments and their associated labor costs. This can be achieved by estimating labor requirements, by labor category, for each of the activities of a schedule. The results are illustrated in Figure 11.5, based upon the Gantt or PERT subnetwork charts shown in Figure 11.1. The person loading, for each of three labor categories, is tabulated [Figure 11.5(a)] and the composite person loading is built up on a time line [Figure 11.5(b)]. The time line shows a starting total of seven people, phasing down to five people after four weeks and then down to three people for the last week. These values are specifically based upon the schedules of Figure 11.1 and graphically show the composite assignments by labor category. Given the cost per week for each category, the costs for each category for each week can be easily generated [Figure 11.5(c)]. The sum by row is then the total cost to the

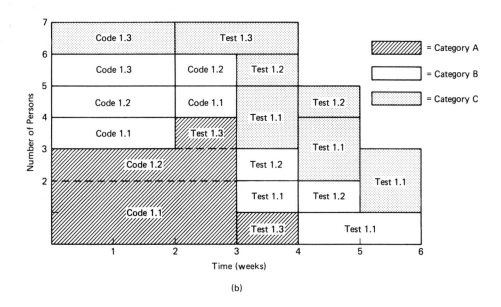

	Number of Persons (by Category)			Person-Weeks (by Category)		
Activity	A	B	C	A	B	C
Coding Unit 1.1	2	1	—	6	3	—
Testing Unit 1.1	—	1	2	—	3	6
Coding Unit 1.2	1	1	—	3	3	—
Testing Unit 1.2	—	1	1	—	2	2
Coding Unit 1.3	—	1	1	—	2	2
Testing Unit 1.3	1	—	1	2	—	2
			TOTALS	11	13	12

(a)

(b)

	Cat. A	Cat. B	Cat. C
Cost Per Week	500	600	700

	Week						
Category	1	2	3	4	5	6	Total Cost
A	1,500	1,500	2,000	500	—	—	5,500
B	1,800	1,800	1,200	1,200	1,200	600	7,800
C	700	700	700	2,800	2,100	1,400	8,400
TOTALS	4,000	4,000	3,900	4,500	3,300	2,000	21,700

(c)

Figure 11.5 Person-loading and cost-tabulation illustration.

project for each category. The sum by column is the weekly cost over all three categories. The sum of both is, of course, the total labor cost.

We note that Figure 11.5 has a tabulation input, a graphical output, and a cost-tabulation output. The tabulations can be associated with spreadsheets, which can be found in an integrated software package together with the graph. Many project management types of software packages will provide the results of Figure 11.5 quite readily. Other variations are likewise available such as the tabulations of cost by category and week, converted to a graphical type of output and the addition of cost elements for materials, computer time charges, and other nonlabor related costs.

11.3.8 Econometrics

There are a variety of cost considerations that are more properly categorized within the field of economic analysis. When the latter is consistently quantitative and uses complex mathematical bases it is sometimes called econometrics. Types of analyses or measures within this field include

- theory of investment
- supply–demand analysis
- preference assessment
- break-even analysis
- capital recovery factors
- service life
- fixed and variable costs
- short- and long-term marginal costs
- return on investment (ROI)
- payoff period
- price per unit utility
- profitability index
- viability index (see Chapter 10)

Numerous economic analysis texts can be referred to for the reader interested in these and other related subjects.

Another particularly interesting form of econometrics may be characterized as input–output analysis [14], which involves a macroeconomic model of the behavior and interaction between various sectors of our economy. Such analyses are sometimes called Leontief models, named after the well-known economist who first developed them. Parameters of such a model may include

- gross national product
- unemployment rate
- consumer price index

- R&D spending
- government spending (total)
- industrial production
- industry growth rates
- inflation rates
- interest rates
- labor productivity

In general, there is not a strong tie between econometrics and systems engineering. The former is included here since it represents a more sophisticated form of economic analysis and, at times, it may be necessary to use econometrics in order to analyze important aspects (e.g., payback period) of a large-scale system. For systems of sufficient size (e.g., national aviation system), impacts may be felt on a macroeconomic level. Thus, systems acquisition decisions can flow through to the national economy. In the military arena, the national economy can likewise be affected in substantial ways by large-scale system procurements (e.g., Star Wars, the Trident submarine). On occasion it is necessary that the systems engineer be prepared to examine such far-reaching economic effects.

11.3.9 Department of Defense Cost References

A very large number of references is available that can be accessed in terms of life-cycle costing and analysis. Table 11.8 provides a set of references cited in a Navy Military Handbook [10], all of which can be used as source materials in developing a system life-cycle cost approach and model.

TABLE 11.8 SELECTED GOVERNMENT COST REFERENCES

OMB Circular A-94—Discount Rates to be Used on Evaluating Time-Distributed Costs and Benefits

OMB Circular A-104—Comparative Cost Analysis for Decisions to Lease or Purchase General Purpose Real Property

OMB Circular A-109—Major System Acquisitions

DoD Directive 4245.3—Design to Cost

DoD Directive 5000.1—Major System Acquisitions

DoD Directive 5000.4—OSD Cost-Analysis Improvement Group

DoD Directive 5000.26—Defense Systems Acquisition Review Council (DSARC)

DoD Instruction 5000.33—Uniform Budget/Cost Terms and Definitions

SECNAVINST 4000.31—Life-Cycle Costing

DoD LCC-1—Life-Cycle Costing Procurement Guide

DoD LCC-2—Casebook Life-Cycle Costing in Equipment Procurement

DoD LCC-3—Life-Cycle Costing Guide for System Acquisitions

NAVFAC P442—Economic Analysis Handbook

TABLE 11.8 *(cont.)*

NAVMAT P5242—Joint Design to Cost Guide

AD 728481—Cost Considerations in Systems Analysis

AD 901477L—Military Equipment Cost Analysis

AD A082273—Naval Material Command Life-Cycle Cost Guide for Major Weapon Systems

AD A083845—Naval Material Command Life-Cycle Cost Guide for Equipment Analysis

Note: AD numbered documents available from the Defense Technical Information Center, Cameron Station, Alexandria, VA 22314.

11.4 CASE APPLICATION SUMMARY

Scheduling and costing for large-scale systems involve a variety of tasks and activities whose efficiency can be enhanced through the use of computer-aided techniques. A selected set of such tasks and activities and related potential means of providing personal computer-based assistance are listed next.

Task/Activity	CASE Application
Gantt schedule chart	Project manager
PERT schedule chart	Project manager
Stochastic networks	Special purpose software
Overall LCCM tabulation	Spreadsheet, integrated package, multitask manager
Overall LCCM construction	Modification of existing LCCM, conversion to a p-c environment using language compiler, spreadsheet, integrated package
Schedule/cost plans and documentation	Word processor
Schedule and cost graphs	Graphics/integrated package
Present value/cost escalation tables	Spreadsheet
Cost uncertainties	Statistical package, spreadsheet
Input-output econometrics	Mathematics package, special purpose software
Schedule and cost coupling	Spreadsheet, integrated package

REFERENCES

1. *Engineering Management,* Military Standard 499A (USAF), May 1, 1974.

2. *Cost/Schedule Control Systems Criteria* (Implementation Guide), DoD Instruction 7000.2, Department of Defense, October 1980.

3. Malcolm, D. G., J. H. Roseboom, C. E. Clark, and W. Fazar, "Application of a Technique for Research and Development Program Evaluation," *Operations Research,* no. 7, pp. 646–669, 1959.

4. Eisner, H., "A Generalized Network Approach to the Planning and Scheduling of a Research Project," *Operations Research,* 10, no. 1, February 1962.

5. Whitehouse, G. E., *Systems Analysis and Design Using Network Techniques.* Englewood Cliffs, NJ: Prentice-Hall, 1973.

6. Pritsker, A. A. B., and C. E. Sigal, *Management Decision Making: A Network Simulation Approach.* Englewood Cliffs, NJ: Prentice-Hall, 1983.

7. Noah, J. W., and R. W. Smith, *Cost-Quantity Calculator.* RM-2786-PR, The Rand Corporation, January 1962.

8. *Software Cost Estimation Study, Volume I: Study Results.* Technical Report No. 151, Rockville, MD: Doty Associates, Inc., February 1977.

9. Augustine, N. R., *Augustine's Laws.* © Norman R. Augustine; all rights reserved, New York: American Institute of Aeronautics and Astronautics, Inc., 1982.

10. *Life Cycle Cost in Navy Acquisitions.* Military Handbook 259 (Navy), April 1, 1983.

11. *LCC FLEX-9E, Navy Material Command Life Cycle Cost Methodology Computer Tape.* AD A115621, Defense Technical Information Center, Cameron Station, Alexandria, VA.

12. *User's Guide for Naval Material Command's Life Cycle Cost [(FLEX) Model],* AD A115622, Defense Technical Information Center, Cameron Station, Alexandria, VA.

13. *Defense Communications Agency (DCA) Cost and Planning Factor Manual.* DCA Circular 600-60-1, Defense Communications Agency, Washington, D.C., March 4, 1983.

14. Intriligator, M. E., *Econometric Models, Techniques and Applications.* Englewood Cliffs, NJ: Prentice-Hall, 1978.

See Table 11.8 for additional references.

PROBLEMS

P11.1. For the PERT chart shown, find
    a. the critical path and its expected time
    b. the standard deviation associated with the critical path
    c. the slack in all other paths

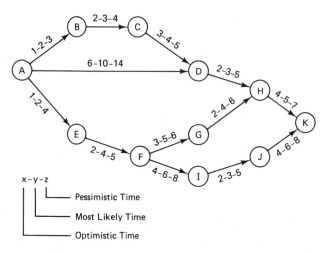

x–y–z

└── Pessimistic Time

└── Most Likely Time

└── Optimistic Time

Figure P11.1

P11.2. Select a set of numbers to demonstrate the log-linear relationship of Equation 11.4 and plot the results. Interpret these results in terms of the change in cost as a function of number of units.

P11.3. Develop and plot the relationship associated with a linear CER using Equations 11.6 and 11.7 and the following data set:

Cost (Y)	Key Cost Driver (X)
10	1
25	2
30	2
35	3
50	3
55	4
80	5
65	5
80	6
90	6

P11.4. Verify Equations 11.6 and 11.7 from Equations 7.1 of Chapter 7.

P11.5. Select values for the constants associated with the equations in the text relative to the following nonlinear forms referred to in the text:
1. geometric
2. exponential
3. parabolic
4. hyperbolic

Plot the results in each of these four cases.

P11.6. Select values for the constants in Equation 11.8. Plot and interpret the results.

P11.7. Select one of the equations in Figure 11.4. Plot and interpret the results.

P11.8. Expand the results shown in Table 11.4 by writing a program in BASIC or Pascal that will print out present values for 15% and 20%.

P11.9. Write a program in BASIC or Pascal that will compute values of D in Equation 11.10 for n from 1 to 20, $D = 1000$, and $r = 8$, 10, 12, and 14%.

P11.10. Plot and interpret the results for the following CERs in Table 11.7 in relation to the types of equipment listed here.
   a. analog LOS radio equipment
   b. parabolic dish antenna
   c. facsimile terminal transceiver
   d. communications-satellite attitude-control subsystem

P11.11. In Figure 11.6, we may assume that there are many persons in labor categories A, B, and C such that the cost per week has a (probability) distribution with the mean values shown in the figure. Assume further that the standard deviation for each category is a fixed 10% of each mean value. Calculate the standard deviation associated with the total cost of the subnetwork shown. Show all steps in the calculation.

P11.12. Design a spreadsheet that will produce all of the data shown in Figure 11.6.

P11.13. Investigate the work of economist Leontief and relate your reading to the discussion of econometrics in this chapter. Discuss how computer tools may be used to handle Leontief's model forms. Explain your rationale.

P11.14. Obtain a copy of one of the references in Table 11.8 and write a five-page interpretation of how the subject of the reference can be addressed by computer tools.

Any mental activity is easy if it need not take reality into account

Proust

PERFORMANCE MODELING AND SIMULATION

The subject of this chapter is performance analysis of large-scale systems. This analysis can be carried out by a *modeling* process, the construction of a set of relationships that represent the key performance variables of the system and how they interact. All quantitative descriptions of a system may therefore be called models of the system or portions thereof; even diagrams, flow charts, and other similar semiquantitative or semiqualitative descriptions can be models. Thus modeling is considered here to be a rather broad generic term.

Simulation is viewed, however, as a particular type of modeling approach. It is quantitative and usable in place of the real system in order to represent the behavior of that system. In that sense, for example, a rapid-transit simulator would run a small railroad through its paces on a digital computer. A way of looking at various kinds of simulations is to categorize them as follows [1]:

- static versus dynamic
- continuous versus discrete
- deterministic versus stochastic

The stochastic type of simulation is often based upon Monte Carlo techniques. These involve sampling from distributions representing key variables in the simulation and then having each sample set constitute a trial or sample run of the simulation. We will make no further significant distinction between modeling and simulation other than the above, and where the latter is referred to it will generally be associated with Monte Carlo methods. Exceptions may be the commercially available packages, many of which are called simulations but are not necessarily Monte Carlo in their basic construction.

Performance modeling and simulation fits directly into the notions of top-level systems engineering, as delineated in Chapter 10. In Figure 10.1, the following major factors involved in top-level systems engineering were described as:

- top-level requirements/specifications
- evaluation criteria
- design alternatives
- MOMs/MOEs
- evaluation framework

The major role of performance modeling and simulation is to be able to compute the MOMs and MOEs for the system alternatives under consideration. Models and simulations that facilitate these computations are also thought of in terms of parametric and trade-off analyses that assist in the optimization of these same measures.

Modeling frameworks and hierarchies have taken many different forms including attempts at defining taxonomies of model types. As an example, one popular text [2] defines four basic model types as physical, analog, schematic, and mathematical. Another author [3] considers the two basic types of physical and mathematical, and then breaks each of these into static and dynamic, and other lower-level elements. We select here only three simple model types that also represent sequential steps in a modeling process

- conceptual model
- configurational model
- flow and computational model

A conceptual model is the broadest and the starting point in developing a model. As the word suggests, it is the formulation of a concept for the system—the purpose of the system, the system requirements translated into a broad conceptual design consisting of the system functions, and how they interact. It is a statement of what the system is to do in distinction to how it is to do it. The configurational model conceives of methods of implementation—blocks of hardware, software, and personnel. It also consists of pictures of the system, which can range from physical sketches of what the system looks like to parts lists that define the physical elements of the system. The flow and computational model describes, in both qualitative and quantitative terms, the true operation of the system in terms of data flow, computation, and how the physical elements are believed to interact in the real world. The latter contains the mathematics of the system and relates most strongly to the tools and techniques of systems engineering as described herein.

As an example, consider the notion of modeling an automobile. The conceptual model would deal with such factors as speed, acceleration, vehicle range, number of passengers, and the definition of functions such as ignition, suspension, braking, and speed control. The configurational model might include a clay mockup, a 3-D computer-graphics representation, a sketch, and a photograph as well as descriptions of the various subsystems or components, for example, fuel pump, carburetor, and radiator. Finally, the flow and computational model might contain curves of performance such as acceleration versus horsepower, stopping distance versus speed, and fuel economy versus speed. It also includes either, or both, flow charts or timing diagrams that, for example, show the sequence of operation of the pistons, valves, and fuel injection. These simple factors for the familiar system of an automobile provide some insight into the difference between the three simple model forms listed.

An important aspect of modeling, given the latter discussion, is what is sometimes called *output analysis*. This is a specific focus on outputs as a starting point, particularly when considering the flow and computational model construction. The notion here is simply that if you are not clear about where you're going, you're not likely to get there. As an example, if false-alarm rate and detection probability are two performance parameters of a radar system to be calculated, then an output analysis starts with their definition and works backward from there by successively asking the question—what does this parameter depend upon? Thus by working backward from the desired results to the means by which these results are obtained, it is considerably more likely that the model will both produce the required outputs and have front-to-back structural integrity. Many of the diagramming techniques shown in Chapter 4 (e.g., PDDs) use this technique implicitly and provide a discipline for working backwards and assuring that all necessary parameters are both defined and interrelated.

A mid-1960s framework for modeling the performance of large-scale systems was the so-called Weapon System Effectiveness Industry Advisory Committee (WSEIAC) effort [4]. This activity focused on the matter of assessing the effectiveness of a system. If indeed there were measures of effectiveness (MOEs), then there clearly needed to be a means of computing specific values of these measures. Effectiveness was construed, as shown in Figure 12.1, as being a function of three key factors

1. availability (A)
2. dependability (D)
3. capability (C)

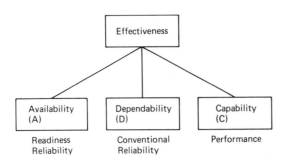

Figure 12.1 Elements of effectiveness.

Availability considers the condition of a system at the start of a mission; dependability refers to the condition during the mission, given its condition at the start of the mission; capability addresses the overall results of the mission, given the system's condition during the mission. Other sources, including this text, associate the dependability measure with reliability. Capability has also been referred to as performance.

This basic effectiveness construct is still being used and represents an important technique for what in this text is generically called *technical performance analysis*. Thus, for purposes of this exposition, technical performance includes system effectiveness. Technical performance measurement, a term used in Mil. Std. 499A [5], thus also includes effectiveness evaluations and a specific means by which the deepest type of technical performance measurement is carried out is called performance modeling and simulation.

12.1 TECHNICAL PERFORMANCE MEASUREMENT

Technical performance measurement (TPM) has been defined [5] as "the continuing prediction and demonstration of the degree of anticipated or actual achievement of selected technical objectives." A distinction is specifically made between a current estimate of a technical performance measure and its achievement to date. The former is the best current estimate of the value of a TPM at the end of the contract period. The latter is the value at the time of the assessment of the TPM. This assessment may be a result of modeling, simulation, or testing, depending upon what is happening at the time of the assessment. Both of these values are to be compared with the specification or requirement for the system. Other requirements relative to TPMs are as listed here

- **a.** technical performance measures (or parameters) are to be key indicators of program success
- **b.** TPM interrelationships are to be depicted by means of tiered dependency trees
- **c.** a time-phased value profile (with a tolerance band) is to be constructed for each TPM
- **d.** the TPM program is to be related to cost and schedule

The first area of activity, given items a–d, is to identify a set of TPMs for a given system. Examples of candidate TPMs for large-scale air transportation and communications systems are listed in Table 12.1. This articulation satisfies item a. above. With respect to item b, dealing with tiered dependency trees, the parameter dependency diagramming technique (see Chapter 4) is ideally suited to the construction of such a tree. Figure 4.16 shows a diagram for which the key TPMs would be the detection probability and the false-alarm probability. The parameters in the diagram upon which the TPMs depend are sometimes called technical performance parameters or TPPs. The notion of a time-phased diagram for a TPM is illustrated in Figure 12.2. This diagram shows a profile projected over time for a typical TPM. The matter of TPMs in relation to cost and schedule as well as an overall program is

TABLE 12.1 TECHNICAL PERFORMANCE MEASURE EXAMPLES FOR AIR TRANSPORTATION AND COMMUNICATION SYSTEMS

Air Transportation Systems
- Availability
- Capacity-to-demand ratio
- Convenience/comfort
- Flexibility
- Human factors
- Induced demand
- International effects
- Intramodal balance
- Maintainability
- Military back-up utility
- Noise
- Pollution
- Reliability
- Safety
- Security
- Service frequency
- Social factors/effects
- Speed
- System capacity
- Technology utilization
- Trip time
- Vehicle capacity

Communication Systems
- Antijam margin
- Availability
- Bandwidth/data rate
- Capacity (number of channels by data rate)
- Computer memory size
- Computer resource utilization
- Computer throughput
- Coverage
- Dynamic range
- Grade of service
- Interface compatibility
- Interoperability
- Maintainability
- Power
- Processing time
- Quality of service (bit/message error rate)
- Range
- Reliability
- Response/set-up time
- Security
- Signal-to-noise ratio
- Software complexity
- Source lines of code
- Switch size/speed
- Speed of service
- Survivability
- Vulnerability

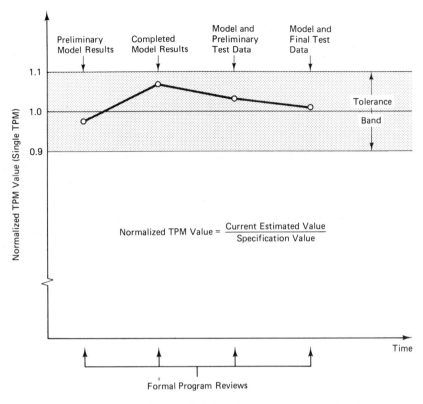

Figure 12.2 Time-phased technical performance measurement profile.

discussed later in this chapter. The following section addresses technical performance measurement by means of several illustrative model examples.

12.2 ILLUSTRATIVE TECHNICAL PERFORMANCE MODELS

In this section we examine several examples of specific performance models. These necessarily have to be rather simple in their construction. It is noted that in Chapter 10, an intercity freight-transportation example was discussed, leading to an evaluation framework for alternative systems. Figures 10.7 and 10.8 contained a number of measures for a comparison of five systems. These measures can be put into a one-to-one correspondence with top-level performance parameters. Clearly, a large number of performance models were constructed and applied in order to develop the numeric values for the measures and parameters of interest. Thus, performance models were implicit and rather complex tools that were needed in order to make the desired computations. Only simplified models can be considered within the constraints of this text, serving to illustrate the nature and form of the modeling process. Examples in this section include

- alternative measurement/monitoring schemes

- the strategic defense initiative
- a search radar
- an illustrative Monte Carlo simulation, based upon the search radar analysis
- an air transportation procedural model

12.2.1 Alternative Measurement Schemes

This simplified model involves a generalized set of alternative measurement schemes [6]. The measurements are being made of one or more parameters that, for example, might be a radar's pulse width, repetition interval, or frequency. The measurement device might be a simple piece of equipment or might represent a large-scale measuring system such as a satellite, aircraft, or ship.

The measurement situation is depicted in Figure 12.3. The parameter to be measured is A and this example shows that there are a variety of surrogate measurement parameters, X_i, that can be measured in order to estimate the A parameter. In particular, the diagram illustrates six alternative ways of estimating A, namely, by measuring

1. X_1, X_2, X_3
2. X_2, X_3, X_4, X_5
3. X_4, X_5, X_6, X_7

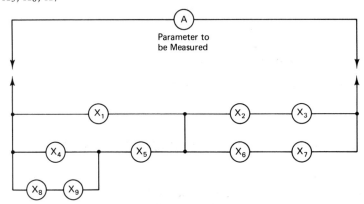

Surrogate-Measurement Parameters

Estimating the Parameter A Through
Six Measurement Schemes

$\hat{A}_1 = f_1(X_1, X_2, X_3)$ with error σ_1

$\hat{A}_2 = f_2(X_2, X_3, X_4, X_5)$ with error σ_2

$\hat{A}_3 = f_3(X_4, X_5, X_6, X_7)$ with error σ_3

$\hat{A}_4 = f_4(X_1, X_6, X_7)$ with error σ_4

$\hat{A}_5 = f_5(X_2, X_3, X_5, X_8, X_9)$ with error σ_5

$\hat{A}_6 = f_6(X_5, X_6, X_7, X_8, X_9)$ with error σ_6

Figure 12.3 Alternative measurement schemes.

4. X_1, X_6, X_7

5. X_2, X_3, X_5, X_8, X_9

6. X_5, X_6, X_7, X_8, X_9

Any path that bridges the A parameter constitutes a legitimate measurement scheme in the construct of the diagram of Figure 12.3. That is, X_1, X_2, and X_3 bridge the parameter and therefore constitute a measurement scheme. The measurement of X_4 and X_5 is equivalent to measuring X_1 and the same is true for the equivalence between $X_2 - X_3$ and $X_6 - X_7$.

The next point is that an error, σ_i, can be associated with each of the surrogate-measurement schemes. This is likewise depicted in Figure 12.3 together with the notion that some functional relationship f_i is required in order to convert from the raw measurements of the X_i to the estimate of A. In Figure 12.4(a) we show examples of

Surrogate Measurement Parameters	X_1	X_2	X_3	X_4	X_5	X_6	X_7	X_8	X_9
Error Variance	1.0	1.2	1.4	1.6	1.8	2	2.2	2.4	2.6
Cost	90	80	70	60	50	40	30	20	10

(a) Surrogate-Measurement Parameters

Measurement Scheme	X_1	X_2	X_3	X_4	X_5	X_6	X_7	X_8	X_9
1	1	1.2	1.4						
2		1.2	1.4	1.6	1.8				
3				1.6	1.8	2	2.2		
4	1					2	2.2		
5		1.2	1.4		1.8			2.4	2.6
6					1.8	2	2.2	2.4	2.6

(b) Error Variances Associated with Measurement Schemes

Measurement Scheme	Overall Error Variance (σ_i^2)	Cost	Signal Value	Signal-to-Noise Ratio per Unit Cost
1	3.6	240	1000	1.16
2	6.0	260	1000	0.64
3	7.6	180	1000	0.73
4	5.2	160	1000	1.20
5	9.4	230	1000	0.46
6	11.0	150	1000	0.61

(c) Figure-of-Merit Calculations for Each Measurement Scheme

Figure 12.4 Surrogate-measurement parameters and figures of merit for measurement schemes.

the error variances for each of the surrogate-measurement parameters, X_i. We make the simplifying assumption, for purposes of this illustration, that the overall error variances (σ_i^2) are the sums of the surrogate-parameter error variances (see Chapter 5). This is a typical assumption made in this type of analysis, but it should be recognized that the individual surrogate-parameter errors actually propagate through the functional relationships, f_i. The costs of measuring each of the surrogate parameters are likewise shown. These may be comparative or absolute values. As previously alluded to, the costs may be associated with a simple measuring device or a large-scale system whose purpose is to make a key measurement.

We now adopt as a figure of merit (FM) for the various measurement schemes the *signal-to-noise ratio* per unit of cost. The signal value is assumed to be some constant. The noise is directly related to the error variance for each of the measurement schemes. The cost of a measurement scheme is the sum of the costs of measuring each of the surrogate parameters. We thus wish to maximize this FM, which is equivalent to looking for a minimum product of error variance and cost. [See Figure 12.4(c)]

The illustrative numeric values in Figure 12.4(c) show measurement scheme four as the best according to the aforementioned FM, with number one not far behind. These are the only two schemes requiring three surrogate parameters; all others require four or more such parameters. In this example, the costs of making the additional surrogate-parameter measurements drive the FM down so that the measurement schemes are not competitive. Despite this, the FM for scheme six is close to that of scheme two, notwithstanding the fact that the former requires five surrogate parameters as compared with four for the latter. This results from the interplay of both the values of error as well as cost.

We may now expand the situation to one in which we use more than one measurement scheme simultaneously in an attempt to replicate measurements and thereby drive down the error variance. The situation is illustrated in Figure 12.5. Figure 12.5(a) shows the three single-measurement schemes and four replicated-measurement schemes, the latter being various combinations of the former. The benefit of a replicated measurement is to reduce the error variance, which for this example will be assumed to decrease inversely with the number of replicated measurements [see Figure 12.5(c)]. Thus, for measurement schemes one, two, and three we have a simple addition of variances to obtain the overall error variance. For the next three schemes, each of which involve two replications, the error is reduced by a factor of two. For the last measurement scheme, the reduction factor is three. Figure 12.5(c) shows the FM calculations for the same surrogate-parameter errors and costs as Figure 12.4.

The costs are aggregated as before and the overall FM results shown in Figure 12.5(c). Measurement scheme three turns out to be clearly superior as a consequence of both its small overall error variance of 1.8 and its low cost of 50. Measurement schemes 1, 2, 1 + 3, and 2 + 3 are all relatively close. The replication does in fact help the latter two schemes in terms of their competitiveness with the single-measurement scheme.

From this discussion we find a relatively simple model that could apply to low-level as well as high-level (i.e., large-scale systems) measurement. The framework

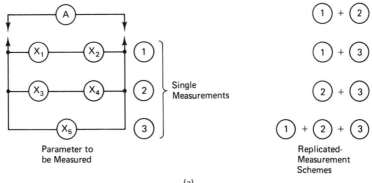

(a)

Measurement Scheme	X_1	X_2	X_3	X_4	X_5
1	1	1.2			
2			1.4	1.6	
3					1.8
1 + 2	1	1.2	1.4	1.6	
1 + 3	1	1.2			1.8
2 + 3			1.4	1.6	1.8
1 + 2 + 3	1	1.2	1.4	1.6	1.8

(b) Error Variances Associated with Measurement Schemes

Measurement Scheme	Overall Error Variance (σ_i^2)	Cost	Signal Value	Signal-to-Noise Ratio per Unit Cost
1	2.2	170	1000	2.67
2	3.0	130	1000	2.56
3	1.8	50	1000	11.11
1 + 2	2.6	300	1000	1.28
1 + 3	2.0	220	1000	2.27
2 + 3	2.4	180	1000	2.31
1 + 2 + 3	2.3	350	1000	1.24

(c) Figure-of-Merit Calculations for Each Measurement Scheme

Figure 12.5 Replicated-measurement-scheme illustration.

for the model is straightforward and assumptions are made for purposes of keeping the model structure simple. The signal-to-noise ratio, an important parameter in most communications systems, was used to develop an FM for choosing among alternatives. Other formulations in this regard include converting to logarithms, resulting in such forms as $\log(1 + S/N)$ or $W\log(1 + S/N)$, where W = bandwidth, which are both recognized as expressions for the capacity of a classical communications channel. This model explicitly included a conceptualization of the measurement problem, a configurational model of how to make the measurements, and a computational model showing relationships among key parameters imbedded in an FM.

12.2.2 Strategic Defense Initiative Example

In this example we select some of the variables related to the strategic defense initiative (SDI) program [7] and illustrate how a simple set of parametric trade-off relationships may be constructed. The scenario is the boost phase during which it is assumed that a missile attack might occur. These missiles are to be engaged by directed energy weapons (DEWs) such as high-power lasers, which are carried aboard a constellation of satellites.

A DEW aboard a satellite, after firing, can be thought of as slewing to another target, settling, and then firing on another target for a certain dwell time. For a total time available during boost, $T(t)$, the total number, N_m, of target engagements per DEW would then be

$$N_m = \frac{T(t)}{t(s) + t(e) + t(d)} = \frac{T(t)}{T(\text{sed})} \qquad (12.1)$$

where $t(s)$, $t(e)$, and $t(d)$ are the slew, settle, and dwell times, respectively. This simple relationship is depicted in parametric form in the top left corner of Figure 12.6. Thus, if the total time available were 300 sec and $T(\text{sed})$ were equal to 6 sec, then at most 50 target engagements could be carried out per DEW.

Given these calculations, we follow the directional arrow counterclockwise to the bottom left corner of Figure 12.6. For a selected number of engagements per DEW and an assumed number of targets, N_t, we calculate the number of DEWs engaging targets, N_d, as

$$N_d = \frac{N_t}{N_m} \qquad (12.2)$$

To illustrate, if we had to engage 1400 targets (N_t) and could handle 50 targets per DEW, a total of 28 DEWs engaging targets would be needed. The parametric form of this simple relationship is shown in the figure.

Knowing the number of DEWs engaging targets (N_d), we recognize that many satellites with DEWs aboard are on the "wrong" side of the earth at the time of engagement and thus cannot be used to engage targets. Thus they are "absent" from the battle. The relationship between DEW satellites in the battle and engaging targets (N_d) and the total number of DEW satellites in orbit (N_o) is sometimes reflected through a parameter called the absentee ratio, R_a. This relationship, shown in the lower right-hand corner of Figure 12.6, is given as

$$N_o = R_a N_d \qquad (12.3)$$

Using the number of DEWs engaging targets as 28 and an absentee ratio of 5, then a total of 140 satellites would be needed.

Finally, knowing the total number of DEW satellites required and the cost per satellite, $C(s)$, a simple product yields the total cost C_T of the constellation of satellites:

$$C_T = N_o C(s) \qquad (12.4)$$

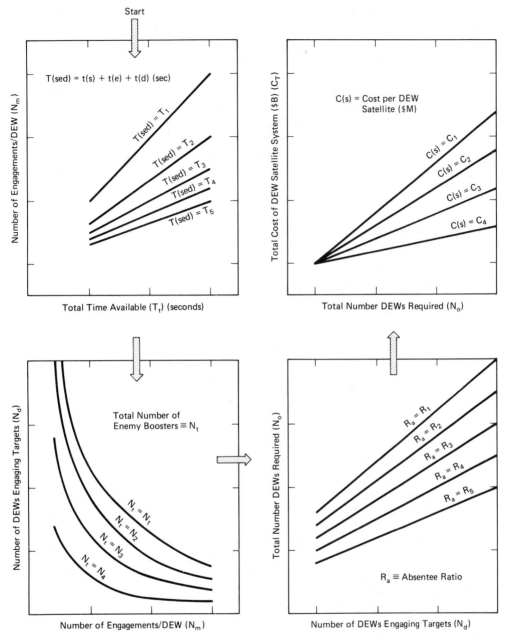

Figure 12.6 Simple strategic defense initiative parametric relationships.

This relationship is plotted parametrically in the top right-hand portion of Figure 12.6.

The parametric analysis example outlined in this section is acknowledged to be an oversimplification of a complex issue. It was constructed, however, to illustrate how a set of large-scale system parameters may be interrelated in parametric form that may be suitable for overview trade-off analyses. The methodology of tracking through the four plots of Figure 12.6 shows some limiting cases. It illustrates, for example, how some insight may be obtained as to the significance of the slew, settle, and dwell times in terms of their ultimate impact on the cost of such a defensive system. Of course, this is just the tip of this particular iceberg since no part of the example relates to the effectiveness of a DEW in terms of its ability to destroy a target. Such considerations and many similar ones would have to be added to develop a complete model of this strategic defense scenario. A lot can be done, however, by exploring system relationships in parametric form in order to get a feel for the key variables of a system and how they vary with one another. An important issue for the model designer, of course, is how to select these key variables. Good designers generally will do the above "back-of-the-envelope" types of calculations in order to try to isolate such variables and understand their significance.

12.2.3 A Search Radar Example

A third example of a model approach and structure utilizes the parameter dependency diagramming (PDD) technique described in Chapter 4. The particular model refers to a search radar for which the principal outputs are detection probability and false-alarm probability [8]. The relevant figures and tables from Chapter 4 are

- Figure 4.16—Parameter dependency diagram (PDD)
- Figure 4.17—Definitions of the illustrative parameters
- Figure 4.18—Functional relationships related to the PDD
- Figure 4.19—Signal-flow block diagram

As indicated in Chapter 4, the lines of a PDD represent parameters of interest and the blocks are functional relationships between input and output parameters. The PDD is itself a model, in this case it is a computational model of the most significant parametric relationships.

Tracking through the PDD in Figure 4.16, block 1 is the familiar radar-range equation. It calculates the signal returned at i-f (intermediate frequency), given such input parameters as output power, target strength, range to target, and gains. Along a parallel computational line, block 2 computes the noise-power spectrum, and when combined with the i-f bandwidth, W, (block 3), produces the noise at i-f. Carrier and noise-power inputs are used (block 5) to determine the post-discriminator noise-power spectrum. Knowing the upper- and lower-filter cutoff frequencies and other parameters, both the output noise (block 6) and maximum output voltage (block 7) can be developed. We then come to blocks 8 and 9 that compute the false-alarm and detection probabilities, respectively. It is recalled that these latter relationships were cited in Chapter 5.

The PDD and its associated figures and tabulations thus are seen to represent a specific modeling methodology. It has front-to-back integrity, is easy to understand, and facilitates the development of trade-off analyses and curves. In principle, it is then possible to relate an output parameter (detection or false alarm probability) back to input parameters such as range to target, R, transmitted power, P_t, or i-f bandwidth, W. A system designer can use such relationships to trade-off between key variables in an attempt to optimize the design. If the system is already built and fielded, it is also possible to use such a model to evaluate suggested engineering change proposals (ECPs). An example of the latter could be an ECP to change the i-f bandwidth, W. By examining the PDD we see immediately the impact of this parameter on the noise at i-f, the post-discriminator noise-power spectrum, and the maximum-output voltage. Detailed exploration of these multiple impacts is necessary before such an ECP is accepted as beneficial to the overall operation of the system.

The discussion in this section is an example of another simple model. It could also be representative of a continuous simulation, as defined by various texts on simulation. It is not, however, a sampling model utilizing Monte Carlo procedures. A way of looking at some of these relationships in terms of Monte Carlo techniques is examined in the next section.

12.2.4 Monte Carlo Simulation Example

We may visualize a simple simulation model as consisting of three sets of variables

1. output variables
2. exogenous input variables
3. endogenous internal variables

We will assume that the relationships (equations, procedures, truth tables, logic diagrams, data flows, and so on) between these variables are well defined and embedded in the model. Unfortunately, if these relationships are extremely complicated, it may be considered impossible to develop a *closed form* set of relationships that define outputs as a function of inputs, parametric in the internal variables. In such a case, practitioners often resort to Monte Carlo simulation procedures. Some situations that are typically known for such an approach include

- war gaming
- assembly-line analysis and balancing
- transit-system scheduling
- transportation-system network analysis
- communications-system network analysis

In a Monte Carlo simulation, some significant number of the variables are normally defined in terms of their probability distributions. Typically, this definition would apply to the input variables but could also apply to the internal variables. For example, for a command, control, and communications (C^3) system the traffic load between two nodes might be an input variable, represented as a probability distribution, whereas the size and speed of a switch might be endogenous internal variables defined as fixed-parametric values or distributions. For all the variables defined in terms of probability distributions, single samples are drawn. These samples are run through the model relationships to compute the values of the outputs. Thus, one point in the input–output space is developed. By replication of this process and judicious design, input and output sample distributions are generated. Sample statistics then characterize these input–output relationships, parametric in whatever variables are selected as such by the designer. There is typically no guarantee, however, that these sample statistics will converge with increased sample size to the true values of the desired variables. A considerable amount of activity has been devoted to issues surrounding the matter of sample size with respect to Monte Carlo simulations [1].

A key element in the example is the specific procedure for sampling from the distribution of each sampled variable. Although random number generation has been studied by various investigators [e.g., 1], it can be carried out by means of a random number generator (RNG), which is a computer language feature (see Chapter 9). That is, with a few simple computer program statements, a sequence of random numbers can be generated, each of which lies in the interval 0 to 1. This is equivalent to sampling from a uniform distribution over that interval and it allows for the development of a sample value from essentially any well-defined distribution, as illustrated in Figure 12.7. The random number (RN) is generated by the program and through the variable's cumulative distribution, a single sample of the variable in question is developed.

To illustrate the process, assume a random number equal to *RN* and a typical exponential distribution. The density and cumulative distribution functions for the exponential are, respectively

$$p(x) = \lambda \exp(-\lambda x) \qquad x \geqslant 0 \qquad (12.5)$$

$$F(x) = 1 - \exp(-\lambda x) \qquad x \geq 0 \qquad (12.6)$$

Formally, the graphical procedure implied by Figure 12.7 is carried out by solving the cumulative probability distribution function (CDF) equation for the corresponding value of the variable, given the random number, as here

$$x = -\frac{1}{\lambda} \ln[1 - F(x)] \qquad (12.7)$$

$$x(\text{sample}) = -\frac{1}{\lambda} \ln(1 - RN) \qquad (12.8)$$

Using some simple numbers, for $\lambda = 1$ and if the random number selected for the exponential distribution in Equation 12.8 were 0.4, then the corresponding sample value from the exponential distribution would be 0.51. It is recognized that the $(1 - RN)$ term can be replaced by (RN) when sampling from a uniform distribution in the $(0, 1)$ interval in order to save one step in the calculation. For each variable of interest, this process is replicated many times over the full set of variables. In this fashion the Monte carlo simulation is implemented.

Conversion from the input sampling procedure, shown in Equations 12.7 and 12.8, to an output set of results can be illustrated by means of another simple example. We take the case of detection of a radar pulse, as developed quantitatively in

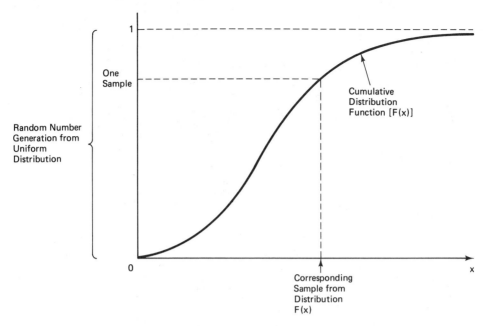

Figure 12.7 Process of sampling from a random distribution.

Chapter 5, and shown also in this chapter. Figure 12.8 illustrates a representative set of output variable samples from the distributions characterized by noise alone and signal plus noise. To simplify the example, the standard normal distribution is used (mean equal to 0 and variance equal to 1), and the pulse magnitude is taken to be equal to 4σ, or in this case numerically equal to 4. From the sample data in Figure 12.8, when there is no pulse present, nine cases (sample values) out of ten were below threshold and thus not detected. One sample value exceeded threshold and therefore was detected, constituting a false alarm. Although no proper simulation would generate only ten samples, for illustrative purposes we show only ten samples and a conclusion that the false alarm probability, with this data, would be estimated as 0.1. Similarly, in Figure 12.8, we show ten samples from the signal-plus-noise distribution. Two such samples happened to fall below the threshold so that the detection probability estimate for this sparse sample set would be 0.8. Although these are not representative values for a true radar pulse detection system, the process of Monte Carlo simulation is illustrated in terms of the output false-alarm and detection-probability variables.

It is noted that there is no need to go into a Monte Carlo simulation when a closed-form solution can be found. Such is clearly the case for the latter example. For the values illustrated in Figure 12.8, and observing that the threshold is at the 2 sigma value, a simple integration of the normal distribution yields the results that $P(fa) = 0.0228$ and $P(det) = 0.9772$. Chapter 5 should be consulted to verify these results in terms of integration of the normal distribution.

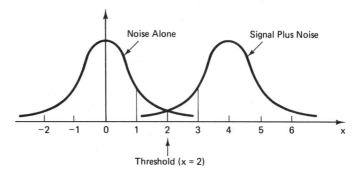

Noise Alone: Standard Normal, Mean = 0; Variance = 1
Signal Plus Noise: Standard Normal, Mean = 4; Variance = 1

| Trial Number | NO PULSE PRESENT (Noise Alone) | | | PULSE PRESENT (Signal Plus Noise) | | |
	Sample Value	Above Threshold ?*	Consequence[†]	Sample Value	Above Threshold ?*	Consequence[†]
1	−0.67	N	ND	+3.47	Y	D
2	+1.35	N	ND	+4.53	Y	D
3	+0.14	N	ND	+3.61	Y	D
4	−0.32	N	ND	+5.18	Y	D
5	+0.81	N	ND	+1.96	N	ND
6	−0.19	N	ND	+3.84	Y	D
7	−0.54	N	ND	+4.12	Y	D
8	+2.13	Y	FA	+4.75	Y	D
9	−0.78	N	ND	+1.89	N	ND
10	+0.26	N	ND	+3.72	Y	D

| Estimated False Alarm Probability → 0.1 | Estimated Detection Probability → 0.8 |

*N = No; Y = Yes
[†] ND = No Detection; FA = False Alarm; D = Detection

Figure 12.8 Monte Carlo simulation example using noise-alone and signal-plus-noise distributions.

12.2.5 National Air-Transportation Evaluation

In Chapter 10 (Table 10.3), we defined a series of alternative air-transportation system concepts. The evaluation of these concepts against a set of criteria can be considered a very large-scale-modeling task. As a consequence, only a suggested procedure for carrying out such an evaluation process is discussed here. Selected steps in the process require submodeling activities in order to develop the necessary performance data comparing the alternatives against one another.

One way to conceptualize such an evaluation is to separate it into two subevaluations, one quantitative and the other qualitative or subjective. A possible procedure for doing this is presented in Figure 12.9. Here Figure 12.9(a) is formulated

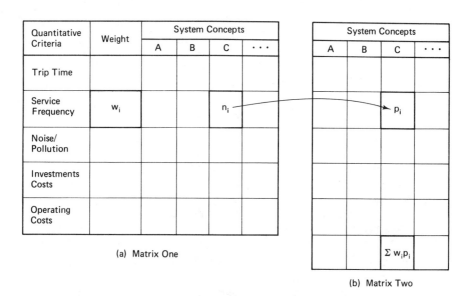

(a) Matrix One

(b) Matrix Two

(c) Matrix Three

Figure 12.9 Quantitative and qualitative top-level matrix evaluations.

w_i = Weights for quantitative criteria
n_i = Numerical evaluation
p_i = Normalized quantitative rating
w_j = Weights for qualitative criteria
r_j = Normalized qualitative rating

based upon a set of quantitative criteria, including

- trip time
- service frequency
- noise/pollution
- investment costs
- operating costs

These represent a limited subset of the criteria listed in Table 12.1, and the latter two criteria are cost rather than performance related. The procedure calls for a numerical evaluation of all system concepts against all the quantitative criteria, which are weighted. Figure 12.9(b) converts this quantitative analysis into a normalized quantitative rating of each system concept against each criterion.

In conjunction with the criteria listed, a set of qualitative criteria are defined, as shown in Figure 12.9(c). The implication is not that it is impossible to find quantitative surrogates for these criteria but rather that in the final analysis the procedure is to allow for the exercise of "expert judgment" on at least some set of qualitative evaluation criteria. In this particular example we take these qualitative criteria to be a set of factors dealing with

- social
- safety and security
- reliability and convenience
- human use of the system
- international/economic
- technology prospects
- special service features
- flexibility

A weighting and rating scheme is postulated for the quantitative criteria.

Given the factors outlined, a procedural flow model for carrying out the quantitative and qualitative evaluations is illustrated in Figure 12.10. This is a procedural model in the sense that it defines the evaluation process steps; the execution of these steps, in many cases, requires the construction and exercising of quantitative submodels. An overview of these steps with respect to the three quantitative performance criteria is provided next.

Trip time. Trip-time calculations can be described as follows:

1. for each of the defined system concepts, establish a set of typical door-to-door trips
2. establish the mix of modes (auto, rail, and air by type) appropriate for each typical trip under each system concept

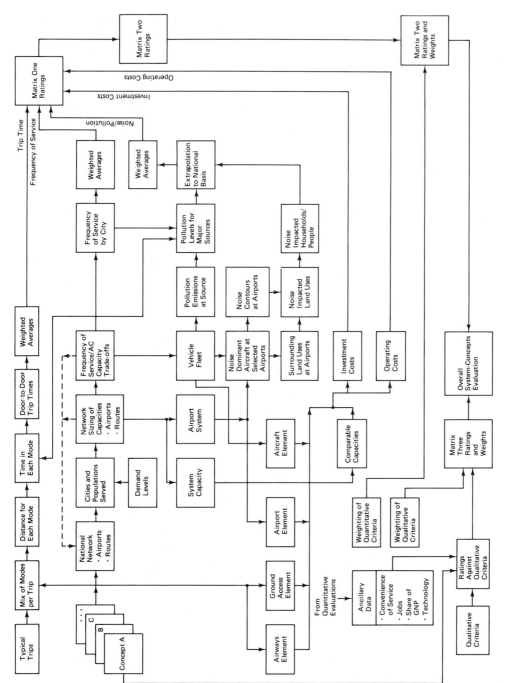

Figure 12.10 Procedural flow model for air-transportation evaluation.

3. define a travel distance for each mode of each typical trip and a door-to-door travel distance

4. calculate the travel time for each mode of each typical trip and a door-to-door travel time, including estimated time consumed for intermodal processing and potential delays

5. carry out the steps 1–4 for passengers and cargo

6. carry out the steps 1–4 for private transport

7. carry out the steps 1–4 for specific future years and compare with present situation

8. aggregate trip time results to calculate weighted average trip times

9. examine changes to all previous calculations under varying demand assumptions

Service frequency.   Illustrative steps for the calculations with respect to this criterion are

1. for each of the defined system concepts, lay out typical national and regional commercial air-service networks with airports as nodes and service routes as links

2. identify the cities served by each airport, the populations served, and the demand levels for each airport

3. size the network according to the demand levels by establishing required airport and route capacities needed to satisfy the demand

4. examine trade-offs between frequency of service and airport types/capacities to satisfy demand; develop recommended service frequencies

5. carry out steps 1–5 for cargo and passengers, accounting for interactions where possible

6. carry out steps 1–5 for discrete future time periods and compare with present situation

7. aggregate results to show typical service frequencies by class of city served and size/type of vehicle used

8. examine changes in all results as a function of alternative demand assumptions

9. modify trip time results to the extent that they are affected by service frequency calculations; rationalize and conform the two sets of calculations

10. tabulate consequences in terms of total airport system, total vehicle fleet, and total system capacity needed to satisfy demand

Noise and pollution.   Quantitative comparisons of noise and pollution effects can be developed through the following steps:

1. for all airports considered and the recommended aircraft mix, determine those aircraft that exhibit the greatest noise levels

2. select from the results of step 1 a representative set of airports and noise-dominant aircraft

3. estimate and overlay noise contours for the noise-dominant aircraft that operate into and out of the selected airports

4. establish the adjacent land uses for the selected airports

5. calculate the land areas impacted at various noise levels (e.g., 20 NEF, 30 NEF) for the selected airports

6. calculate the number of households and people impacted at the different noise levels

7. extrapolate the results of steps 1–7 to a larger population of airports

8. carry out steps 1–7 for each alternative system concept, by time frame

9. estimate the pollution emission levels, for various pollutant types, at the source for all pollution-dominant aircraft in the fleet

10. extrapolate these results to a national level through the number of aircraft by type in the fleet and the number of hours of flight per year

11. tabulate pollution results by alternative system concept over time periods of interest

Note the various interplays between all of the 11 steps in the evaluation procedure for the aforementioned performance criteria, the cost criteria, and both the quantitative as well as qualitative criteria, as shown in Figure 12.10.

We observe from the examples in this section that performance modeling and simulation is not a simple task although we have had to resort to relatively simple examples to demonstrate some of the inherent processes. Where simple models can be used to gain insight into a problem, they certainly should be. Indeed, there are substantial schools of thought among well-known engineers and scientists that argue persuasively for "back-of-the-envelope" analysis. This author supports that point of view whenever it is feasible and particularly insofar as it sheds light on the nature of the problem, its principal variables, and their key interactions. Too often, very complex models and simulations are developed when the analysts do not have the insight, skill, or perseverance to address simpler and more effective approaches. As with many aspects of systems engineering, it is often the case that simple, elegant, and correct solutions can be found by the talented practitioner.

12.3 SETTING UP A TECHNICAL PERFORMANCE MEASUREMENT PROGRAM

The previous two sections considered the requirements for a TPM program and examples of technical performance models that might be part of such a program. In this section we will briefly cite the various steps in setting up a TPM program for a large-scale system development. A defense systems engineering guide provides further information on TPM programs [9].

The first step is to review the system requirements and specifications in order to isolate the key performance drivers as articulated by this documentation. The requirements document, in particular, is a statement of what performance characteristics are important to the user. TPMs, in general, should therefore be keyed to user needs in this respect. Thus the next step is to define the program TPMs. Experience with large-scale defense programs [10] would suggest that of the order of a dozen top-level TPMs be considered (see Table 12.1 for lists of candidate TPMs). A third step is to develop tiered dependency trees, or parameter dependency diagrams, that interrelate the TPMs. This step will necessarily lead to the definition of parameters upon which the TPMs depend. These technical performance parameters (TPPs) can be structured in a set of dependency diagrams together with the TPMs. At this stage, all TPMs and TPPs should also be placed in a requirements-traceability matrix to relate them specifically to the existing requirements and specification documents. The latter should define, at least in principle, the required values of each of the TPMs and TPPs.

The fourth step is to develop modeling and simulation activities that support and are in effect part of the TPM program. The notion is that in order to predict the values of some of the TPPs and TPMs it will be necessary to run various models and simulations. The latter should also support other program requirements for system synthesis, analysis, and trade-off studies. Thus the performance-modeling effort, if properly designed, should provide a strong input to several of the necessary systems engineering tasks and activities. These same models, ranging from relatively simple sets of equations to full-scale simulations, will provide preliminary estimates of the key TPPs and TPMs. These values may be estimated numbers and can be continually compared with required values as defined in the requirements and specifications. Both are to be tracked and plotted on a time-phased profile, as illustrated in Figure 12.2.

As the program progresses, estimated values from the models are modified and updated as real-world hardware and software is being developed. As this occurs, test and evaluation (T&E) data is used, together with continuing modeling results, for this updating process. In principle, it should be possible to better understand, for each key parameter, whether or not the design is leading closer to the required values. If performance appears to exceed the requirement, such an area may possibly be relaxed and traded off against one in which the performance versus requirement appears to be at risk. A typical example is a performance-weight trade-off in a weight-constrained environment such as a satellite system.

Finally, in conjunction with the steps outlined in this section, the process of meeting or failing to meet requirements and specifications should be tied to both cost and schedule. The time-phased TPM profile is schedule oriented but should be made explicit by relating the TPPs and TPMs to the approved schedule that governs the overall development. Required and estimated performance needs to be cited on the master PERT, Gantt, or other schedule-control mechanism. Verification of these performance values is essential either by design data, model estimation, or test information. Since program costs are normally tied to the master schedule, a relationship between performance, cost, and schedule should be a natural product at all key program milestones. Cost and schedule reporting requirements (e.g., military standards) will dictate the format of these interrelated reports.

If properly designed, the TPM program will help both the customer as well as the developer of the system. If fragmented it may be treated as an off-line requirement and will not significantly affect the design approach and implementation. A key element is to closely tie the design team to the performance-modeling team and have both share responsibility for the TPM and TPP estimates as well as the overall TPM program. For complex programs a mature computer-aided modeling and simulation capability is essential in order to be able to keep pace with the design team and thus hopefully have a positive influence on the design.

12.4 MODELING–SIMULATION PACKAGES

A large number of computer software packages that can be used for modeling and simulation exists. A selected sample of such packages is listed in Table 12.2. Software known to be available at the p-c level is so indicated. If not available for micros, the others can be found in operation at the mini and mainframe levels.

TABLE 12.2 SELECTED SIMULATION PROGRAMS/LANGUAGES

• ACES*	• GPSS*
• ACSL	• ISIM*
• CAPS	• MICRONET*
• CSMP	• Q-GERT
• CSSL	• SIMAN*
• DARE	• SIMLAB
• DEMOS	• SIMPAS
• DSL	• SIMSCRIPT*
• DYNAMO*	• SIMULA
• EASY5	• SLAM*
• GASP	• TUTSIM

*Known to be available for the personal computer.

The systems engineer with a technical performance analysis and measurement problem can thus look to the possibility of using an existing software package to help work the problem. If only a p-c is available, then of course only those packages running at the microcomputer level are usable. As with other types of packages, increasing use of microcomputers for simulation purposes can be expected over time as the software is modified for that environment.

The matter of selecting a package to solve a particular problem is not a simple one. A primary issue is whether or not the package is suitable, from a technical point of view, to do the job. In general, this requires an investigation into the features of the packages in relation to the problem at hand. It may be necessary to refer to a book about the language first in order to make that judgment. Source materials for the more popular simulation languages are readily available. This is not necessarily the situation for the lesser-known simulations.

If the simulation language is suitable for application to the problem at hand, the next issue is one of availability of both the package as well as the host machines. Thus, availability of both the package and the computer resources become a second criterion for the selection of a simulation approach.

Cost is generally a third criterion for most users and applications. The systems engineer operating in an industrial setting normally has greater ability to spend more for a package than does, for example, the student. Simulation packages are not inexpensive, at least in relation to the more conventional tools such as spreadsheets and DBMSs. For more complex simulations the user may have to consider the cost of installation and maintenance.

The quality of the simulation package, the first criterion mentioned, may not have a lot to do with the quality of its documentation. A potential user should consider this issue in some detail, especially when there are no reasonable textbooks or other references available on the open market. This may not be easy to do prior to making a purchase. Knowledge of a simulation language is important, of course, but there can be a big difference between the language in the abstract and how it is implemented in a particular software package.

Another criterion that should be made more explicit is the relationship of the particular package to the available computer resources. Assuming basic compatibility between the package and the computer still leaves such matters as speed of operation and storage requirements as factors to be analyzed. Some packages require large amounts of RAM storage, which could be a limiting factor in certain applications, meaning that the user may simply run out of internal storage. Processing speed can be a critical factor in simulation since a large number of runs are normally made in order to generate a parametric set of trade-off curves. Where Monte Carlo procedures are used the situation may be even worse since this technique is based upon replication. Coprocessors may have to be used in order to amplify the processing power and efficiency.

The aforementioned items appear to be the principal criteria to be examined when considering the selection of a particular simulation package. Other factors may also come into play, such as

- degree of familiarity with computer/simulation languages
- time available to solve the problem
- flexibility and ease of operation of the package
- portability of the package to other environments
- possible future needs for other classes of problems

The potential user thus has much to explore in order to apply these computer packages to current as well as projected future problem areas. A decision diagram is documented and may be examined in regard to the selection of a simulation language in relation to various application areas [11].

The various model-simulation packages cited in Table 12.2 can be categorized as discrete or continuous or containing the characteristics of both. Figure 12.11 below shows a simple breakdown, representing a way to think about these two forms.

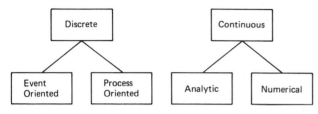

Figure 12.11 Discrete and continuous modeling–simulation packages.

The discrete approaches step through time, which is represented in a discrete manner. For the event-oriented approach, characteristic events are identified and routines are written that describe how the state of the system changes when each event occurs. The events are stored on a time-oriented list and executed as they come up in increasing order of their time of occurrence. When a particular event is considered, no actual time elapses at that instant. Changes of state that are caused by that event are examined and new events and their times of occurrence are generated and stored at their proper time on the event list. The GASP package is an example of an event-oriented discrete approach. For the process-oriented approach, a particular time-ordered process is tracked, but in respect to, for example, a single customer or class of customers. Thus the entire time history of a customer evolves in a process subroutine. Therefore, time passage is implicit in the process itself. Interactions between customers are considered as well, when they occur. This is a very natural modeling framework and is used, for example, by GPSS. Some simulation languages design have both an event and a process orientation (e.g., SIMSCRIPT and SLAM).

As the name implies, for the continuous approach, time is normally a continuous variable. The model–simulation will usually contain a variety of differential equations that describe the rates of change (with respect to time) of the key variables of the system. If the solutions to the equations can be developed analytically, the results are stored in the simulation. If not, numerical integration techniques are used. In either case the situation is viewed as one that changes continuously as a function of time. The continuous system modeling program (CSMP), advanced continuous simulation language (ACSL), continuous system simulation language (CSSL), and All-Purpose Continuous Equation Simulator (ACES) are examples of packages embodying a continuous approach. We will briefly cite some of the main features of the more popular of the simulation packages listed in Table 12.2.

12.4.1 General Purpose Simulation System

The general purpose simulation system (GPSS) is a very popular simulation whose roots go back into the 1960s. It was first developed by IBM but other companies have brought other versions into the market. It is process oriented and its basic design makes it an excellent tool for investigations of various kinds of queueing problems. In this context, one can conceive of a series of customers proceeding through some type of service situation. However, it may be applied to many other queuing problems such as

- the flow and execution of instructions through a computer-based system
- the servicing of constellations of satellites by a variety of ground stations and terminals
- the flow of packets through the nodes of a packet-switched telecommunications network
- the handling of aircraft traffic into the hubs (airports) of our air-transportation system

Three versions of the language—GPSS/360, GPSS V, and GPSS/H—are in use today, the latter being the most advanced and apparently also the most expensive (supported by Wolverine Software in Annandale, VA). GPSS has also been adapted to a micro environment [12]. Several good references are available [3, 13] to further investigate the features of this language.

GPSS is perhaps the most popular of the simulation languages. It is easy to learn and to use. Defining a simulation exercise can be thought of first in terms of constructing a flow- or process-oriented diagram of nodes and connecting branches. In general, nodes define some operation upon the flow of an entity (e.g., customer) and there are more than 40 such node types in GPSS. As with other process-oriented simulation languages, from the network structure it is a straightforward matter to write the language statements of the simulation. Thus, there is a coupling between the node-diagram methodology and the instructions of the language.

GPSS has been known to have at least two limitations. In general, it has had relatively long execution times, although the most recent version (GPSS/H) apparently runs about five times faster than earlier versions. Also, the GPSS clock has been integer valued, limiting the granularity of the recording of time. There are scaling workarounds for this kind of problem but it is at least a significant inconvenience.

As one of the first languages that received widespread use, GPSS set the stage for the development of other languages by demonstrating their utility in solving a variety of systems engineering problems.

12.4.2 SIMSCRIPT

The initial development of SIMSCRIPT is generally attributed to H. Markowitz (working with P. J. Kiviat and R. Villanueva), who first described this work while at the Rand Corporation and then later when he was at a company by the name of C.A.C.I. [14]. The C.A.C.I. company now plays a major role in providing this language and related services in the open market. Indeed, the micro version of this language is made available through this company.

SIMSCRIPT is provided in its own language. The syntax is relatively free form and simple, making it easy to understand, check out, and document various routines. It is discrete and event oriented and therefore events play a major role in the language. However, it also admits of a process orientation. As with other simulations, events occur at specific points in time (i.e., they do not occur over more than a particular instant of time). The times between events are called activities. This nomenclature is similar to that previously described in PERT networks in regard to events and activities. Unlike PERT, however, SIMSCRIPT also uses the explicit notion of processes, which are collections of related events. Processes are established by declarations resulting in a specific process routine.

Other important aspects of SIMSCRIPT include the use of such items as entities, attributes, and sets. These can be defined as

- entities—structured data items for an element in the simulation, similar to a subscripted variable
- attributes—values that characterize the entities
- sets—collections of entities that have one or more common properties

A typical run of a SIMSCRIPT simulation, as with several other event-oriented simulations, would contain the following steps:

- initial declarations
- initializing of all appropriate variables, entities, and sets
- the scheduling of initial processes and events
- starting of the simulation
- development of output statistics
- ending the simulation by appropriate end statements

SIMSCRIPT has a variety of built-in functions that are similar to those provided by many other languages (see Chapter 9). For example, in order to sample from a random variable, a simple SIMSCRIPT statement will suffice to provide a sample value from a built-in set of distributions. The same is true for output statistics; simple function statements will provide computations of means, variances, and the like. This is a particularly strong feature of the language.

As a generalized simulation language, SIMSCRIPT is very useful, popular, and powerful. Available at the p-c level, it is an excellent candidate for use by the practicing engineer. However, its cost is relatively high in relation to most of the other commercially available packages cited in this text. This is not unusual for simulation languages especially with the extensive capability of SIMSCRIPT.

12.4.3 GASP

GASP apparently had its origins at U. S. Steel, and then was built upon at Arizona State University by A. A. B. Pritsker, a leader in the field of simulation. In 1969, Pritsker and P. J. Kiviat documented GASP II [15] and described both the nature of

GASP as well as application areas such as queueing, inventory systems, and networks. Since GASP is Fortran-based it is usable only on machines with a Fortran compiler. GASP has the following basic functional areas of operation:

1. initialization of system states
2. event control
3. program monitoring and error reporting
4. statistical analysis and report preparation
5. an executive to control items 1 through 4

GASP is event oriented but is able to handle discrete, continuous, and combined discrete-continuous simulation constructions. It contains entities, their related attributes, and files of common entities. Its most recent and capable version, GASP IV [16], is well documented and available through Pritsker and Associates, Inc., West Lafayette, IN. GASP is considered less powerful than SIMSCRIPT but it is also less expensive.

12.4.4 Simulation Language For Alternative Modeling

The simulation language for alternative modeling (SLAM) is relatively new, containing most of the features of GASP but with greater flexibility and power. It is therefore viewed as a language that will eventually supersede GASP. It provides the capability to deal with both event- and process-oriented simulations, with the event orientation similar to that in GASP. The process orientation is facilitated by the use of network symbols and techniques. A set of standard symbols is used to construct a network of nodes and branches. Such a network is process oriented, representing a flow of customers, commodities, or items whose state and performance one wishes to track and analyze. The language's formal program statements can be developed from the overall network structure. In this sense it is similar to Q-GERT, which is cited in Table 12.2 but whose discussion is reserved for Chapter 13.

SLAM was developed by C. D. Pegden and A. A. B. Pritsker and is available through Pritsker and Associates, Inc. (as are other simulations). Their book on SLAM [17] shows a wide variety of application areas including series work stations, inventory-system operation, inspection systems, a PERT network, traffic light operations, and a world (system) dynamics model.

SLAM is also available at the microcomputer level [18] with a particularly flexible capability (SLAM II, a registered trademark of Pritsker & Associates, Inc.). This implementation allows for modeling through any one or a combination of network, discrete event, and continuous approaches. The core of the p-c version is a network of nodes and branches. The nodes can be of the order of 20 different types, representing what it is that happens in the flow of entities through a process. These nodes handle such operations as selection, matching, waiting, assigning, and gating. Routing alternatives allow for deterministic, probabilistic, and variable-related approaches. Simulation statements are based upon the network representation and embody the implementation code.

If the capability of the network and associated features are exceeded, user-defined Fortran subroutines can be written to allow for more complicated discrete-event handling. At the same time, a third capability can be added by means of Fortran-based continuous variable modeling and analysis. Thus, three capabilities are combined, which substantially extends the power and scope of the simulation.

Typically, for the aforementioned implementation, three processors are needed, dealing with input, execution, and output modes. Operating at only the network level, one would need a p-c with a floppy disk and at least 320K of RAM. The addition of a second disk and a coprocessor will extend the storage and speed capabilities. If user-defined Fortran subroutines are used, a Fortran compiler is necessary. In a comparison with the VAX 11/780 [18], a sample simulation at the p-c level, in terms of CPU time, ranged from 21 to 56 times faster at the mini level. This places quite an effective tool in the hands of a p-c user when considering both cost and performance. Expected price–performance improvements in the micros will enhance the p-c as a selection for limited production application areas.

12.4.5 DYNAMO

DYNAmic MOdels (DYNAMO) is a series of models based upon the system-dynamics analysis methodology developed by J. W. Forrester [19]. It was formulated and applied to a variety of systems, including social systems. It was Forrester's conviction that social systems could be modeled only by means of simulation techniques that utilized very complex numerical solutions. Coupled to that was the notion that purely analytic solutions were not feasible. In addition to his theory of systems [19], Forrester documented his views of industrial dynamics [20], urban dynamics [21], and world dynamics [22].

Although DYNAMO was explained by Forrester [19], additional focus and model developments were addressed by A. L. Pugh III [23]. Further, specific versions and applications are supported by Pugh through his company, Pugh-Roberts Associates, Cambridge, MA.

One of the central themes in DYNAMO is the notion of feedback loops of which there are both negative and positive types. Negative feedback is well known to the control-systems engineer. Positive feedback normally leads to system instability in that the output can grow without bound. Time, of course, is the usual independent variable. The dependent variables are of two basic types

1. level variables—describing the states of the system
2. rate variables—describing actions that cause changes in the level variables

Paths through the system model display alternation of these two variable types. A simple-rate equation in DYNAMO might start with

$$OR = \frac{1}{AT}(DI - I) \qquad (12.9)$$

where, for an inventory or logistics support system, OR is the order rate, DI is the desired inventory, I is the (current) inventory and AT is the adjustment time.

DYNAMO basically combines these and more complex equations end-on-end, together with an integration capability, to yield a powerful set of simulation packages.

12.4.6 Other Packages

As suggested by Table 12.2, simulations other than those described are available for the micro, mini, and mainframe. For example, a continuous-system simulation language (CSSL) called interactive simulation interpreter for microcomputers (ISIM) exists for the micro [24]. This relatively small simulation is written in Fortran and basically contains the capability of solving certain types of differential equations and providing monitor and plotted output. Another modeling package is called Micro PASSIM and it handles both discrete-event and continuous models [25]. It is written in Turbo Pascal and is menu driven, which allows for user interaction and control. The SIMAN simulation language, provided by the Systems Modeling Corporation, was developed by C. D. Pegden who coauthored the work on SLAM [17]. This simulation will run on micros with a Fortran compiler. Key features available in SIMAN include

- a block diagram modeling approach (with about 40 basic blocks)
- discrete and continuous orientations
- new functions that can be added using Fortran subroutines
- an interactive output processor
- built-in statistical-analysis functions
- interface capability with respect to other p-c graphics packages

Other p-c based packages referred to in the literature [26] are: micro-DYNAMO provided by Addison-Wesley; a discrete simulation called MicroNET available through Pritsker & Associates; and a continuous simulation known as ACES, from the Modulo 2 Company. Other selected continuous simulation packages [26] include: CSSL-IV available through Simulation Services in Chatsworth, CA; a package called EASY5 from Boeing Computer Services; DARE from the University of Arizona's Electrical Engineering Department; and ACSL available from Mitchell and Gauthier Associates in Concord, MA.

Workstation systems, such as those provided by Apollo Computer, are continuously moving toward making available a full range of simulation as well as other packages (e.g., decision support systems, and software development tools) to their users. For example, Apollo provides at least the following three simulations at their workstations [27]:

1. advanced continuous simulation language (ACSL)
2. continuous system simulation language (CSSL-IV)
3. DYNAMO

These three continuous simulations are embedded in an extremely powerful workstation environment for the practicing engineer with access to the appropriate machines and associated software. This capability is not readily available to the student or systems engineer who wishes to purchase and retain these tools for their own personal use. Various microbased packages, by name and vendor, can be found in Appendix A.

12.5 OTHER SELECTED MODELING AND SIMULATION ACTIVITIES

This section addresses a few modeling and simulation activities that are related to computer-aided techniques, either directly or indirectly. Areas to be briefly considered are

- mechanical computer-aided engineering
- flight simulation
- military applications
- algorithms

The final words of this section highlight some additional sources of information concerning modeling and simulation.

12.5.1 Mechanical Computer-Aided Engineering

Mechanical Computer-Aided Engineering (MCAE) is a simulation activity that is being applied to large-scale, expensive, or very complex mechanical systems. Industry is becoming increasingly involved in these applications in order to prove the integrity of mechanical designs on the computer in distinction to building and testing real physical models [28]. A case in point is the crash testing being performed by the automobile manufacturers, which could be replaced by MCAE. With MCAE, the vehicles can be crashed on the computer instead of in real life. In addition to saving considerable sums of money in the long run, observations can be made in the simulated environment that are difficult to deal with in real crash tests. For example, the mechanical elements that take both the least and most stress can be isolated in the simulation. Thus, in addition to beefing up the highly stressed members, it is also possible to lighten the less-stressed members.

Some of the industry MCAE activities that are cited [28] include

- minimization of noise and vibration for automobiles
- design of flexible moorings for oil platforms in the North Sea
- design of a device that will hold artificial lenses in the eye
- reduction of optical distortion in large mirrors and lenses such as those carried aboard the Space Telescope

Market analysts have predicted a very large increase in industry's use of these MCAE techniques. In a sense, this is a second wave of application, the first being the acceptance of conventional CAD/CAM/CAE methods. In response to these trends, software specialists have been working on improving computer simulation packages whose applications are focused on mechanical systems of all types.

12.5.2 Flight Simulation

We remark briefly on a software simulation package known as Flight Simulator (produced by Microsoft Corporation). This simulation is available for the p-c for around $50 and is remarkable in its price–performance capability [26]. Graphics appear on the screen that are split into two parts: a windshield display that shows the surrounding view outside the aircraft and a lower screen display of a control panel that shows airspeed, altitude, as well as turn, heading, and rate-of-climb indicators. Other displays are available making the "pilot's view" extremely realistic. Flying the aircraft is actually controlled through the keyboard so that a lot of practice is necessary in order to be able to handle the aircraft. Various preset modes are built in to the package, allowing flight in the four basic areas of Los Angeles, Seattle, Chicago, and New York/Boston.

Another product similar to the above is one known as Jet (produced by Sub-LOGIC Corporation). In this simulation, options are available to fly F-16 or F-18 aircraft in military missions that include engagement with other aircraft (MiG-21, MiG-23) and the delivery of bombs to specified targets. A free-flight mode allows the user to become proficient in the use of the keyboard. The screen display includes such features as altitude and altitude indicators, airspeed, thrust, magnetic heading, fuel remaining, a range to target indicator, and information regarding the status of weapons carried. Taking off can be performed from a carrier or a fixed ground-based hangar. The displays are extremely realistic and the overall simulation very challenging. This author has yet to successfully land the aircraft back on its home-based carrier.

The significance of these types of packages is clearly not in terms of their immediate value in a systems engineering context. However, they do indicate what can be done at the micro level for a minimum investment. We can thus extrapolate this price–performance to a more serious systems engineering set of problems. The one that comes to mind, of course, is flight simulation, but in a mode that can be complementary to the large-scale flight simulators that cost millions of dollars. Further extrapolations can be imagined, the point being that the p-c with state-of-the-art software can be a significant adjunct to, and in some cases a replacement for, programs that cost thousands of times more than the microbased package.

12.5.3 Military Applications

The military has been an arena in which numerous models and simulations have been developed. It is likely that thousands of such tools are in use today, some based upon the simulations discussed in this chapter and others starting from scratch with standard programming languages. Examining just one single source [29], we note the existence of models and simulations in the following application areas:

- general purpose forces

- theater nuclear forces
- evaluation, training, and operations
- strategic forces
- strategic recovery and civil defense
- cost
- C^3
- logistics

In the C^3 area, a fundamental notion is that of a strategic network that is minimally vulnerable to attack and destruction. An example shown is related to the Minimum Essential Emergency Communication Network (MEECN) and its associated MEECN System Simulator known as MSS. This model simulates the threat that might be imposed upon the MEECN and assesses the degree to which the network will be able to carry out its functions in the presence of such a threat. The model actually contains several submodels, each of which addresses a different subfunction of the network operation. One such submodel, for example, is called the MEECN Dynamic Simulator (MDS); another is the MEECN Attack Allocator (MAA). The latter utilizes Monte Carlo techniques. A second series of models in the C^3 area is called STRAT COMMAND. These models address the performance of the C^3 links to our strategic bomber force. Other classes of models with respect to C^3 consider the so-called *multiplier effect*, the presumed enhancement that a C^3 system provides, for example, to a defensive weapons system, if the former operates properly.

With respect to logistics, a variety of models are cited [29] that evaluate alternative schemes for providing logistics pipelines to various forces. Logistics support is being recognized as a more significant and often neglected area of planning and analysis with respect to assuring the readiness of our defense systems. Further discussion of this area can be found in Chapter 15.

Simply stated, massive amounts of effort have gone into the development of models and simulations in support of military analyses. Often, these are unclassified and can be accessed through the Defense Documentation Clearinghouse and other references in the open literature. The serious student and practitioner can therefore tap a vast resource of information in order to assist in the development of new models or the adaptation of old ones.

12.5.4 Algorithms

Computational algorithms, of course, represent the central core of all quantitative models and simulations. In this regard, every now and then a special algorithm is formulated that is recognized as a breakthrough. When this is the case, such an algorithm is used again and again in various models and computational routines. One such algorithm is the so-called Kalman filter (see also Chapter 16). This filter is able to accept measurement data and provide estimates of the position, velocity, and so on of a body in a more efficient manner than had been achieved before. Another such algorithm is the Fast Fourier transform (FFT), a procedure that has improved previous methods for calculating these types of transforms. This technique is used extensively in the arena of signal processing. Another recent example is the newly developed Karmarkar algorithm [30], a procedure for improving upon current methods (i.e., simplex methods) of solving linear-programming problems.

In general, when a new and more efficient algorithm comes upon the scene, it is quickly brought into the modeling and simulation domain. The model developer must be tracking the literature in order to look for possible improvements in computational techniques. As with all fields, the obligation of the practitioner is to keep current with the most applicable and efficient procedures as tools of the trade. With respect to systems engineering application areas, two special information sources are of particular importance. The first is represented by the various documents and conferences associated with the Institute of Electrical and Electronics Engineers (IEEE). The overview IEEE *Spectrum* magazine as well as several of the IEEE transactions (e.g., *Systems, Man, and Cybernetics*) should be reviewed for the latest advances in a wide variety of fields. In regard to modeling and simulation, conferences and proceedings are available (for example, see Reference 6) every year to help in keeping up with the latest developments in this field.

12.6 CASE APPLICATION SUMMARY

In relation to this chapter, CASE deals with the ways in which the computer can be of assistance in terms of the tasks of performance modeling and simulation. Since modeling concepts were defined in a very broad context, it follows that many of the computer-based tools discussed both in this chapter as well as earlier chapters appear to be relevant. Table 12.3 cites the computer tools that can assist in performance modeling and simulation, together with an example of a particular model application area.

TABLE 12.3 COMPUTER-BASED TOOLS AND ILLUSTRATIVE PERFORMANCE MODELING AND SIMULATION APPLICATION AREAS

Basic Computer Tools (Chapter 3)
- Spreadsheets/integrated packages—modeling the power budget trade-offs in a satellite-communications systems

Diagramming Techniques (Chapter 4)
- Signal-flow block diagrams—describing the signal flow through a radar system
- Hierarchical decomposition diagrams—functional decomposition of a manufacturing system
- Data flow diagrams—describing the flow of data through a multiserver inventory-control and logistics system
- Functional flow diagrams and descriptions—breakdown of a configuration-management system, including the flow of data between system elements
- Hierarchical input-process-output diagrams—hierarchical description of inputs and outputs of a large-scale, air-traffic-control system
- Warnier–Orr, Michael Jackson, action diagrams—computational data flow and algorithms for an air-defense command and control system
- Sequence and timing diagrams—signal flow through a high data rate signal processor and correlator

TABLE 12.3 (*cont.*)

- Parameter dependency diagram—definition and interaction between key parameters of a command, control, and communications network system
- Logic flowcharts, Nassi–Shneiderman charts—logical computation sequence associated with a large-scale simulation of a rapid-transit system
- Decision-network diagrams—decision alternatives with respect to the acquisition parameters of a space-surveillance system and its ground-based support system

Specialized Computer Tools (Chapter 6)
- Multitask applications managers—combining spreadsheet and DBMS data developed by two different systems engineering groups working on the same program
- Project management aids—tracking the progress of a TPM program
- Alternatives and preference Evaluators—laying out subsystem alternatives for a worldwide telecommunications system
- Decision support systems—providing quantitative trade-off data for the latter telecommunications system

Mathematical and Engineering Tools (Chapter 7)
- Statistical—estimating the error inherent in a Monte Carlo simulation
- Mathematical—developing equations of motion for a constellation of satellites
- Engineering—computing the power spectral-density function and bit-error rate for a communications system
- Reliability—estimating the reliability of an automated manufacturing line
- Mathematical programming—development of cost-material usage trade-offs for an automobile assembly line
- Curves and plotting—automatic presentation of results for a variety of model–simulation outputs
- CAD/CAM/graphics—graphical representation of a space-station docking procedure
- Workbenches—provision of diagramming tools in an automated environment

Programming Languages (Chapter 9)
- BASIC/Pascal/other languages—Development of a model–simulation without benefit of existing packages; use of selected programming language

Modeling and Simulation Tools (Chapter 12)
- Modeling/simulation packages—Adaptation of existing package (e.g., Simscript, GPSS) to address queuing in a large-scale computer complex

REFERENCES

1. Law, A. M., and W. D. Kelton, *Simulation Modeling and Analysis*. New York: McGraw-Hill, 1982.
2. Blanchard, B. S., and W. J. Fabrycky, *Systems Engineering and Analysis*. Englewood Cliffs, NJ: Prentice-Hall, 1981.
3. Gordon, G., *System Simulation*. Englewood Cliffs, NJ: Prentice-Hall, 1969.
4. Weapon System Effectiveness Industry Advisory Committee, *Chairman's Final Report*, AFSC-TR-65-6, U. S. Air Force, Air Force Systems Command, January 1965.
5. *Engineering Management*, Military Standard-499A, Department of Defense, May 1, 1974.

6. Eisner, H., *A Systems Engineering Evaluation Model for Selecting Among Alternative Measurement Schemes,* Fifteenth Annual Pittsburgh Conference on Modeling and Simulation, University of Pittsburgh, Pittsburgh, PA, April 1984.

7. *Defense Against Ballistic Missiles: An Assessment of Technologies and Policy Implications,* and *The Strategic Defense Initiative: Defensive Technologies Study,* Department of Defense, April 1984.

8. Eisner, H., "The Assessment of Military Systems' Performance Using Parameter Dependency Diagram Techniques," The Third Annual Israel Conference on Operations Research, Tel Aviv, Israel, 1969.

9. *System Engineering Management Guide.* Defense Systems Management College, Fort Belvoir, VA, October 3, 1983.

10. "Technical Directions, DSM: Satellite Command and Control," IBM-Federal Systems Division, Bethesda, MD, 10, no. 3, 1984.

11. Shannon, R. E., *Systems Simulation: The Art and Science.* Englewood Cliffs, NJ: Prentice-Hall, 1975.

12. Cox, S., and A. J. Cox, "GPSS/PC: A User Oriented Simulation System," *Modeling and Simulation on Microcomputers: 1985.* ed. R. G. Lavery, The Society For Computer Simulation, La Jolla, CA, 1985.

13. Schriber, T. J., *Simulation Using GPSS.* New York: John Wiley, 1974.

14. *SIMSCRIPT II.5 Programming Language.* C.A.C.I, Los Angeles, CA, 1983.

15. Pritsker, A. A. B., and P. J. Kiviat, *Simulation With GASP II, A FORTRAN Based Simulation Language.* Englewood Cliffs, NJ: Prentice-Hall, 1969.

16. Pritsker, A. A. B., *The GASP IV Simulation Language.* New York: John Wiley, 1974.

17. Pritsker, A. A. B., and C. D. Pegden, *Introduction to Simulation and SLAM.* New York: Halsted Press, 1979.

18. Lilegdon, W. R., and J. J. O'Reilly, "SLAM II for Microcomputers," *Modeling and Simulation on Microcomputers: 1985.* ed. R. G. Lavery, The Society For Computer Simulation, La Jolla, CA, 1985.

19. Forrester, J. W., *Principles of Systems.* Cambridge, MA: Wright-Allen Press/MIT Press, 1968.

20. Forrester, J. W., *Industrial Dynamics.* Cambridge, MA: Wright-Allen Press/MIT Press, 1961.

21. Forrester, J. W., *Urban Dynamics.* Cambridge, MA: Wright-Allen Press/MIT Press, 1969.

22. Forrester, J. W., *World Dynamics.* Cambridge, MA: Wright-Allen Press/MIT Press, 1971.

23. Pugh III, A. L., *DYNAMO User's Manual.* Cambridge, MA: MIT Press, 1970, 1973, 1976.

24. Crosbie, R. E., and J. L. Hay, "ISIM for MSDOS and Other 16 Bit Machines," see reference [8].

25. Barnett, C. C., "Micro PASSIM: A Modeling Package for Combined Simulation Using Turbo Pascal," see reference [8].

26. *BYTE Magazine,* Simulation Issue, 9, no. 3, March 1984.

27. *Catalogue of Applications for Domain Systems.* Apollo Computer, Inc., Chelmsford, MA, 1985.

28. "Tests By Computer Make Trial-and-Error Old Hat," *Business Week,* June 17, 1985.

29. Battilega, J. A., and J. K. Grange (eds.), *The Military Applications of Modeling*. Air Force Institute of Technology Press, Wright-Patterson Air Force Base, Dayton, OH.

30. "Karmarkar's Algorithm: A Threat to Simplex?" IEEE *Spectrum*, December 1985.

PROBLEMS

P12.1. Find an effectiveness model in the literature, such that effectiveness is defined in terms of availability, dependability, and capability. Explain and discuss the structure and algorithms that make up the model.

P12.2. Develop lists of a dozen TPMs for each of the following types of systems:
   a. a stereo system
   b. an automobile
   c. a telephone system
   d. a large-scale computer complex
   e. a computer-aided networked workstation system

P12.3. Reevaluate the parameter-measurement example in the text for costs that are twice the values listed for surrogate measurement parameters $X1$ to $X9$ listed in Figure 12.4.

P12.4. Referring to the SDI example in Figure 12.6, fill in all ordinates, abcissas and parametric curves with representative numerical values. Based upon this, trace one numerical example through each of the four graphs and interpret the results.

P12.5. Construct a parameter dependency diagram that reflects the parametric relationships represented in the SDI example and the curves in Figure 12.6.

P12.6. Select five random numbers from a Rayleigh distribution. Detail your steps.

P12.7. Referring to Figure 12.8, calculate the detection and false alarm probabilities if
   a. the threshold is moved from 2 to 3
   b. the threshold remains at 2, but the signal increases to 5

P12.8. Develop a numerical example by choosing your own values for the matrices in the top-level evaluation in Figure 12.9.

P12.9. Referring to the literature, demonstrate a simple GPSS application to the level of the program instructions. Explain all the steps in detail.

P12.10. Do problem P12.9, replacing GPSS with SIMSCRIPT.

P12.11. Do problem P12.9, replacing GPSS with GASP.

P12.12. Do problem P12.9, replacing GPSS with SLAM.

P12.13. Do problem P12.9, replacing GPSS with DYNAMO.

P12.14. Investigate a p-c based simulation model other than those in the problems here. Discuss its characteristics in a three-page evaluation.

Betwixt the devil and the deep sea

Erasmus

RISK AND DECISION ANALYSIS

In the broad popular meaning of the term, *risk* is associated with a set of conditions or factors that, in a large-scale system-development sense, could result in lack of success for the system and its development or operation. Such factors or conditions might be categorized as related to

- cost
- schedule
- technical performance
- administrative issues

However, in a context larger than that of system development, it may be necessary to deal as well with societal risk. In this area, it is assumed that the program is successful with respect to all risk factors and conditions with the possible exception of some of the key technical issues and that if such is the case, disastrous societal effects might follow. Examples from previous programs have included

- nuclear power plant accidents and the possible release of radioactive materials
- inadvertent release of noxious chemicals and gases from various chemical processing plants

- depletion of portions of the ozone layer as a result of operating a fleet of supersonic transports, and the consequent increase in skin cancer incidence as well as long-term climatic effects

Both types of risks, namely program development and societal, will be explored in this chapter.

Decision analysis may be thought of as a set of procedures that provides information to decision makers that hopefully allows for the execution of better decisions. In a sense, most of the systems engineering tasks and activities support the goal of improved decision making at all levels of design and analysis. However, there are certain broad tools and techniques that can be used to assist in decision making and an overview of some of these tools and techniques is provided in this chapter.

It should be noted that there are linkages between risk and decision analysis from at least two points of view. The first is that proper decision analysis should take into account the risk factors associated with any large-scale system. If significant program-development or societal risks are present, the decision analysis must highlight these risks and attempt, to the maximum extent, to mitigate them. It is an obligation of the decision analyst to do so, whatever tools and techniques are being employed. Second, it turns out that several important tools and techniques are essentially the same for both the fields of risk analysis and decision analysis. Thus, if these methods are mastered, they can be applied to both analysis arenas. Specifically, various probabilistic modeling and decision network or tree analysis procedures are applicable to both subjects. These are highlighted in the text due to their broad applicability.

Finally, we return to the subject of computer aids by means of a brief reexamination of commercially available decision analyzers and support systems. The question here is: Given the risk and decision analysis problems, issues, and formalisms, to what extent can such available software packages provide assistance to the systems engineering analyst and the decision maker? This question is explored in Section 13.3.

13.1 RISK ANALYSIS

Program-development risks can be present with respect to cost, schedule, technical performance, and administrative matters. The risk in regard to cost is simply that the system development may overrun budgeted costs, a rather common occurrence for large-scale complex systems. If a life-cycle cost model (LCCM) has been applied, with good input data, estimated costs can readily be compared with budgeted costs by cost category and by year. Uncertainty in the cost elements can be estimated in several ways, one of which is suggested in Chapter 11, Equations 11.11 and 11.12. This procedure allows for probabilistic cost estimating, yielding an estimate of the variance of the individual and composite cost elements. If the cost uncertainty can be identified, at least some attempt can be made to minimize the risk associated with out-of-budget costs, by priority area.

In terms of schedule, a technique is available that is similar to that alluded to in the preceding paragraph about costing. If we assume implementation of a PERT-like scheduling system, through the three estimates for activity times, probabilistic time estimates for the various subsystems and the overall program are readily obtained. Schedule risk is related to these time estimates in a straightforward manner.

Technical performance risk is the most complex area, depending upon the intent, scope, design, and operation of the system. As such, it is the primary focus of the risk-analysis discussion of this chapter. Administrative risk refers to factors associated with the acquisition of the system such as the interactions between developer, procurement agent, and user and the rules and regulations under which the acquisition process is administered. Many system developments run afoul in regard to such administrative matters as

- inability to flexibly negotiate scope, requirements, and specifications changes
- improper fee and cost-control incentives
- improper form as well as terms and conditions of contract

Developing a large-scale system is a two-way street and if a solid working relationship is not established, the overall risks can increase dramatically.

13.1.1 Technical Performance Risk

Technical performance risk is present when the satisfaction of key performance parameters is in serious question and the consequences of such a failure could be considered disastrous. Situations of this general type can be illustrated by the following kinds of system problems:

- failure of a rapid-transit signal and braking system, leading to the crash of one or more trains
- failure of an air-defense system to properly distinguish between friendly and enemy aircraft, leading to fratricide
- failure of a space-shuttle rocket, leading to the death of our astronauts

The mitigation or elimination of risks of these types is the objective, in part, of government standards such as Mil. Std. 499A [1], which identifies the following program risk analyses to be carried out:

- continuing assessment of cost, schedule, and technical parameters
- identification of critical areas
- investigation of methods for system or hardware proofing, prototyping, testing, and backup development
- assessment of key test requirements, TPM parameters, and critical milestones

The latter risk analyses, as well as common sense, suggest a linkage between program-risk considerations and technical-performance measurement. If the latter defines and tracks the key technical performance parameters of a system, then risk must at least in part be related to the possibility that the desired levels of these parameters might not be achieved. In short, given a TPM program (see Chapter 12), a simple risk analysis can be keyed to such a program as outlined next.

We assume the existence of a TPM program, which includes the definition of TPMs as well as a continuing evaluation of the actual TPM values in relation to the required or specified values (see Figure 12.2). We may divide the TPMs into *criticality* categories, reflecting their importance toward achieving key mission objectives. Thus, for example, we may have four criticality categories, ranging from most to least critical. The TPM program provides, for each TPM, a ratio of actual achievement to desired achievement, as

$$R(i) = \frac{A(i)}{D(i)} \tag{13.1}$$

where $A(i)$ is the actual assessed value for the ith TPM, $D(i)$ is the desired or specified value, and $R(i)$ is the ratio. A simple index of the extent to which technical risk is present, for the jth criticality category, is relatable to the average value of the ratio in Equation 13.1 expressed as

$$I(j) = 1 - \sum_i \frac{R(i,j)}{n(j)} \tag{13.2}$$

where $n(j)$ is the number of TPMs in the jth category and $I(j)$ is the risk index for the jth risk category. Lower values, of course, are indicators of less risk, using these definitions. The following example shows how this might work. Assume three criticality categories ($j = 1, 2, 3$), and the $R(i, j)$, and $I(j)$ values listed

j	$R(i, j)$	$I(j)$
1	0.8	
1	0.9	0.2
1	0.7	
2	1.1	
2	0.9	
2	0.7	0.1
2	0.9	
3	0.5	
3	0.7	0.4
3	0.6	

These sample numbers show category three with the highest risk index (0.4) and category two with the lowest (0.1). This simple measure is straightforward and provides a sense of priorities regarding risk such that a category of TPMs considered critical with a high risk index requires additional management attention. It is further noted that if all ratios $R(i, j)$ are greater than unity for a given criticality category, the risk index will drop below zero. Negative values become indicators of possible areas for trade-off analysis, as suggested in Chapter 12. Other more complex measures are possible if it is desired to spread the risk index over a greater numeric range. Whatever the situation, simplicity is urged as is a direct tie in to the TPM program, given that such a program has been put in place.

Another quantitative approach to risk assessment includes both the possibilities of failing to meet specification as well as the consequences of such failures [2]. In broad terms, the top-level construction may be written as

$$RF = 1 - (1 - FI)(1 - CI) \tag{13.3}$$

where RF is defined as a risk factor, and FI and CI are failure and consequence indices, respectively. The previously described risk index may be considered to be a surrogate for the FI, and the consequences of the possible failures become associated with the more significant failures.

Both FI and CI take on values between 0 and 1. If both values are low, then from Equation 13.3 the risk factor would tend toward zero. Conversely, for values of FI and CI near unity, the risk factor approaches one. An illustrative implementation might call for FI taking on the following discrete magnitudes with their associated interpretations:

- 0.1—not conducive to failure
- 0.3—moderate failure likelihood when stressed
- 0.5—moderate failure likelihood even when not heavily stressed
- 0.7—high failure likelihood when stressed
- 0.9—high failure likelihood even when not heavily stressed

The system in question is broken down into its functional elements and into their subsystems and judgments are made regarding propensity toward failure. A suggested specific approach is to focus on the hardware, software, and personnel-related configuration items and rate them with respect to failure likelihood. Thus, for a particular configuration item, the three ratings might be averaged to calculate FI as

$$\frac{Ph + Ps + Pp}{3} = \frac{0.3 + 0.7 + 0.5}{3} = 0.5$$

where Ph, Ps, and Pp refer to failure indices relative to hardware, software, and personnel. This value then becomes the failure index for that configuration item and may be compared with estimates for other configuration items. These values can then be aggregated with other configuration items, if desired, into higher level functional elements.

A similar approach is suggested [2] for the consequence index, CI. Staying with a particular configuration item, the consequences of a failure, should it occur, might be related to technical performance, cost, and schedule. An analogous approach to that described in the previous discussion for the failure index leads to a consequence index defined as

$$CI = \frac{Ct + Cc + Cs}{3} \tag{13.4}$$

where Ct, Cc and Cs are indices related to technical performance, cost, and schedule. Using a measure similar to the failure index, a consequence index might be calculated numerically as

$$CI = \frac{0.1 + 0.3 + 0.1}{3} = 0.167$$

leading to an overall risk factor, from Equation 13.3

$$RF = 1 - (1 - 0.5)(1 - 0.167) = 0.58$$

The overall indices can be set up such that the final risk factors are interpreted as follows:

- $0.0–0.3$ = low risk
- $0.3–0.7$ = medium risk
- $0.7–1.0$ = high risk

Based upon this scaling, the example with the result of 0.58 would be classified as a medium-risk situation.

The aforementioned risk analysis techniques are structured procedures and are based largely upon a series of judgmental ratings. Other risk-analysis methods, using fault and event trees, are considered under the topic of societal risk. Some of the vagaries of judgmental or subjective estimation are discussed in Section 13.2.5.

13.1.2 Risk-Abatement Activities

The first rule of risk abatement is that it is necessary to acknowledge that risk exists in a program and that a formal attempt should be made to assess the levels and priorities of risk, using methods such as those described in the previous section. Without such acknowledgment and risk-program structure, it is unlikely that serious risk-abatement considerations will surface.

The focus for risk abatement, given a risk-analysis program, becomes the items (configuration items, subsystems, etc.) with relatively high failure tendencies and significant consequences should failure occur. Failure in this context includes both failure to meet spec as well as failure on a reliability (so-called random failure) basis. Risk-abatement methods, in general, may be different for technical-performance risk, schedule risk, and cost risk. Also, the existence of one of these types of risk will likely contribute to the others. Risk-reduction activities for those items considered to be high in risk include the following suggestions:

TECHNICAL PERFORMANCE RISK

- engagement of best design tiger team
- modeling and simulation of key technical parameters
- use of mature computer-aided systems engineering tools
- parallel developments for high risk items
- incentives for success

- use of outside specialists for brainstorming and review
- extensive developmental test and evaluation

SCHEDULE RISK

- formal master-schedule development and adherence
- focus on critical and near-critical paths
- best worker assignments on time-critical tasks
- overtime incentives
- maximum shift to parallel versus serial activities
- early project organization and careful staffing plans

COST RISK

- identification of and focus on key cost drivers
- special low-cost design alternative reviews and assessments
- proof of performance through modeling and simulation
- maximum use of commercial-off-the-shelf (COTS) equipments
- early breadboarding, prototyping, and testing
- design to cost approaches

Another area that has been developing whose purpose is related to risk reduction is that of preplanned-product improvement (P^3I). A formal program in this regard addresses the early definition and description of methods of improving a particular product (system). Instead of a broad "we can put that capability in later" response to design alternative issues, a P^3I program would capture all the potential areas for further product improvement and formalize them by means of documented plans that include schedules, costs, and performance assessments. The attraction of such an approach is that it separates the current program from a future program and tends to force the former into meeting immediate performance, schedule, and cost specifications and constraints. In addition, if future P^3I enhancement possibilities are defined early, current design alternatives can be selected so as to facilitate these future upgrades. A properly conceived P^3I program can be an important tool for the system's program manager who experiences the relatively normal but very real pressures of cost and schedule growth, together with increases in performance uncertainty.

13.1.3 Societal Risk

Societal risk, alluded to earlier in this chapter, is a subject of continuing investigation, especially by federal government agencies. These agencies are charged, at least in part, with the responsibility of assuring that risks taken by the citizenry at large are not too great. This becomes a very difficult set of issues that involves both assessments of the nature of such risks as well as the judgment as to when a set of risks is considered to be too high. Unfortunately, it often takes a serious accident to precipitate action. Notwithstanding these types of responses to catastrophes, ongoing efforts attempt to deal with risk issues on almost a day-to-day basis.

Two issues stand out with respect to matters of public risk, particularly as they also relate to the rights of citizens to take independent action. One has to do with cigarette smoking and cancer. This issue has been studied for a long time and after much analysis and industry lobbying, a standoff of sorts has been reached. Cigarette makers are now required to let potential users know about the potential hazards of cigarette smoking on the packages and smoking is restricted in various public areas. Another issue is that of airbags for automobiles. This subject has also been publicly discussed and studied for a while in terms of public and private interest. It is known through testing that air bags can save lives in crashes of automobiles. However, they are viewed by many as an intrusion into their personal freedom and an unnecessary cost. Many of the issues of public risk involve balancing the same kinds of matters

- the proven decrease in hazard that can be achieved
- the increases in costs to be incurred by the potential users
- the possible intrusion into personal and individual freedom of action and choice

The systems engineer working on a program with possible societal hazard increases or reductions should at least be aware of the need to assess the potential hazard changes and some of the aspects of costs, especially those related to the cost of producing the system itself. Ideally, the systems engineer should help to assure that such assessments are carried out as carefully and completely as possible.

One study sponsored by the National Research Council, in response to a legislative mandate, examined risk assessment in the federal government [3] in an attempt to strengthen the reliability and objectivity of scientific assessment that underlies policies applicable to carcinogens and other public health hazards. One major conclusion was that the data that support risk analyses are often inadequate. This should be noted as well by the systems engineer who may be in a position, at early stages of product development, to call for the development of hazard-related data, if possible. Some of the recommendations for improvement in public-risk assessment can be summarized by the following points:

- a comprehensive risk assessment should be carried out and made accessible
- the comprehensive risk assessment should be reviewed by an independent science advisory panel
- uniform inference guidelines should be used by agencies in their risk-assessment processes

These and other suggestions reinforce the importance of societal risk analysis and represent a challenge to the systems engineer to contribute whenever possible to the improvement of these types of analyses.

A relatively early and significant contribution to the matter of dealing more effectively with societal risk was set forth in the context of the design and evaluation of nuclear systems' safeguards [4]. The essence of safeguards relates to the possible willful misuse of nuclear materials or facilities on the part of an adversary, where such willful misuse constitutes an event. Risk in this context is defined in terms of the following measurement units:

$$\frac{\text{consequences}}{\text{unit time}} = \frac{\text{attempts}}{\text{unit time}} \times \frac{\text{events}}{\text{attempt}} \times \frac{\text{consequences}}{\text{event}}$$

and the corresponding relationship

$$R = F P C \tag{13.5}$$

where R is the risk, F is the frequency of attempts to produce an event, P is the conditional probability that the attempts will be successful, and C are the consequences of the events that are produced. This definition corresponds to the functions of a safeguards system, which have been articulated as

- reduction of the probability that misuse will be attempted
- reduction of the probability that any willful misuse attempt will be successful (i.e., result in an event such as a nuclear explosion)
- reduction of the consequences of any successfully perpetrated event

If the possible events are subscripted with the index i and the attempts defined as adversary-action sequences, the index j, the following formula sums up the risk elements over these events and attempt sequences as (see also Chapter 5, Equation 5.9)

$$R = \sum_{i,j} F(i,j)P(i,j)C(i) \tag{13.6}$$

Thus the quantification of risk in this application depends upon being able to estimate the individual terms in this equation. This is not a trivial task, but it has been addressed in considerable detail in the literature [4]. The following section examines some of the aspects of a risk-analysis technique.

13.1.4 Fault-Tree and Event-Tree Methods

The fault-tree methodology is a diagramming technique that has been used effectively in carrying out risk analyses [4, 5, 6]. Such analyses have largely been at a detailed level, dealing with specific events within and between subsystems. The technique utilizes a series of gate symbols and events. The gate symbols embody the logic inherent in sequences of states of the system; the events describe the states themselves. An illustrative set of gate symbols is shown in Figure 13.1. We note some of the more familiar logic elements such as the AND gate, the OR gate, and the EXCLUSIVE OR gate. The events can also be coded to distinguish between different types of events.

Referring to Figure 13.2, we see a simple example of a portion of a fault tree. In this example, a particular failure is assumed and the fault tree delineates the possible combinations of sequences that could have caused the particular failure. We as-

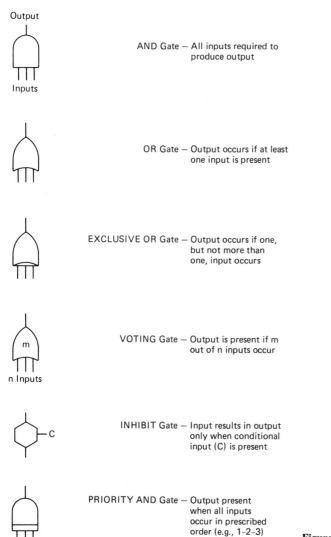

AND Gate — All inputs required to produce output

OR Gate — Output occurs if at least one input is present

EXCLUSIVE OR Gate — Output occurs if one, but not more than one, input occurs

VOTING Gate — Output is present if m out of n inputs occur

INHIBIT Gate — Input results in output only when conditional input (C) is present

PRIORITY AND Gate — Output present when all inputs occur in prescribed order (e.g., 1–2–3)

Figure 13.1 Selected gate symbols for fault-tree analysis.

sume intrusion into an alarmed area and that the failure is defined as the alarm system not ringing. This could have been caused by a failure in the alarm box itself or loss of power to the alarm system. The latter, in turn, could have been caused by either (a) loss of main power, (b) failure of the circuit breaker, or (c) a break in the wiring for the system. The probabilities of each of these causes can be added to the tree and aggregated according to the conventional rules of event probabilities (see Chapter 5). Thus the fault tree emphasizes a given failure and works backward from the assumed failure to define all the possible causes of the failure. Adding probability estimates overlays a quantitative dimension allowing for the assessment of the likelihood of failure as a whole and of the chains of sequences leading to various

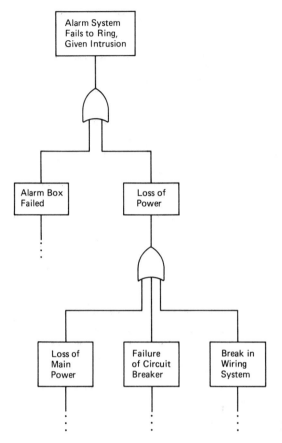

Figure 13.2 Portion of a fault tree.

events. For example, the AND, OR, and EXCLUSIVE OR probabilities, for two inputs (A and B), are developed in the usual way (see Chapter 5):

- AND Relationship

$$
\begin{aligned}
P(\text{output event}) &= P(A \text{ and } B) \\
&= P(A \mid B)P(B) \\
&= P(B \mid A)P(A) \\
&= P(A)P(B) \text{ if } A \text{ and } B \text{ are independent}
\end{aligned}
$$

- OR Relationship

$$
\begin{aligned}
P(\text{output event}) &= P(A \text{ or } B) \\
&= P(A) + P(B) - P(AB) \\
&= P(A) + P(B) \text{ if } AB = \varnothing \text{ (null set)}
\end{aligned}
$$

- EXCLUSIVE OR Relationship

$$
\begin{aligned}
P(\text{output event}) &= P(A \text{ or } B, \text{ not both}) \\
&= P(A) + P(B) - 2P(AB)
\end{aligned}
$$

All gate symbols should be representable, in one way or another, by some type of probability relationship, including rather complex compound statements.

The event-tree methodology, which is basically the same as a decision tree that is addressed in Section 13.2.1, builds upon an initiating event and the consequences, rather than the causes, of that event. A classical example of an event tree is contained in the well-known and documented Reactor Safety Study of the Nuclear Regulatory Commission [7]. Figure 13.3 shows an example of a general event tree with three events, A, B, and C. Event A is the initiating event, which can occur with probability $P(A)$. The next failure event, B, can occur with probability $P(B)$. Successive failure events are shown to the left and success events to the right. Lack of independence between brances can be incorporated in the tree, as appropriate. Numerical applications of this method can be lengthy but are basically straightforward.

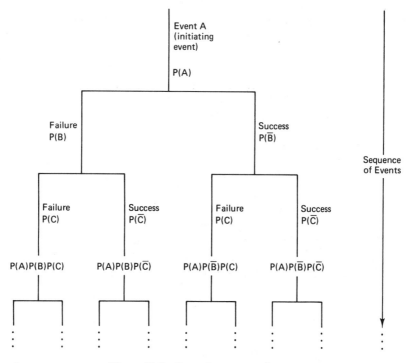

Figure 13.3 General event-tree diagram.

As a footnote to the matter of fault-tree construction and analysis, it has been observed that fault-tree development is not unique, is complicated and thus difficult to check for accuracy, and is quite time consuming [5]. An attempt has been made to develop computer-aided techniques to help to systematize the process and solve some of these related problems. A *decision table* approach has been used to this end, requiring the identification of all input and output events for all system components with a table that defines their connectivities. As with other diagramming techniques (see Chapter 4), it is certainly possible to represent the information in a diagram by a decision table, a truth table, a matrix of inputs and outputs, or some other multidimensional array. As pointed out in Chapter 4, this is often the key to developing a true computer-aided representation of a process. In regard to fault trees, however, some early computer-aided steps have been taken. It can be expected that as other investigators work on this problem, a computer-aided fault-tree software package will become available, leading eventually to a usable microbased commercial product. Indeed, it appears that such a package can be developed through specialized overlays upon existing integrated-package software.

The fault-tree and event-tree methods discussed in this section are tools that have been used by the analyst to assess the risks associated with large-scale systems, which could have disastrous consequences if they fail in certain ways. They are considered microanalytic tools in distinction to the procedures that develop risk indices at a larger macrosystem level. The latter are generally applicable at the top-level performance, cost and schedule levels; the former tend to be very detailed tools that examine specific logic sequences built in to the system's hardware, software, and operating procedures. Although there are numerous other variations on these themes (e.g., relative accident probability (RAP) methods), the methods discussed in this section tend to span the spectrum of analytic tools that are used to perform risk analyses. By analyzing the problem at hand, the reader can examine these tools and decide which ones are most appropriate to the given application area.

Whatever the method selected, it is noted that risk analysis has at least one inherent difficulty. For the top-level formulation, the procedure requires subjective estimation as to the risk associated with the various elements of the system. For the detailed-level fault-tree and event-tree formulations, very low probability events are to be estimated, at least in the quantitative versions of these methods. Data bases are generally not readily available from which to make these judgments. Essentially, for both kinds of subjective estimation procedures, although they are in different arenas, so-called expert judgments are called for. These are limitations, but in the final analysis a considerable amount of good systems engineering is based upon expert judgment. These judgments, however, can be embedded in a systems approach and some degree of formalism and can be excellent if properly applied. In addition, recognizing these issues, considerable attention has been paid to the matter of obtaining expert judgment in a clear and consistent manner. Some of these subjective estimation techniques are discussed later in this chapter.

13.1.5 Other Risk-Related Areas

We note that there are many other risk-related items of interest to the systems engineer but which will not be explored further in this book. Some of these items deal with the matters of

- differences in risk perception
- risk aversion profiling
- checklists for hazard and risk analysis
- group interactions for risk assessment
- failure modes, effects, and criticality analysis (FMECA)

- relative accident probability analysis
- preference analysis
- utility assessments
- regret modeling
- importance analysis

A noteworthy overview of methods of assessing risk and hazard, touching upon those just listed and other risk-related areas, is provided in the literature [8], coauthored by A. P. Sage, a significant contributor to the field of systems engineering in general. This overview highlights

- the nature of risk
- risk-assessment methodologies
- the formulation of risk issues
- risk-impact analysis and interpretations
- needs for a theory of hazard and risk assessment
- biases that might affect risk and hazard assessment
- a descriptive model of risky choicemaking
- fundamental limits of risk and hazard analysis

A total of 250 references are discussed and cited in Sage's survey and status report. The reader with a further interest in risk and hazard analyses is urged to consider this overview article and then move on to more specialized studies in areas of particular interest. Some well-known investigators and their work are cited as references to this chapter [9, 10, 11].

13.2 DECISION ANALYSIS

Decisions, in the context of large-scale systems engineering, are presumed to be made with the following two ingredients present:

1. a defined set of alternatives
2. a set of attributes related to these alternatives

The presence of alternatives, although apparently obvious, is sometimes not explicit. One of the precepts of systems engineering is to assure that alternatives are made explicit in a balanced manner throughout the systems engineering process. Often it is the task of the lead systems engineer to force this issue so that the alternatives as well as their attributes are clearly articulated. These attributes may fall into the categories of cost, schedule, technical performance, and related uncertainties. It is an implicit task in many but not all aspects of formal decision analysis to quantify the uncertainty. Thus the subject of decision analysis is often viewed as a matter of selection of alternatives in the presence of uncertainty.

Unfortunately, many of the constructions of decision analysis are not posed in such a way that the systems engineer can relate them to real-world issues. Some of these issues can be posed as questions, such as [12]

- Which system requirements and specifications are being satisfied by the system design and which are not?
- What is the minimum cost-design approach to meeting all of the system requirements/specifications?
- Do we have alternative system/subsystem designs, all capable of meeting specifications?
- How do we trade between higher performance and lower life-cycle costs?
- What are the developmental risks and how can they be reduced?

It is seen that we have bridges yet to be built in this arena—constructs that will bring the formalisms of decision analysis closer to the problems that are faced by the systems engineer. We do not expect to develop such constructions here but attempt in the main to point out how some aspects of decision analysis are formulated and how they may be applied to systems engineering. A subsequent section deals with computer-aided techniques that are further attempts at providing formal decision analysis tools.

13.2.1 Decision Networks and Trees

This discussion is based upon a construction by this author [13] that is sufficiently general to subsume both decision networks and trees which, in any case, are very similar. We demonstrate these notions by means of an illustrative example, as shown in Figure 13.4. The figure shows a simple decision network consisting of

- a starting node
- decision nodes
- end nodes
- activities that connect the various nodes

In general, interim event nodes may be part of the network but are omitted in this example.

The decision network can be viewed as an R&D process in which uncertainty exists and is explicitly identified. It can also be seen as an articulation of design alternatives in a decision-tree format [14]. In the former context, after initiation we come to node 1 which defines two subsequent alternatives. Estimates of the probabilities of ultimately selecting each of these alternatives are provided which in this case are 0.3 for *no* and 0.7 for *yes*. This probability estimation procedure is required at all decision nodes. One path leading from decision node 1 leads to decision node 5 and the other leads to decision nodes 2 and 3. The latter implies that both paths leading to nodes 2 and 3 will be taken if the 0.7 (*yes*) alternative is chosen. The merging of activities shown as X and Y indicates an alternative such that both X and

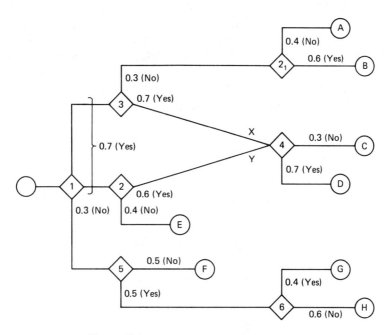

Figure 13.4 An illustrative decision network.

Y are required in order to proceed to decision node 4. This notion was originally identified as a conjunctive path-merge point. Another special network construction is the conjunctive path dependency, shown by decision node 2_1, which is a repetition of node 2. This represents the idea that activities following the *no* alternative from decision node 3 depend upon the outcomes of decision node 2. Hence node 2 is repeated as node 2_1.

When all the paths are followed, it is seen that there are eight possible outcomes for this illustrative network, with the related probabilities listed next [13]

Outcomes	Probability Components	Numerical Probabilities
A and E	$P(A)P(E\|A)$	0.084
B and Y	$P(B)P(Y\|B)$	0.126
C	$P(C)P(X)P(Y)$	0.0882
D	$P(D)P(X)P(Y)$	0.2058
X and E	$P(X)P(E\|X)$	0.196
G	$P(G)$	0.06
H	$P(H)$	0.09
F	$P(F)$	0.15

Thus we have a complete set of alternative outcomes and estimated probabilities of occurrence for the outcomes. This network formulation is immediately applicable to the definition and evaluation of alternatives in a large-scale systems engineering program. These alternatives might represent fundamental design choices, such as frequency-division versus time-division multiplexing approaches in a communications system, or success at achieving various performance levels, such as response time, for a multiuser centralized computer complex.

If we further associate a set of gains of losses with each of the alternative outcomes, the decision-network structure shown also incorporates a classical representation of a decision tree. An example might be

Outcomes	Gain Or (Loss)
A and E	(250)
B and Y	(500)
C	(750)
D	(1000)
X and E	250
G	500
H	750
F	1000

There are several ways of looking at this illustrative data. At one level, the decision network or tree can be compared with another such network or tree on the basis of expected gains or losses. The latter are calculated in the conventional expected value sense, from chapter five, as

$$\text{Expected value} = \sum (\text{outcome gain or loss}) \, \text{Prob (outcome)}$$

Staying within the confines of one network or tree, the results can be ordered by the product on the right hand side of the above equation, namely, the expected-value element. For the illustrative example, this leads to the following ordered list of possible outcomes:

Outcomes	Expected Value Element
F	150
H	67.5
X and E	49
G	30
A and E	(21)
B and Y	(63)
C	(66.15)
D	(205.8)
	(59.45) = Overall Expected Value

A further sensitivity analysis can be performed for the decision maker who wishes to explore variation possibilities in the probability estimates and the expected-value calculations.

It is further noted that other schedule, cost, and performance information can be overlaid upon the basic decision-network structure. The gain and loss data shown in the previous table implicitly represents such an overlay. With respect to scheduling, we observe in Figure 13.5 a set of expected time estimates for each of the activities and the resultant decision network on a time scale. This network structure is thus seen as rather broad in its ability to handle a variety of program-oriented decision representations.

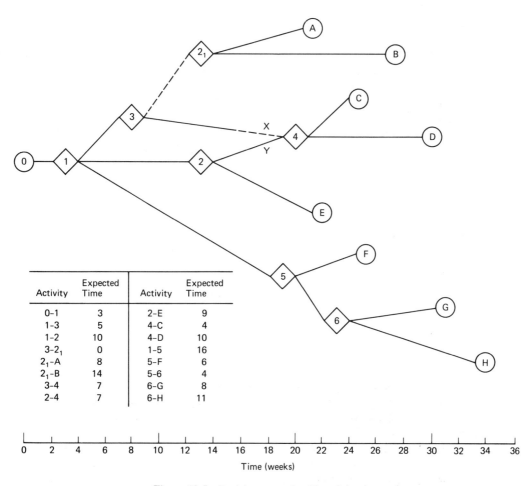

Activity	Expected Time	Activity	Expected Time
0-1	3	2-E	9
1-3	5	4-C	4
1-2	10	4-D	10
3-2_1	0	1-5	16
2_1-A	8	5-F	6
2_1-B	14	5-6	4
3-4	7	6-G	8
2-4	7	6-H	11

Figure 13.5 Decision network with activity time estimates.

There are numerous variations and extensions of the aforementioned fundamentals that can be found in the decision-theory literature. From a systems engineering point of view, however, progress toward better decisions can be made by using even these simple elements that involve

- the formal definition of alternatives
- probability estimation
- gain or loss expectations
- overall alternative assessment

13.2.2 Q-GERT

Building upon the decision-network approach described in Section 13.2.1, a technique known as Q-GERT has been developed. We touch upon its characteristics since it represents both a significant extension as well as a computer-aided state-of-the-art simulation for a wide class of decision-related and other types of problems. A. A. B. Pritsker and C. E. Sigal [15] as well as G. E. Whitehouse [16] appear to be leading proponents of this technique. Q-GERT is a process-oriented network language consisting of branches and nodes. The branches represent activities that expend time or can inject a time delay. Transactions flow over the branches and can be designated as information flows, entities, or any other item that is to be processed. There are seven basic types of nodes that represent milestones, events, routing decisions, resource allocations, and storage points. Four of the seven nodes relate to the transaction flows and are designated as

1. basic
2. selector (S-node)
3. queuing (Q-node)
4. matching (Match node)

Three node types involve resources and are defined as

1. allocate
2. free
3. alter nodes

As a simulation, Q-GERT generates basic transactions and processes them through the network. Routing can be deterministic, probabilistic, and conditional. After all or a representative set of transactions have been processed, outputs are generated automatically. Other features include the ability to add user-developed Fortran subroutines and access to a variety of built-in functions. A special graphics package known as Q-GRAF has also been developed [15] so that the user can build the network model, show operations on the network, and display the results on a monitor and obtain hard copy graphics. Areas of application include

- throughput analysis for a computer or communications system, or network
- reliability evaluations

- queuing for various transaction processors
- logistics support
- risk assessment
- project planning
- decision network and tree representations

Q-GERT is a full simulation language and system in the context of those discussed in Chapter 12. It is cited here since it is particularly useful in relation to risk-analysis and decision-analysis applications.

13.2.3 A Decision Theoretic Approach to Signal Detection

Other elements of decision theory can be seen by examining a particular signal-detection example. In Chapter 8, we characterized a communications channel by means of a conditional matrix of probabilities, $p(y_j|x_i)$, the probability that y_j was received, given that x_i was transmitted. The full matrix is depicted in Figure 13.6 (note Figure 8.2 in Chapter 8). Also shown in the figure is a decision and cost matrix. If we assume that y_j is received, then D_i is the decision that x_i was sent, and the C_{ij} are generalized costs associated with the two decisions, given the two transmitted signals. There are two possible decisions in this example, and each has a cost associated with it, framed in probabilistic terms. These decisions and related costs are sometimes also defined in terms of risks, such that the risks of choosing decisions D_1 and D_2 may be expressed as

$$\text{Risk}(D_1) = C_{11} p(x_1|y_j) + C_{21} p(x_2|y_j) \tag{13.7}$$

$$\text{Risk}(D_2) = C_{12} p(x_1|y_j) + C_{22} p(x_2|y_j) \tag{13.8}$$

An intuitively appealing decision rule, given the above, is simply to decide D_1 if the

Received Signals (y_j)

$$
\begin{array}{c}
& \begin{array}{cc} y_1 & \quad\quad y_2 \end{array} \\
\begin{array}{l}\text{Transmitted} \\ \text{Signals } (x_i)\end{array}
\begin{array}{c} x_1 \\ x_2 \end{array}
& \left[\begin{array}{cc} p(y_1|x_1) & p(y_2|x_1) \\ p(y_1|x_2) & p(y_2|x_2) \end{array}\right]
\end{array}
$$

(a)

Decisions

$$
\begin{array}{c}
& \begin{array}{cc} D_1 & \quad\quad D_2 \end{array} \\
\begin{array}{c} x_1 \\ x_2 \end{array}
& \left[\begin{array}{cc} C_{11} & C_{12} \\ C_{21} & C_{22} \end{array}\right]
\end{array}
$$

(b)

$C_{ij} = \text{Cost}$

Figure 13.6 (a) Channel matrix of conditional probabilities, and (b) decision/cost matrix for decision theoretic-detection problem.

$\mathrm{Risk}(D_1) < \mathrm{Risk}(D_2)$, and vice versa. Thus we have a situation in which costs, risks, and decision criteria are all combined in a single formalism.

It can be shown (see problems at end of chapter) that the aforementioned decision criterion can be translated into the following terms:

Decide on decision D_1 if

$$\text{Likelihood ratio} = \frac{p(y_j|x_1)}{p(y_j|x_2)} > \frac{p(x_2)}{p(x_1)} \frac{C_{21} - C_{22}}{C_{12} - C_{11}} \qquad (13.9)$$

This is sometimes called a Bayesian approach to decision analysis [14]. It uses Bayes' Theorem and calls upon the decision maker to make a minimum-risk decision. It also uses the likelihood ratio, a commonly used criterion to discriminate between two possible hypotheses. This ratio has a central role in the field of statistics known as *hypothesis testing*.

The example may be continued by interpreting the original two transmissions as

$$x_1 = \text{signal plus noise, and}$$
$$x_2 = \text{noise alone}$$

and simplifying the cost parameters as

$$C_{11} = C_{22} = 0$$
$$C_{12} = C_m = \text{cost of a miss, and}$$
$$C_{21} = C_f = \text{cost of a false alarm}$$

The decision rule can then be rewritten as

Decide on decision D_1 if

$$L = \frac{p(y_j|x_1)}{p(y_j|x_2)} > \frac{p(x_2)}{p(x_1)} \frac{C_f}{C_m} = \alpha \qquad (13.10)$$

otherwise, decide on D_2.

This decision criterion may be further illustrated by selecting particular distributions for the likelihood ratio. If the numerator (signal present) is normally distributed with mean m and standard deviation σ, and the denominator (noise only) is normally distributed with mean zero and the same standard deviation σ, then the likelihood ratio can be found to be

$$L = \exp\left(\frac{2y - m}{2\sigma^2}\right) \qquad (13.11)$$

Further, using the threshold, α, from Equation 13.10, the decision boundary can be set as

$$\frac{2y - m}{2\sigma^2} = \ln \alpha \qquad (13.12)$$

Solving for the values of y at the decision boundary, we have

$$y = \frac{m}{2} + \sigma^2 \ln \alpha \qquad (13.13)$$

A numerical example will help to clarify the meaning of this result. We will assume the following values:

$$p(x_1) = 0.4$$
$$C_m = 2$$
$$C_f = 4$$
$$m = 10 \text{ volts}$$
$$\sigma = 2$$

We then calculate α as

$$\alpha = \frac{p(x_2)}{p(x_1)} \frac{C_f}{C_m} = \frac{0.6}{0.4} \frac{4}{2} = 3$$

Thus the value of y at the threshold boundary is

$$y = \frac{10}{2} + (4) \ln 3 = 9.39 \text{ volts}$$

This means that the decision should be made that a signal was sent if the received signal exceeds the value 9.39. Otherwise, it should be decided that no signal was sent.

13.2.4 Air-Transportation-System Concept Evaluation

As a final example of a decision framework, we carry forward previous examples, cited in Chapters 10 and 12, of air-transportation-system concept evaluations. Building upon the notions described in Chapter 10, Table 10.3, and the evaluation criteria listed in Chapter 12, Figure 12.9, we have the results of a real-world rating and weighting exercise, as shown in Figure 13.7. The weights were developed by averaging weight estimates from a total of nine participating evaluators, as listed in Figure 13.8. The scoring system for the ratings was based upon a maximum score of 100. Both weighted and unweighted results are shown in order to get some idea as to the sensitivity of results to this process.

It is observed that concept E is marginally preferred on the basis of the listed evaluation criteria whereas the choice shifts to concept A when the various implementation factors are considered (concept A represents a minor extension of the current air-transportation system). If both the criteria and factors are put together, concept D is marginally preferred, with E and A not far behind.

The latter example represents a real-world implementation of the weighting and rating decision framework. Although this is a very simple notion, it has yet to find its way into the mainstream of systems engineering decision analysis and decisionmaking. It is, of course, ultimately tied to subjective judgments and estimating.

Evaluation Factors		System Concept					
I. Evaluation Criteria	Weight	A	B	C	D	E	F
Social	11	65	60	54	62	74	68
Environmental	16	30	24	44	72	67	61
Service Quality	16	41	61	52	62	72	72
System Capacity	14	30	59	54	67	73	73
Human Factors	6	68	73	57	53	56	57
International/Economic	9	55	63	67	65	70	66
Investment Costs	13	56	42	50	65	54	45
Operating Costs	14	45	42	46	65	58	58
Sum I – Unweighted (%)	99*	49	53	53	64	66	63
Sum I – Weighted (%)		45	50	51	64	65	62
II. Implementation Factors							
Financing	21	69	39	53	49	42	28
Management/Administration	13	67	39	53	42	42	39
Community Acceptance	25	31	19	47	81	67	75
Legislation	16	86	67	67	39	36	25
Regulation	14	78	61	64	47	44	39
Technological Risk	11	72	44	50	58	61	53
Sum II – Unweighted (%)	100	67	45	56	53	49	43
Sum II – Weighted (%)		64	42	55	55	50	45
Overall Average[†] – Unweighted (%)		58	49	55	59	58	53
Overall Average – Weighted (%)		55	46	53	60	58	54

*Column does not add to 100% due to rounding errors.

[†] Average of sum I and sum II.

Figure 13.7   Evaluation of air-transportation-system concepts.

Evaluation Criteria Category	Evaluator									Average*
	1	2	3	4	5	6	7	8	9	
Social	5	10	15	10	15	5	10	21	8	11.0
Environmental	20	40	10	15	20	5	15	8	12	16.1
Service Quality	20	10	10	15	20	15	15	19	18	15.8
System Capacity	10	10	10	15	20	20	15	15	13	14.2
Human Factors	5	10	10	5	5	10	5	1	6	6.3
International/Economic	5	10	5	10	5	15	10	10	13	9.2
Investment Cost	15	5	20	15	5	15	15	12	14	12.9
Operating Cost	20	5	20	15	10	15	15	14	16	14.4

*Column does not add to 100% due to rounding errors.

Figure 13.8   Weights for evaluation criteria.

For this reason, we address such matters in the next section. In addition, the thrust of several decision-related software packages is in this direction. As these packages are both used and improved, it may be expected that these evaluation structures will find their way more and more into systems engineering developments and programs.

13.2.5 Subjective Estimation and Decision-Making Research

The matter of subjective estimation has come up in several areas in this text, including at least the following:

- probability estimation in decision networks and fault trees
- risk-factor analysis
- system assessment versus a set of evaluation criteria and factors

There are no easy roads or definitive solutions to these subjective estimation problems. However, various approaches have been taken in an attempt to make progress in this area. For example, the conventional view of probabilities is in terms of the likelihood of an event occurring, given the possibility of occurrence of a large number of events. For repeated trials, this *frequency of occurrence* approach yields an estimate of the desired probability. In a systems engineering context, however, it is often difficult to make the mental leap to repeated trials in that it is not really possible to truly replicate, for example, a technology development. The issue at hand is most often related to similar experiences and relative likelihoods or odds as the framework for subjective probability estimation. These steps involve

- the definition of alternatives
- estimation of relative likelihoods or odds
- normalization to unity

Emphasizing the notion of relative likelihood, if a success event is twice as likely as a failure event, then the subjective estimates are $P(\text{success}) = 2/3$ and $P(\text{failure}) = 1/3$. With this type of simple framework further attempts at refinement have used multiple estimators as well as consensus-seeking techniques. In the former case, independent estimates from different people are used; in the latter, people gather together to try to reach a consensus after first examining the issue individually.

An example of the above consensus-seeking approach is the so-called Delphi technique [17]. This technique applies to a wide class of subjective-estimation problems, with probability estimation and combining expert judgments [18] representing only one special set of cases. The Delphi technique was developed initially at the Rand Corporation in the early 1960s and deals with techniques for obtaining a consensus of opinion from a group of experts. Judgments are rendered by these experts through a series of questionnaires and interviews. The formal questionnaires are normally augmented by feedback to the experts regarding interim results. Through such a process, an attempt is made to converge to a consensus view in the most reliable way. Numerous experiments have been run with both matters of fact and matters of subjective opinion to try to improve and refine this technique. Notwithstanding the popularity of such an approach it has always been recognized that even a panel of experts can reach consensus incorrectly on matters of objective fact. However, judgments can be acknowledged as issues of opinion and convergence to a consensus can be extremely valuable if used properly. Many systems engineering decisions come down to weighing a large number of incommensurable combinations of both fact and judgment. Thus, the Delphi technique, and others like it, have their place and can contribute in an important way to the resolution of systems engineering questions.

An interesting application of the Delphi technique to the field of personal computers is reported in the literature [19]. The forecasts that were made were compared to actual experience and indicators. This comparison showed that the Delphi panel consensus was almost completely accurate. Factors of interest included use of the personal computer by companies, networking, software pirating, encryption, home application areas, and others.

Returning to the general subject of subjective estimation, it has been noted that given "perfect" information regarding event probabilities, payoffs and losses, further human biases creep into the decision-making process. Leaders in this field, sometimes called psychological decision theory, have been D. Kahneman and A. Tversky [20]. Some of their investigations have centered around the following notions:

- regression to the mean
- representativeness and availability
- loss avoidance

Regression to the mean refers to a tendency to let down after a superb performance and improve after a dismal one, and related manifestations. *Representativeness and availability* both allude to the setting up of a mental model based upon prior experience, possibly neglecting current facts and apparent likelihoods. *Loss avoidance* is the tendency of people to avoid sure losses even when *expected value* losses may be greater. The latter, for example, might influence a decision maker to avoid decisions that are perceived to result in certain schedule slippage or cost overruns in favor of even riskier decisions that have some upside potential, however small, for meeting schedule and cost goals. One implication of the Kahneman and Tversky work is that even with "perfect" decision theoretic formulations, additional biases are present that defy a so-called rational approach to decision making. This could lead to different decisions by two different people examining the same decision model and results. Undoubtedly these areas will be under study for many years to come. As a practical matter the systems engineer must move ahead with the decision process while trying to use research results when feasible. However, the pace and inherent engineering complexity of large-scale programs unfortunately often precludes the use of even modest decision-theoretic frameworks.

The list of other important contributors to decision theory is extremely long and, regrettably, only three other avenues of investigation will be noted here. The first is the work of Keeney and Raiffa [21], and others [e.g., 22], dealing with multiple-objective and criteria decision analysis. This is related to the issue of commensurability of objectives and criteria. Incommensurability leads to difficulty in aggregating results across a set of objectives and criteria so as to produce a clear preference for one alternative over another. Several examples in this book have addressed such a problem by giving weights to the objectives and criteria (e.g., Figure 13.7).

The reader interested in more formal and rigorous relationships is referred to the references cited at the end of this chapter.

A second area of inquiry is the analytic-hierarchy process, formulated by T. Saaty [23]. This is a particularly interesting hierarchical-decomposition approach with systematic-subjective judgments applied to the various levels of the hierarchy. It is also noteworthy that this approach is specifically embodied in a decision-related software package called Expert Choice, cited in Appendix A under the category of alternatives and preference evaluators.

In regard to future research directions, a group of leading researchers in the field of decision analysis participated in a workshop in April 1984 and attempted to define further research needs and the phenomena of decision making and related operations. Presentations were prepared and documented [24] in the following categories:

- descriptive science and psychological decision theory
- analyzing decisions using value analysis
- decision analysis and fuzzy mathematics
- acquiring and using information in decisionmaking
- organization theory and design
- organizational form for command and control
- modeling organizational planning
- research needs in marketing
- decision support systems

Scanning the list, it is observed that some emphasis was placed on organizational issues, the inference being that it is important to properly organize if decisions are to be made appropriately in a large-scale systems context. The last area cited above, decision support systems, is receiving a good deal of emphasis as well, and various software packages are being given that generic name. This latter issue is explored in greater detail in Section 13.3 which follows. The reader with an interest in further aspects of decision theory and analysis can consult the above reference as well as the vast literature available on this significant subject. As an example, A. P. Sage has examined the design of information systems and processes for planning and decision support [25]. This broad review of decision rules and processes cites 415 references, an indicator of the degree of interest in the field of decision analysis.

13.3 COMPUTER-BASED TOOLS

In discussion in Section 13.2, the matter of decision support systems (DSSs) was raised. This is a relatively new phrase for an old concept. Decision makers have been using decision support systems, perhaps by a different name, for a long time. Indeed, the management-information system (MIS) appears to be the phrase that has been replaced by the new decision support system, and some investigators have attempted to sort through and define these various terms. It is an area receiving considerable attention, as demonstrated by the following topics examined at the fifth and sixth conferences (DSS-85 and DSS-86) on decision support systems:

- introduction to DSS and DSS tools
- DSS futures
- DSS practices
- DSS methodologies
- DSS products and services
- executive information systems
- DSS in the public sector
- international DSS

Clearly, DSSs appear to be here to stay, at least for a while.

In a previously cited reference [24], it is suggested that DSSs are basically software packages. For purposes of this text, with a particular interest in computer packages in relation to the tasks and activities of systems engineering, this is a good point of departure. It is further noted that the elements of a DSS, defined as a software package, are

1. a database
2. a model base
3. a software interface allowing
   - entering and retrieving data in real time
   - creation and retrieval of models in real time, and
   - asking questions and obtaining answers in real time

In Chapter 6, we examined several classes of software packages that could be of assistance in various aspects of systems engineering. These packages included

- alternatives and preference evaluators
- decision support systems

Both of these packages apply to the matter of decision analysis. As a means of clarifying the definition of software packages available to support various decisionmaking and analyses, we will use the term decision support software (DSSW) here as a generic term for these packages, which will subsume the two listed categories of alternatives and preference evaluators as well as all classes of software-based decision support systems. Thus we will address DSSW in this section, exploring the computer-based tools that are available to the systems engineer to carry out decision-related systems engineering tasks and activities. Specific commercial packages of decision support software are listed in Appendix A. We briefly examine here the generic characteristics of some of these packages [26].

Several of the available DSSW packages utilize the matrix weighting and rating notions discussed in this and other chapters. In such a scheme, the alternatives are defined and rated against a set of criteria. The criteria are weighted and an overall weighting/rating score is developed for each alternative. Some of the packages focus attention, in addition to the basic weighting and rating, on such features as

- setting goals and objectives
- ranking adverse and unacceptable alternatives
- defining musts versus desirables
- adding logical rules
- adding new variables
- examining limiting high and low cases
- formal decision hierarchies and trees

To the extent that the software listed depends upon subjective estimates, it suffers from the problems previously discussed. However, formalizing the definition and analysis of alternatives is a positive step forward and, as noted, many systems engineering decisions of necessity require the use of expert judgment.

A portion of the advertised DSSW is more modeling oriented, adding features to a basic spreadsheet format to yield a more focused capability. These features include

- special functions such as linear programming and regression equations
- procedural languages
- decision-framework menus
- problem segmentation
- special output statistics
- graphics

Whatever special features are added, it should be noted that spreadsheets and integrated packages (see Chapter 3) represent a powerful modeling and decision-analysis capability as do the modeling and simulation packages previously discussed (Chapter 12). In short, the potential decision maker or analyst has software options that go considerably beyond the DSSW category, depending upon the specific application area as well as the framework for the consideration of decision alternatives.

Microcomputers are becoming increasingly important in terms of decision support systems, especially those that address top-level decision frameworks (as in Chapter 10). It is often the case that such top-level considerations do not require great processing speed, number crunching, or the storage of massive amounts of data. Instead, they need to assist in forming the decision alternatives and their key attributes, and other software is used to develop specific values for those attributes or key performance parameters. At the same time, however, the speed and storage capacity of the p-c have been increasing at a rapid rate.

Some notion of both the interest in microbased DSSW as well as the broad view currently held as to what constitutes a decision support system may be found in a compilation of papers [27] that also cite available software applicable to

- artificial intelligence languages and systems (discussed in Chapter 16)
- aids for military intelligence analysis
- command and control
- managerial efficiency
- networks

The bottom line appears to be that we have both a large number of available DSSW packages as well as a need to further distinguish between various subclasses of decision support systems.

13.4 CASE APPLICATION SUMMARY

This chapter has demonstrated several formulations for risk and decision analysis. These are both fields with extensive backgrounds and supporting literature, as well as a wide variety of application areas inside and outside the scope of systems engineering. Specialized computer aids are available but must still be considered in their early stages of development and application, particularly to the tasks and activities of systems engineering. Table 13.1 lists ten risk or decision-related application areas together with an indication of the software package classes that might be brought to bear in such areas. It may be expected that with time, both the computer aids as well as some of the formalisms of risk and decision analysis will be more widely accepted as mainstream systems engineering tools.

TABLE 13.1 RISK-ANALYSIS AND DECISION-ANALYSIS AREAS AND APPLICABLE SOFTWARE

TPM-related risk-index calculations for categories of risk
   Spreadsheets
   Integrated packages
Risk factor calculations
   Spreadsheets
   Integrated packages
Fault trees and event trees
   Diagramming packages (workbenches)
   Special graphics packages
Probability string constructions
   Spreadsheets
   Integrated packages
   Modeling-oriented DSSW
Decision-network constructions
   Diagramming packages
   Special graphics packages

TABLE 13.1 (*cont.*)

Rating and weighting decision techniques
  Alternatives and preference evaluators
  Selected DSSW
  Spreadsheets

Applications of Delphi technique
  Spreadsheets
  Integrated Packages

Broad definition and evaluation of decision alternatives
  DSSW
  Multitask applications managers

Likelihood ratio calculations
  Statistical packages
  Spreadsheets
  Integrated packages

Complex and detailed modeling for design decisions
  Q-GERT
  Other simulation packages
  Computer languages for bottoms-up construction of models/simulations

REFERENCES

1. *Engineering Management*. Mil.-Std. 499A, Department of Defense, May 1, 1974.

2. *System Engineering Management Guide*. Defense Systems Management College, Fort Belvoir, VA, 3 October 1983.

3. National Research Council, *Risk Assessment in the Federal Government: Managing the Process*. Washington, DC: National Academy Press, 1983.

4. Bennett, C. A., W. M. Murphey, and T. S. Sherr, *Societal Risk Approach to Safeguards Design and Evaluation*. ERDA-7, U. S. Energy Research and Development Administration, Washington, DC, June 1975.

5. Henley, E. J., and H. Kumamoto, *Reliability Engineering and Risk Assessment*. Englewood Cliffs, NJ: Prentice-Hall, 1981.

6. Bunn, D. W., *Applied Decision Analysis*. New York: McGraw-Hill, 1984.

7. *Reactor Safety Study*. WASH-1400, U. S. Nuclear Regulatory Commission, Washington, DC, 1975.

8. Sage, A. P., and E. B. White, "Methodologies for Risk and Hazard Assessment: A Survey and Status Report," *IEEE Transactions on Systems, Man, and Cybernetics*, SMC-10, no. 8, August 1980.

9. Arrow, K. J., *Essays in the Theory of Risk Bearing*. New York: Elsevier, 1974.

10. Fishburn, P. C., *Utility Theory for Decisionmaking*. New York: John Wiley, 1970.

11. Keeney, R. L., "Evaluating Alternatives Involving Potential Fatalities," *Operations Research*, vol. 28, January 1980.

12. Eisner, H., "CASE: Computer-Aided Systems Engineering for C^3I Systems," *SIGNAL Magazine*, Armed Forces Communications and Electronics Association (AFCEA), July 1984.

13. Eisner, H., "A Generalized Network Approach to the Planning and Scheduling of a Research Project," *Operations Research,* 10, no. 1, February 1962.

14. Holloway, C. A., *Decision Making Under Uncertainty: Models and Choices.* Englewood Cliffs, NJ: Prentice-Hall, 1979.

15. Pritsker, A. A. B., and C. E. Sigal, *Management Decision Making: A Network Simulation Approach.* Englewood Cliffs, NJ: Prentice-Hall, 1983.

16. Whitehouse, G. E., *Systems Analysis and Design Using Network Techniques.* Englewood Cliffs, NJ: Prentice-Hall, 1973.

17. Dalkey, N., and O. Helmer, "An Experimental Application of the Delphi Method to the Use of Experts," *Management Science,* 9, no. 3, April 1963.

18. *Management Science,* 32, no. 3, March 1986.

19. Gray, P., and J. M. Nilles, "Evaluating a Delphi Forecast on Personal Computers," *IEEE Transaction on Systems, Man, and Cybernetics,* SMC-13, no. 2, March/April 1983.

20. Kahneman, D., P. Slovic, and A. Tversky, eds. *Judgment Under Uncertainty: Heuristics and Biases.* Cambridge, MA: Cambridge University Press, 1982.

21. Keeney, R. L., and H. Raiffa, *Decision With Multiple Objectives: Preferences and Value Tradeoffs.* New York: John Wiley, 1976.

22. Zeleny, M., *Multiple Criteria Decision Making.* New York: McGraw-Hill, 1982.

23. Saaty, T. L., *The Analytic Hierarchy Process.* New York: McGraw-Hill, 1980.

24. Cohen, M. D., et al. "Research Needs and the Phenomena of Decisionmaking and Operations," *IEEE Transactions on Systems, Man, and Cybernetics,* SMC-15, no. 6, November/December 1985.

25. Sage, A. P., "Behavioral and Organizational Considerations in the Design of Information Systems and Processes for Planning and Decision Support," *IEEE Transactions on Systems, Man, and Cybernetics,* SMC-11, no. 9, September 1981.

26. Spezzano, C., "Decision Support Software," *Popular Computing,* 4, no. 12, October 1985.

27. Andriole, S. J., ed. *Microcomputer Decision Support Systems: Design, Implementation and Evaluation.* Wellesley, MA: QED Information Sciences, 1986.

PROBLEMS

P13.1. Develop your own illustrative example of a risk-index calculation approach based upon Equation 13.2 for five risk categories. Demonstrate how this index may become negative and the significance of such a situation.

P13.2. Develop your own illustrative example of a risk-factor approach based upon Equation 13.3.

P13.3. Write a five-page discussion of a societal risk issue, clearly delineating the pros and cons of both sides of the issue.

P13.4. Construct a fault tree related to a situation of your own choice with the tree containing all the gate symbols in Figure 13.1 and at least ten gates or events. Add probabilities to the tree and calculate the probabilities of the various events that you have constructed.

P13.5. Extend to events D and E and assign numeric probabilities to the event tree shown in Figure 13.3. Calculate all the outcome probabilities. Assure that you have a complete probability scheme.

P13.6. There are standard computer programs that will traverse binary trees. Find or construct such a program in BASIC or Pascal. Modify the program such that probabilities will be assigned at each branching point and such that final outcome probabilities will be automatically computed at the ends of the tree branches. Show the completed program with comments.

P13.7. Formulate your own decision network for a situation that is relevant to you. Develop at least 16 alternative outcomes and calculate the probabilities of occurrence for these outcomes.

P13.8. Assign numerical payoff values to each of the outcomes in P13.7. Place priorities on each of the outcomes and explain your method of prioritization.

P13.9. Obtain a good reference on Q-GERT and write a five-page discussion of the characteristics of the simulation with emphasis on how it may be applied to systems engineering decision-analysis or risk-analysis issues.

P13.10. Prove the likelihood-ratio decision criterion shown in Equation 13.9.

P13.11. Prove that the likelihood ratio for the normal-distribution example is as shown in Equation 13.11. Select another distribution of noise only and signal plus noise and calculate the likelihood ratio. Select numerical values and find the threshold boundary value in a manner similar to that shown in the text.

P13.12. Find and discuss, in a total of five pages, the loss avoidance issue articulated by Kahneman and Tversky (Ref. 20) and mentioned in the text. Show a simple but specific example of how loss avoidance might work and some implications in terms of systems engineering decisions.

P13.13. Obtain specific product information on at least three commercial decision support software (DSSW) packages. Write a five-page discussion and critique of the packages you have examined.

P13.14. Write a five-page discussion of the similarities between risk analysis and decision analysis. Show examples.

All things which we clearly and distinctly conceive are true, only observing, however, that there is some difficulty in rightly determining the objects which we distinctly conceive

Descartes

SOFTWARE DEVELOPMENT AND ANALYSIS

Software has become a crucial part of most large-scale systems. Some systems for which the lines of code are in the one million and greater range have software development, maintenance, and improvement as the single largest life-cycle cost component. As a consequence of the pervasiveness, complexity, and cost of software, several notable efforts have been initiated in order to either reduce cost or improve productivity, or both. A Software Productivity Consortium (SPC) has been formed just outside of Washington, D.C. with the principal focus of cost reduction. Major U. S. defense contractors make up this consortium and their initial priorities appear to have been reusable software and software prototyping. The results are to be equally shared by all SPC members. Another activity is the DoD sponsored Software Engineering Institute (SEI) set up at Carnegie-Mellon University, oriented to immediate software applications. A third is the Microelectronics and Computer Technology Corporation (MCC) whose emphasis appears to be computer architecture, very large-scale integrated circuits (VLSI), semiconductor technology, and software technology [1]. Within the latter, the so-called Leonardo project (low-cost exploration offered by the network approach to requirements and design optimization) was defined to focus on the design process with special attention to formal requirements analyses for software.

Thus we see a recognition of the complex process of synthesis and its many implicit alternatives as well as an approach toward a solution that addresses the front-end requirements-definition area. This direction in terms of software development is especially noteworthy and fully in consonance with several premises of this text.

The DoD's Software Technology for Adaptable, Reliable Systems (STARS) [2] was the forerunner of and, in some cases, impetus behind the three aforementioned efforts, and others, in the software area. Further, the DoD probably represents the most significant single purchaser and user of software. For this reason we will next examine the way in which the DoD looks at software development from the perspective of a DoD software-development standard.

14.1 SOFTWARE DEVELOPMENT PROCESS

A key standard set forth by the DoD for software development is DoD Std. 2167 [3], superseding an older DoD Std. 1679A. Other software standards can be expected to represent, primarily, a variation on the general theme of this DoD standard. The broad requirement areas of this document are shown next with the detailed requirements areas listed in Table 14.1.

TABLE 14.1 DETAILED REQUIREMENT AREAS FOR SOFTWARE DEVELOPMENT

TOP-LEVEL REQUIREMENTS AREAS
1. Requirements analysis
2. Preliminary design
3. Detailed design
4. Coding and Unit Testing
5. Computer Software Component (CSC) integration
6. Computer Software Component Item (CSCI) Testing
7. Configuration Management (CM)
8. Quality Evaluation
9. Project Planning and Control

SECOND-LEVEL REQUIREMENTS AREAS
1. Activities
2. Products
3. Formal Reviews
4. Baselines
5. Developmental Configuration

OTHER SECOND-LEVEL AREAS
1. Audits (for CSCI Testing and Configuration Management)
2. Software Acceptance (for CSCI Testing)
3. Installation and Checkout (for CSCI Testing)
4. Independence (for Quality Evaluation)

- software-development cycle
- computer-software organization
- software quality

- use of commercially available, reusable, and government-furnished software
- subcontractor control
- nondeliverable software, firmware, and hardware
- firmware
- development methodologies
- security
- deliverable data

Table 14.1 provides an indication of the scope of requirements that are considered important in the standard. We will focus here on a handful of requirements areas that are of special interest, namely

- the software development cycle
- software-item structure
- key products
- selected approaches

The software development cycle, normally paralleled by equivalent hardware development steps, is illustrated in Figure 14.1. The key sequential elements are shown in the figure as

- requirements analysis
- preliminary design
- detailed design
- coding, unit testing, and CSC integration
- CSCI testing
- system integration and testing
- operational testing and evaluation
- production and deployment

This cycle fits into the overall system-development cycle shown at the top of the figure as reflected also in an earlier discussion (Figure 2.6 of Chapter 2). From a computer-aided point of view, it is of interest to be able to support the development elements shown in Figure 14.1 to the maximum extent, consistent with increases in productivity and reductions in cost.

The software development hierarchy has been defined as shown in Figure 14.2. The overall system contains segments, which in turn contain hardware-configuration items (HWCIs) and computer-software-configuration items (CSCIs). From each of the latter, the flowdown structure is

- TLCSCs (top-level CSC)
- LLCSCs (lower-level CSCs)
- Units

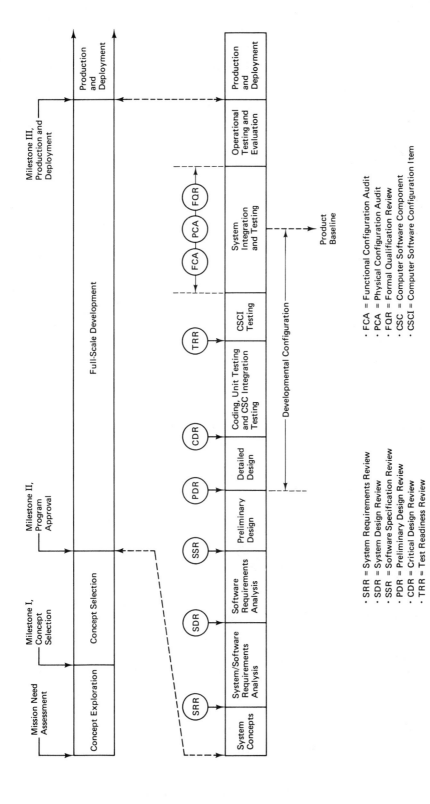

Figure 14.1 Software development cycle.

- SRR = System Requirements Review
- SDR = System Design Review
- SSR = Software Specification Review
- PDR = Preliminary Design Review
- CDR = Critical Design Review
- TRR = Test Readiness Review

- FCA = Functional Configuration Audit
- PCA = Physical Configuration Audit
- FQR = Formal Qualification Review
- CSC = Computer Software Component
- CSCI = Computer Software Configuration Item

365

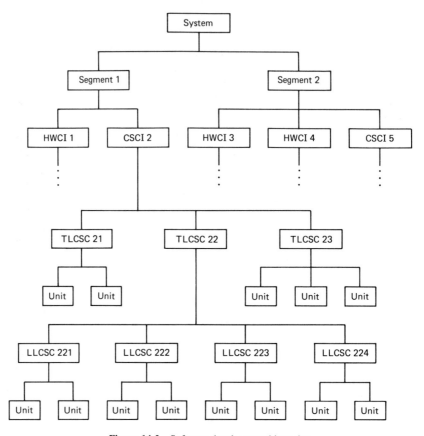

Figure 14.2 Software development hierarchy.

The unit, by definition, is the *lowest level item in the hierarchy*. A CSC is a generic name for a *computer software component*.

A variety of specific written products are called for, with four of the more important ones cited as

1. software development plan (SDP)
2. software standards and procedures manual (SSPM)
3. software configuration-management plan (SCMP)
4. software quality-evaluation plan (SQEP)

The titles of the latter list are self-explanatory in terms of the desired contents. In addition, supporting documentation that represents data requirements includes the following:

- system-segment specification
- software-requirements specification
- interface-requirements specification

- software top-level design document
- software detailed-design document
- interface-design document
- software user's manual
- computer-system diagnostic manual
- software programmer's manual
- firmware-support manual
- operational-concept manual
- computer-resources integrated-support document
- configuration-management plan
- engineering-change proposal
- specification-change notice

The data requirements in this list, along with the more highly featured SDP, SSPM, SCMP, and SQEP, represent a set of products that might be called out for delivery on a typical large-scale DoD program. Since the military standards are often used as a model for systems development in the nondefense government sector, similar requirements for software-related documentation may be expected in these sectors. Such systems pertain, for example, to various aspects of transportation, telecommunications, health and human services, and commerce. Whatever the sector application, there are strong forces calling for detailed and coherent documentation of all aspects of the software development process. Mil. Std. 2167, in addition to citing these typical data requirements, correlates them with the point in the software development cycle at which they are typically needed by the user or procurement agent for the system.

By way of selected approaches, three points in the standard are especially noteworthy. One calls for the use of structured-requirements analysis tools and techniques. Thus the standard acknowledges and emphasizes the importance of such tools and techniques as part of the software development process. The subject of computer aids and packages in relation to requirements analysis is considered in greater detail later in this chapter. A second area in the standard indicates mandatory use of a program design language (PDL) [4] or another top-level design-description tool or methodology. Here again we note the focus on tools, in this case either a specific coding-oriented language or a formal tool that assists in describing the software design at the top level. Many of the diagramming techniques discussed earlier (Chapter 4) are in fact such design description methodologies. Some are available for use in a software package; those that are not should be available soon due to the ever-increasing power of commercially available graphics software.

Finally, the standard calls for a top-down approach to design, coding, integration, and test. This top-down orientation implies a hierarchical decomposition from the system level downward (see Figure 14.2) and implicitly points toward so-called modern structured techniques for software development. A brief overview of such techniques follows since they occupy an important place in today's approach to software development.

14.2 STRUCTURED ANALYSIS

Structured analysis (SA) is a formalized top-down set of procedures for analyzing a system. Since software has become a particularly difficult area, SA has been applied extensively to software developments and systems. Generically, it also can and has been used in systems analysis and engineering, especially as it applies to such areas as functional analysis and decomposition, requirements analysis, and data/information flow analysis. Since SA is normally top-down, in broad terms it can be envisioned as an orderly method for hierarchically decomposing a system and then describing the interactions between the decomposed elements at all levels. In that sense, several of the diagramming techniques presented in Chapter 4 (e.g., HIPO, Warnier-Orr, Action Diagramming) are specific implemention mechanisms of SA.

An important investigator and proponent of SA is D. T. Ross whose early work [5, 6] set the stage for further examination and development by others. About the same time period (1974–1977), E. Yourdon [7] was looking at structured programming, top-down design, and related subjects. Ross appeared to focus on the development and application of SADT (SADT is a trademark of SofTech, Inc.), the structured analysis and design technique. Yourdon seems to have oriented his further developments to building more formal methods of software engineering. Some key aspects of the aforementioned investigators as well as others are discussed next.

In 1985, Ross summarized some of the applications and extensions of SADT [8]. The technique fundamentally consists of a box–arrow diagramming construct and a design approach associated with that construct. The diagram is sometimes called a structured-analysis box (SAB) and consists of an input, an output, a control (input) arrow, and a mechanism (input) arrow. Input, control, output, and mechanism are also referred to by the acronym ICOM. The control is thought of as the *why* input and the mechanism as the *how* input. A full complement of SABs is intended to fully document the architecture of large-scale systems. It can also be thought of as a model of such a system.

The Yourdon thrust, supported in direction by such persons as L. Constantine [9] and T. DeMarco [10], has the data flow diagram (DFD) as a central element. This is a diagramming technique that is discussed in Chapter 4. As with Ross, a key aspect of the DFD is its power as a communication device, that is, a means by which two engineers can convey information to one another. Its graphical format conveys that information concisely and facilitates further analysis and design. As an example of this feature, we show in Figure 14.3 a simple DFD as well as a matrix that embodies the same information, but in a different format. The DFD shows four basic entities

1. terminators
2. files or data stores
3. transforms
4. data flows

Referring to the figure, terminators are the source and sink nodes that define the boundary of the DFD. Since the DFD is a model of the flow and processing of data

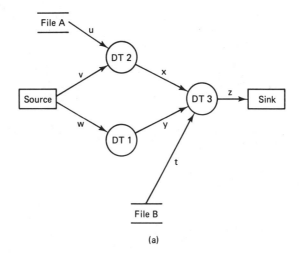

(a)

From (Rows)	To (Columns)						
	Source	File A	File B	D-T 1	D-T 2	D-T 3	Sink
Source	—			w	v		
File A		—			u		
File B			—			t	
D-T 1				—		y	
D-T 2					—	x	
D-T 3						—	z
Sink							—

(b)

Figure 14.3 (a) Data flow diagram and (b) an equivalent matrix representation.

within a larger system context of data flow, the boundaries of this model must be defined. The terminators serve this purpose and provide an explicit interface to other portions of the system data flow. The files or data stores are also represented as file A and file B. The transforms are the circles designated as DT 1 (data transform one), DT 2, and DT 3. The transform nodes operate in some fashion (e.g., add, multiply, convolve) on the input data to produce the output data. Finally, the data flow itself is shown on the diagram arrows by the letters t, u, v, and so on.

A matrix showing the same information as in the DFD is drawn in Figure 14.3(b). All nodes are listed in both the row and column entries with the data flow from the row nodes to the column nodes. Although this is not the only way to develop a matrix representation, it clearly shows the data flow between all nodes and in the proper direction. The DFD is easy to read and understand; the matrix with the same information content is more difficult to interpret at a glance. However, apart from the value of the DFD versus the matrix as a communication medium, the matrix format is immediately conducive to capture of the information content in a computer format through a spreadsheet, DBMS, or other type of software. Thus it would appear that the matrix is a better link to an automated-information capture and processing environment than is the graphical DFD. This need not be the situation if further internal mapping of the screen graphics is performed in a software package. Whatever the vagaries of DFDs in relation to other information formats, however, the former remain as an integral part of SA.

The two other elements normally associated with the DFD are the data dictionary and a set of specifications. The data dictionary describes the data flow and associated files (sources, sinks, or files) in more precise form. For example, in Figure 14.3. the process carried out by node DT 2 in terms of the data flow in and out of it might be conveyed in a data dictionary as

$$\text{Cost}(X) = \text{Hours Worked}(V) * \text{Hourly Salary}(U)$$

In other words, the DT 2 node transforms the *hours worked* and *hourly salary* inputs into a *cost* output through, in this case, a simple multiplication process. In general, the transformation can be much more complex. The reader is referred to the parameter dependency diagram (PDD) and related text in Chapter 4 to see more complex forms of a transformation process. The similarity between the PDD and DFD is further noted, with the former more oriented to analog modeling and the latter to digital or data modeling.

The last element in this general SA process is the specification (spec). The spec elaborates on the data transforms in a fuller language that might take the form of structured English or even a type of pseudocode or decision table. In the aforementioned example for DT 2 in Figure 14.3, a spec might read

> For each employee in file:
> Compute cost for week
> Accumulate costs over all employees
> Print total cost for week

Thus the specification expands the explanation of the transform by giving more complete information to the programmer as to the steps required in order to ultimately develop proper code for the data flow and processing.

Another important aspect of SA to be kept in mind is that the DFD is embedded in a larger system context. The means by which that context is developed and displayed, in general, is known as hierarchical decomposition (see Chapter 4). This decomposition successively breaks a system down, maintaining traceability between all elements. For example, in Figure 14.2, we show a decomposition of software into subordinate elements. Figure 4.5 depicts a hierarchical functional decomposition. For the DFD, this type of decomposition is sometimes also called *leveling* or developing *parent–child* relationships. The process is essential in order to assure that all elements of the system are accounted for and that the relationships (e.g., data flows) can be formulated between all elements, at whatever level they show up in the system.

From a computer-aided point of view, it should be noted that data-dictionary software packages are available in the commercial market [11]. These packages provide the necessary computer-based adjunct to DFDs within the SA process in order to capture all the manually developed analyses in a computerized data base. This is a crucial part of providing a true computer-aided software engineering capability.

The last element of SA to be touched upon here is structured programming or coding. In general terms, structured programming (SP) completes the structured analysis process by filling in the necessary steps to move from the DFD and related methodologies to executable code for a particular computer system. This may be a one-step procedure or may involve several formal steps. For example, one might go from the DFD data dictionary and specification to a program-design language [4], and from there to a structured computer language such as Pascal, C, or Ada. As structured programming constructs, these languages embody the precepts of structured coding (e.g., top-down, GOTO-less, Bohm and Jacopini [12] constructs) as well as some of the more modern programming notions (e.g., case, repeat-until, do-while, information hiding). There is extensive literature available to further explore this subject [13, 14]. Computer-aided software engineering environments are being developed on a continuing basis to provide tools that facilitate structured programming and coding on programmers' workbenches and software engineering workstations. Tools that may be used in these environments are considered in sections 14.3 and 14.4.

14.3 DEVELOPMENT TOOLS FOR THE MINICOMPUTER AND MAINFRAME

Many tools have been constructed and are commercially available for minicomputers and mainframes. In a market calling for more computing power with the ability to afford greater expenditures, tool developers have been plentiful and the successful ones have brought their products to a variety of users. A generic name for tools in this area has been *software development tools*. A considerable amount of attention has been paid to the matter of defining and sorting out these software-development tools. As an example, the National Bureau of Standards [15] has examined features of these tools and defined a taxonomy of the features of such tools. This taxonomy is illustrated in Table 14.2. In principle, each tool on the market can be described by this characterization. The developer of this taxonomy at the NBS has also formulated an interesting and useful profile of these tools [16].

Another way of examining the areas in which tools are available is to relate them to the software development cycle discussed in Section 14.1. The cycle can be boiled down to the following categories:

- requirements-analysis tools
- software-design tools
- coding and testing tools
- documentation tools
- configuration-management tools
- project-management tools

TABLE 14.2 TAXONOMY OF FEATURES
FOR SOFTWARE DEVELOPMENT TOOLS

I. Input
  1.1 Subject
      1.1.1 Text—Natural Language Form
      1.1.2 VHLL—Very High Level Language
      1.1.3 Code—Coded Input
      1.1.4 Data—Character String

  1.2 Control Input
      1.2.1 Commands—Procedural Operators
      1.2.2 Parameters—Associated With Functions

II. Function
  2.1 Transformation
      2.1.1 Editing
      2.1.2 Formatting
      2.1.3 Instrumentation
      2.1.4 Optimization
      2.1.5 Restructuring
      2.1.6 Translation

  2.2 Static Analysis
      2.2.1 Auditing
      2.2.2 Comparison
      2.2.3 Complexity Measurement
      2.2.4 Completeness Checking
      2.2.5 Consistency Checking
      2.2.6 Cost Estimation
      2.2.7 Cross Reference
      2.2.8 Data Flow Analysis
      2.2.9 Error Checking
      2.2.10 Interface Analysis
      2.2.11 Management
      2.2.12 Resource Estimation
      2.2.13 Scanning
      2.2.14 Scheduling
      2.2.15 Statistical Analysis
      2.2.16 Structure Checking
      2.2.17 Tracking
      2.2.18 Type Analysis
      2.2.19 Units Analysis

  2.3 Dynamic Analysis
      2.3.1 Assertion Checking
      2.3.2 Constraint Evaluation
      2.3.3 Coverage Analysis
      2.3.4 Resource Utilization
      2.3.5 Simulation
      2.3.6 Symbolic Execution
      2.3.7 Timing
      2.3.8 Tracing
      2.3.9 Tuning

III. Output
  3.1 Machine Output
      3.1.1 Data

TABLE 14.2 (*cont.*)

3.1.2 Intermediate Code
3.1.3 Object Code
3.1.4 Prompts
3.1.5 Source Code
3.1.6 Text
3.2 User Output
 3.2.1 Computational Results
 3.2.2 Diagnostics
 3.2.3 Graphics
 3.2.4 Listings
 3.2.5 Text
 3.2.6 Tables

Potential users of the kinds of tools listed can therefore examine their operating environment, i.e., computer, operating system, and so on, and select a tool that is compatible with that environment. However if many tools are to be chosen, the user may have both a compatibility and an interoperability problem. Each selected tool may function quite well if it is compatible with the user's environment, but the tools may not "play together." For example, it may not be possible to exchange data between a requirements-analysis tool and a software-design tool. This was the essential purpose of the multitask manager in the p-c arena, as discussed in Chapter 6, namely, to contain an "overlay" program that would allow for the swapping of data between existing packages. Thus, it is a general premise that the user who needs a full range of capabilities is looking for a fully integrated set of tools—one that allows for compatibility with present computer assets as well as one that is fully interoperable. This is sometimes called an integrated software environment. In addition, the specific user location for such an integrated environment may be called an integrated workstation. If the latter has a wide variety of powerful and interoperable tools, as well as raw computer power, extensive memory, and other attributes (e.g., graphics, networking) it may be viewed as a high-performance workstation. Examples of such systems are produced by Apollo Computer and Sun Microsystems.

The matter of selecting a set of tools in order to construct a compatible and interoperable system can be visualized as a design process similar to that described in Chapter 10. The six tool categories listed previously can be likened to system functions and each such function can be implemented, in general, by several tools. The design problem is thus to select a set of tools (presumably one in each category) such that the criteria (compatibility, interoperability, cost, and so on) are best satisfied. Tool developers and workstation producers are continually mindful of the potential users' needs and evaluation criteria.

Although many of the specific tools cited in this section are not currently available for the p-c, the more successful ones are likely to migrate to a micro environment as the latter's price–performance capabilities continue to improve. In any case this is a small sampling of some tools available for a more powerful setting and is intended to give the reader a sense of what the future may bring for the p-c.

14.3.1 Requirements-Analysis Tools

A set of so-called requirements-analysis tools is listed in Table 14.3, together with the producers of these tools. Tools of this general type, if offered for use, are typically available for a license fee ($\sim$\$40,000) and a yearly maintenance charge ($\sim$\$2,000). Of particular significance is the work of D. Teichroew [17] who is generally given credit for problem statement language/problem statement analyzer (PSL/PSA) and related computer-based tools available through his company (ISDOS, Inc.). Requirements analysis is a first step in the systems engineering process. Thus it is also a starting point in the software development process. The notion is that the software engineer begins with a requirements document from the user. This document is in its usual textual form and is to be captured in a computer data base so that it can be analyzed and manipulated as early in the process as possible. Thus, for most requirements-analysis packages one looks for the capability to go from the user's requirements document to a computer database format. The power in the package then relates to the ability to naturally move through the later steps in the software development process. The ultimate objective is to be able to stay in a computer-based format in an efficient manner all the way through to machine-readable code. In addition, the tools are to provide such functional capabilities as requirements traceability through verification of requirements by means of the test and evaluation (T&E) process.

TABLE 14.3 SELECTED REQUIREMENTS-ANALYSIS TOOLS

- PSL/PSA—Problem Statement Language/Problem Statement Analyzer, (ISDOS, Inc.)
- SEM—System Encyclopedia Manager, (ISDOS, Inc.)
- SREM—Software Requirements-Engineering Methodology, (TRW, Inc.)
- IORL/TAGS—Input-Output Requirements Language/Technology for the Automated Generation of Systems, (Teledyne Brown Engineering)
- RPL/RSI—Report Procedure Language/Report Specification Interface, (University of Michigan/ISDOS, Inc.)
- RLP—Requirements-Language Processor, (GTE Laboratories)
- RSL—Requirements-Statement Language [18]
- RTRL—Real-Time Requirements Language [18]
- CSDL—Conceptual Schema-Definition Language [19]

The requirements analyzers or languages listed in Table 14.3 are only some of the packages that are available. The reader is referred to the considerable literature in this area for additional information on this particularly important class of software packages [18].

14.3.2 Design Tools

The earlier cited Leonardo project appears to be focused on the matter of moving from requirements into the difficult and critical area of design. Indeed, a key objective of the project is to provide a set of algorithms that allow for design optimization, certainly an ambitious task. Thus the Leonardo project seems to be addressing requirements and design along with the issue of building a coherent bridge between the two, even to the point of attempting to optimize the design. In this context, most of the requirements-analysis tools are also considered to be design tools. Other tools advertised as specifically oriented to design include the Performance Oriented Design (POD, from BCS, Inc.), the SDL/SDA (from ISDOS, Inc.) and a framework for design such as the Total System Design (TSD) [19]. In addition, essentially all of the computer diagramming techniques and tools (Chapter 4) are viewed as design tools. These would include the following:

- Warnier-Orr
- HIPO
- F^2D^2
- Action diagramming

Indeed, a key claim of the last item is that it can be used for both top-level analysis all the way through to detailed coding. As noted earlier, a crucial issue with respect to diagramming techniques is that of capturing the information content of the diagram in computer-based form. The potential user should examine this matter carefully before committing to a particular diagramming technique.

14.3.3 Coding and Test Tools

In this area one is specifically interested in having the capability to write code or assist in the specific development of code. The packages available address code all the way from a type of pseudocode to a PDL and on into a specific language (e.g., Fortran, C) that can be readily compiled for execution. Also included is a class of packages known as software generators [20]. Other code and test-oriented tools have been catalogued and documented by the Bureau of Standards [21], a sampling of which is listed in Table 14.4.

14.3.4 Documentation

Most of the software tools have an internal report generator that serves as documentation of the results of the process embedded in the tool. The format is therefore closely tied to the format of the tool itself. For example, an automated HIPO package provides documentation in the HIPO format, focusing on input, process, and output (IPO) documentation. However, as seen earlier with respect to Mil. Std. 2167, there are many specific documents that are not easily handled in this fashion. These include such documents as the SDP, SSPM, SCMP, and SQEP. An internal word processor within a tool will assist in meeting such needs. A more conventional approach is simply to move to a separate word processing system for these types of documentation needs.

In addition to the aforementioned, there are text formatting systems available

TABLE 14.4 SELECTED CODING AND TEST TOOLS

ADA-ATOM	EFFIGY	SAP
ADS	ENFORCE	SARA-U
AMPIC	EXPEDITER	SELECT
ASSIST-I	FACES	SPTRAN
ATA-FASP	FAST	SSA
ATA-SAI	FAVS	STRUCTURING ENG
ATDG	FORAN	STRUCT
ATTEST	FTN ANALYZER	SUBCRS
AUDIT	GENTESTS	SURVAYOR
AUDITOR	INSTRU	SUS
BSC	JAVS	SYDOC
CADA	JIGSAW	SYMCRS
CASEGEN	JOVIAL/VS	SYSTEM MONITOR
CAVS	JOYCE	SYSXREF
CCA	LOGIC	TAFIRM
CCREF	MENTOR	TATTLE
CCS	MONITOR	TCAT
CENSUS	NASA-VATS	TDEM
CGJA	NODAL	TEST PREDICTOR
CHILL TRANS	NUMBER	TEVERE-I
COBOL/DV	OPTIMUS	TFA
COGENT	PACE	THE ENGINE
COMSCAN	PET	TIMECS
CORE	PROGLOOK	TIMER
COTUNE-II	RADC/FCA	TPT
CPA-ADR	REFLECT II	TRAILBLAZER
DAVE	REFORM	TSA/PPE
DRIVER	REL MEAS MODEL	UCA
DYNA	RISOS TOOLS	XPEDITOR
EAVS	SADAT	YACC

Source: [21]

on the larger computers that can be used to produce large amounts of very professional-looking documentation. These include SCRIPT (IBM) and NROFF (Digital Equipment Co.) [18]. Of course, in order to use these types of systems, immediate availability of larger scale computer resources is necessary. If such is the case, then these tools are preferable to the tools of the stand-alone p-c, but not necessarily the imbedded word processing capability of a specialized software-development tool.

14.3.5 Configuration-Management Tools

Several producers of software have focused on the matter of configuration management (see also Chapter 15), because they have recognized the significance of this area in software development. Tools in this area usually have a DBMS as the central core of the package. In this same context, many developers have obtained a general

DBMS (see Appendix B) and used it to deal with configuration management by writing an overlay set of code (macro) to adapt the general package to that application. More specific aspects of configuration management are considered in Chapter 15.

14.3.6 Project-Management and Other Tools

Tools in this arena are plentiful and mirror the tools of the same subject listed in Chapter 6. In a mini and mainframe environment, however, these tools are more powerful and extensive. They normally have PERT and Gantt routines for scheduling and a variety of database and even spreadsheet capabilities. A good source listing of such available packages is that produced by the Datapro Company under the category of project management and control.

Notwithstanding the aforementioned categorization of tools, it should be recognized that several of the tools span more than one category. That is, several can be used for more than one software-development function such as requirements analysis, design, and coding and test. A reviewer of software "development support systems" [22] has examined a variety of tools as listed next

- AIDES
- CADES
- DREAM
- EPOS
- IORL
- MASCOT
- PCSL/ESPRESO
- PDL
- PSL/PSA
- RDL
- SREM (RSL/REVS)

These tools have been mapped against elements of the software development process, verifying that most of the tools apply to more than one aspect of the process. The potential user thus has many choices and therefore a difficult trade-off analysis and decision as to which tools are best for particular application areas.

In addition to the tools described, a framework for considering the support of large-scale software engineering projects has been articulated under the name *integrated project software environments* (IPSEs). Various practical and experimental systems can be identified as falling within the IPSE category [23], such as

- DSEE (available on Apollo workstation)
- Cedar
- Flex
- SMALLTALK
- MENTOR
- ASPECT
- ECLIPSE
- MULE
- SPRAC

For further information regarding such systems, the reader is referred to the cited compilation and other references to IPSE-type packages and techniques.

Another class of IPSEs is implicit in the development of the Ada language (Ada is a registered trademark of the United States Government, Ada Joint Program Office) and its set of related tools. The Ada language development has been supported by the DoD and is finding its way more and more into DoD systems, especially mission-critical systems. Associated with the Ada language is the Ada programming support environment (APSE) and indeed efforts can be identified whose purpose it is to promote the portability of data and tools between APSEs. As of 1985, the first integrated software-development environment actually written in the Ada language became available.

In a sense, if one generalizes from the IPSEs mentioned to a broader class of computer aids for both hardware and software development, we come to the world of high-performance workstations, as referred to earlier in this chapter. Such workstations represent a considerable upgrade from the p-c, with capabilities that ap-

---

TABLE 14.5 CATEGORIES OF SOFTWARE PACKAGES AVAILABLE AT HIGH-PERFORMANCE WORKSTATIONS*

Architectural Engineering
Artificial Intelligence
Chemical Engineering
Civil Engineering
Communications
Computer-Aided Publishing
Earth-Resources Engineering
Electrical Engineering
Electronic Design Engineering
General Professional Support[†]
Manufacturing Automation
Mechanical Engineering
Plant Engineering
Power-Systems Engineering
Software Engineering[‡]

---

*Examples listed are provided by Apollo Computer, Sun Microsystems, and others

[†]General professional support

• DBMS	• Simulation
• Decision support systems	• Mathematics
• Graphics	• Spreadsheets
• Logistics	• Statistics
• Project management	• Word processing

[‡]Software engineering

• Cross-development tools	• Languages
• General-development tools	• Utilities
• Editors	

proach and in some areas exceed the minicomputer. Special attention in such workstations is paid to high-resolution graphics (e.g., of the order of 1000 by 1000 pixels), networking, and the provision of third-party software of various kinds that can be used at the workstation to address systems engineering and software engineering design and analysis problems. Listed in Table 14.5 are some of the categories within which software packages are made available to the workstation user. This is truly a powerful computer-aided environment for systems and software engineering and should be considered seriously by all industrial users interested in providing computer-aided capabilities. Such workstations, however, are normally outside the price range of the student or individual practicing systems or software engineer. Further considerations of this important area can be found in Chapter 17.

Finally, as alluded to earlier, software development and analysis is sometimes called *software engineering*. For situations in which computer tools such as those discussed in this section have been developed and applied to software engineering, the overall subject, in some portions of the literature, has been termed computer-aided software engineering [24]. Thus, to the extent that one finds the CASE acronym in various references, it is very likely that it refers to the growing subject of computer-aided software engineering. This is considered to be a subset of the overall subject area of this text since software engineering is normally viewed as a subset of systems engineering.

14.4 DEVELOPMENT TOOLS FOR THE MICROCOMPUTER

The emphasis of this text has been the p-c in systems engineering applications; this chapter addresses the subset of systems engineering having to do with software development and analysis. Thus we come specifically to the matter of what tools may be available at the p-c level that can be applied to the tasks and activities of software engineering. In broad terms, these tasks and activities are the same as those cited in Section 14.3 in relation to minis and mainframes (e.g., requirements analysis, software design, and coding and testing). However the application areas are defined, we envision a software engineer, analyst, or programmer working at a p-c workstation, and we are concerned with the tools that can be made available to facilitate the software-development process.

Three previous chapters (3, 6, and 7) discussed several classes of p-c software packages that can be used to assist in the systems engineering process. In general terms, a good many of these packages are applicable as well in a software-engineering environment. For example, the project-management packages are immediately usable for planning and monitoring progress on a software-development program. Likewise, the multitask managers that overlay spreadsheets, DBMSs, and other software can also be brought to bear on software-development tasks. In addition, the mainline tools cited in Chapter 3 (spreadsheets, DBMSs and so on), although not specifically designed for software development and analysis, can be readily adapted to many of these tasks. We cite some illustrations of this latter point here.

Basic Software Packages	Related Development Application Areas
Spreadsheets	• Lines of code and effort estimation/tabulation • Person loading/time/cost tabulations • Software project evaluation
DBMS	• Requirements traceability • Module/program tracking • Configuration management
Word Processor	• Software development plan (SDP) • Software standards and procedures manual (SSPM) • Software configuration-management plan (SCMP) • Software quality-evaluation plan (SQEP)
Graphics	• Diagramming (as per Chapter 4) • Schedule charts • Bar, pie, histogram charts (in above SDP, SSPM, and so on; reports)

The p-c based tools to be considered here, in addition to those mentioned in this section and in earlier chapters, can be thought of as more specifically relevant to the software development process. A listing of representative packages is provided in Appendix A in the following five categories:

- SA tools
- languages
- general support tools
- utilities
- micro-to-mainframe (MTM) links

The SA tools tend to parallel the general SA methodology, as previously discussed. Thus, one can look for a graphics capability oriented to a data flow diagram (DFD) format. High resolution is desired as is the ability to enter and edit this type of diagram quickly and efficiently. Often a mouse or tablet is used for ease and speed of diagram and data entry. Supporting the diagramming function are such capabilities as

- pop-up menus
- a data dictionary
- a specification writer
- a data file manager
- reshaping and "rubber-banding"

- zooming
- data hierarchies (parent–child)

As with all commercial software, the potential user should examine the features considered most important (e.g., resolution, size, command structure, and input devices) and map those against the offerings represented by the packages. At the p-c level, of course, capabilities are necessarily limited but can be more than adequate for smaller software-development jobs. In any case, it is noteworthy that specific SA-oriented tools are in fact available. In general, prices for such tools are considerably greater ($1000–$8000) than those for general-purpose integrated or multitask applications manager packages.

The specific languages listed in Appendix A, and others, are all available for the p-c. Thus it is possible for the computer programmer to work on the stand-alone p-c at home or in the office and develop code that can be tested and checked out with a limited data set. With multi-megabyte storage capabilities, even limited data sets are becoming rather extensive. Each available language is itself embedded in a package whose purpose it is to facilitate the programming process. The user needs to be able to edit the code, check for errors in syntax, store and access data files, compile where necessary, and run the programs. Each available package handles these and other needed functions differently. In addition, various versions of the languages are available (e.g., UCSD Pascal) and the user should assure that the language provided has all the desired features. In the case of the languages, prices start in the $50 range. Often, a language such as BASIC is built in to the initial p-c hardware system purchased from the manufacturer.

General support tools, as illustrated in Appendix A, represent an assemblage of nonhomogeneous packages that can broadly assist in the software development process. Selected characteristics represented by such tools include such items as

1. compiler optimizers
2. application generators
3. program generators
4. prototyping environments
5. so-called 4GLs (fourth-generation languages)

A variety of support packages are available that might be called *fast compilers* or *cross compilers/assemblers*. These tend to speed up operations and are designed for particular types of processors. A class of products known as *programming aids* offers such functions as language primitives and operators, graphics primitives, full-screen editing, help facilities, and scrolling. Some packages use a *shell* to facilitate user interfacing, which in effect establishes a structured programming environment. Another common function is to define a command sequence and relegate it to a particular key in a simplified "macro" type of operation. Some products represent multi-window debuggers that are used to assist the programmer in more rapidly modifying a program that simply will not run. Packages can be found that will allow for the writing of a program in English or a near-English pseudocode. These and other programming aids assist in what is called *rapid prototyping*. The notion here is to test and verify the overall design and compatibility of various program modules. Rapid prototyping is one of the main thrusts of advanced projects at all levels of computing whose purpose it is to increase productivity and assure overall program integrity as early as possible.

Additional functions provided under the general-support tool category include data security, maintenance, operating system overlays, fast editors, display managers, and report writers. In broad terms, this class of tools tends to focus on assisting the user in setting up a programmer's workbench or workstation environment for the p-c.

Utilities, as a set of tools (see Appendix A), can overlap the general support tools. The intent here is to suggest that the latter are more systems oriented and thereby more focused on providing an environment within which the programmer can function with increased productivity. Utilities also increase efficiency but tend to be more narrowly developed and applied. For example, utilities available in commercial p-c packages provide the following kinds of capabilities:

- recovering data from "crashed" disks
- retrieving files that may have been inadvertently "erased"
- access to and rearrangement of directory files
- control of and access to hidden files
- hexidecimal and ASCII editing and display
- pictorial overview of structure and contents of a disk
- rapid copying and reformatting of disks

The list shows that this class of tools tends to focus on access, manipulation, and retrieval of information on disks, and allows for improvements in scope and flexibility of use.

Not only do all formal support tools and utilities have to be compatible with the computer's operating system, the latter, of course, contains support capabilities that facilitate general computer operation. A case in point is the increasingly popular UNIX operating system for the larger computers, also available in a p-c environment. An example is the Xenix system, a version of UNIX for the micro, which is also enjoying increasing popularity, especially in situations involving multitasking and multiusers. The reader interested in operating systems for the p-c should consult the extensive literature contained in various texts as well as almost every issue of the personal computer magazines.

Another category addressed in Appendix A is micro-to-mainframe packages. The concept is both important and straightforward. Through such a linkage, the p-c can be used as a stand alone and the results (e.g., sections of code) are then uploaded to a mainframe via such a link. Conversely, data, programs, and so on, can be downloaded from a mainframe to a p-c for local operation. The major reasons for this, of course, are to provide extended processing power and storage to the micro and also expand the utility of the mainframe. Since one is now involving two different machines with different operating systems, in general, a key issue to address is the specifics of the MTM link in terms of compatibility with both the hardware and the software. Among the issues to consider in detail are compatibility of code structure (EBCDIC versus ASCII) and synchronous versus asynchronous operation. Also, the downloading of data files may be a problem due to the limited micro-storage capability as compared with the mainframe file. Thus the downloading may have to be segmented such that portions of the mainframe file are downloaded sequentially, with the micro transferring each segment from main memory to several floppy disks. Once the link is established, however, it can be a significant productivity enhancer since whatever is done and checked out on the p-c is known to be operative and can be placed in the more powerful domain of the mainframe. Considerable literature is building in this area, which can be accessed by the reader interested in additional information.

In the aggregate, the five categories of tools discussed provide a rather extraordinary capability to assist in software development and analysis. When added to the other general purpose tools (discussed in Chapters 3, 6, and 7) it is clear that a great deal of software engineering can be performed at the p-c level. It is also noted that even these capabilities are not exhaustive. For example, further expansion is possible by means of *networking* many p-cs together, which allows for the transfer and sharing of information among many users. Here again, this capability is extensive and is an ever-increasing segment of the market. However, due to limitations in scope for this text, these and other p-c-oriented systems will not be discussed further in any significant detail. Extensive references are available to the user with respect to networking and other (e.g., data communications) areas.

14.5 MEASUREMENT AND ESTIMATION

An extensive amount of activity is currently going on with respect to measurement and estimation of both the attributes and potential problem areas in software. On one hand, it is desirable to understand in a more complete manner what it is that constitutes good software. Another side of that issue is to be able to recognize and if possible measure poor software. Significant efforts toward these objectives have been undertaken and will be discussed briefly in this section. At the same time, the practitioners, in distinction to researchers, move ahead on a daily basis with the estimation of programming tasks. They need to generate code in response to a particular user or sponsor and therefore have to estimate the effort required (person–days, person–months), the time necessary to do the job, and associated costs. This practical problem is visited essentially every day across the software services industry in which code and program delivery and productivity are key issues in satisfying a customer while maintaining profitability.

Clearly it is not possible here to address all aspects of these far-reaching and complex subjects. However, a brief overview of some quantitative areas is offered below, starting with the research-oriented software science and ending with the practical matter of effort, time, and cost estimation for software development tasks.

14.5.1 Software-Science Concepts

The new field known as *software science* represents a significant attempt to understand the fundamentals of software, especially from a quantitative point of view. As a science, most investigators address this field by trying to define and assess measures or *metrics* related to the software. An interesting approach by Kolence [25] leaned upon measures from physics and applied them to software. These included measures of software work, power, and storage realization. *One unit of software work is performed on a storage medium when one byte of that medium is altered;* this is one example of Kolence's measures. Substantial work by M. H. Halstead has provided a firmer foundation for the field of software science [26]. We will touch here on some of the notions set forth by this investigator.

Halstead formulated an initial hypothesis that a count of operators and operands in a program would be highly correlated with the number of bugs discovered in that program. He demonstrated that this was indeed the case and went on to study and develop a variety of software metrics. Several of these are drawn by analogy from the field of information theory (see Chapter 8), using the fundamental logarithmic properties of the theory in relation to the numbers of events of various types. To illustrate, six selected software-science measures are

1. program length
2. program volume
3. program level
4. language level
5. effort
6. time

Program length, for example, is related to the number of operators (n_r) and number of operands (n_d) in a program and can be estimated by

$$N = n_r \log n_r + n_d \log n_d$$

where N is the length measure. We note the logarithmic definition, which is also used in the program volume metric as

$$V = N \log (n_r + n_d)$$

where V is the program volume, a measure of the number of bits required to specify a program. Program level, L, refers to the level of abstraction of a program and is related to the volume measure in a kind of conservation law that states: The product of program level and volume is a constant. Language level is basically equal to that product and attempts to measure the limitations imposed by the language. Based upon the metrics just discussed, Halstead came to the matter of measuring effort. Using the notions that programming difficulty appears to vary directly with the program volume and inversely with program level, the measure of mental effort necesary to create a program was formulated as

$$E = \frac{V}{L}$$

With respect to programming time, it was reasoned that time, T, should be proportional to effort, so that

$$T = \frac{E}{S}$$

where S represents the speed or rate of the programmer.

The aforementioned six software-related metrics equations have been tested through several empirical studies, lending credence to their formulations and support for the overall concept of a software science [27]. Other metrics have also been hypothesized [28] in order to make further progress in this infant but important area of investigation. A potential goal of software science is to have a set of metrics that can be computed automatically by running the software in question through some type of validated measurement program, thus obtaining an immediate measure of the metric for that software. As further empirical data is obtained, this computed value could then be used to assess other related factors such as the expected number or probability of bugs remaining in the software.

Two further points should be noted in regard to the software-science activities discussed in this section. The first is that there are numerous detractors that argue that such measurement schemes do not work and that statistical correlation techniques are fraught with the usual dangers of inappropriate implications of causality. Such reactions are not without merit and to a certain extent are to be expected. However, research-oriented software-science activities continue and it is conjectured that the gap between the science and the practice of software engineering will continue to decrease with time and further investigation. In broad terms, the second point is that the common meeting ground of software science and practice might well be in the domain of what is called *software reliability*. We can observe the "Halsteadians" moving toward software reliability metrics and analyses; the practitioners will be doing the same, but they will do it from the point of view of probabilistic modeling and characterization. Numerous mathematical structures (e.g., Bayesian, Poisson, binomial, exponential) are being hypothesized and tested in at least the following classes of models [29, Part I]:

- times-between-failure models
- failure-count models
- fault-seeding models
- input-domain-based models

The reader desiring further information in this area may refer to some of the papers that are regularly published in the arena of software reliability.

14.5.2 Effort Estimation

The matter of estimating the effort required in order to perform a software-development task may appear to be simple but is actually rather complex. As a consequence, considerable attention has been devoted to this issue and it is expected that this emphasis will continue indefinitely, particularly as coding techniques improve and as computer-assisted tools become increasingly available.

F. Brooks, Jr. was one of the early investigators in this area [30] and his essays on software engineering have done much to raise the consciousness of practitioners in this field. His main point was that using the person–month as a means of estimating the size of a software job is improper and may lead to significant problems. This can be illustrated by looking at the following conventional sequence of steps:

a. p = number of lines of code producible by one person in a day

b. pt = number of lines of code producible by one person in t days (t person–days)

c. ptn = number of lines of code producible by n persons in t days (nt person–days)

d. M = number of total lines of code required for coding task

e. M/ptn = number of person–days required for coding task

The calculation in (e) is the effort, in person–days, required to do the overall coding job. Brooks' main point is that whereas (e) is a correct formulation, one cannot trade people for time without considering the details of the coding task, especially the constraints imposed by the sequential nature of subtasks as well as the need for internal communications between programmers. The person–time trade thus has definite boundaries and limits and for any given type of coding activity there may be an optimum combination of people and an associated time for completion. Brooks goes on to make the point even more strongly—adding additional people to a task already late may make it even later. The reasoning is simply that the new people will use up productive time of the people on the job in order to come up to speed and understand what they need to do to become productive members of the team. Thus, according to Brooks, a constant effort (in person–months) may be associated with a software task as long as the team is operating in the proper domain of time and numbers of people. At extreme values of time or person assignments, the effort in person–months will not remain a constant but will tend to increase as time and person-loading change.

As a guideline for scheduling a software development, Brooks offers a rule of thumb suggested by his own experience

- 1/3 planning
- 1/6 coding
- 1/4 component/early-system test
- 1/4 system test

The one-third planning and one-half testing factors are to be noted, particularly in relation to the remaining one sixth for the actual coding. This may be compared with the estimates in Table 14.6, gleaned from a variety of other sources. In relation to the list, Brooks reviews early results with respect to effort estimation, providing an appropriate background both in philosophy of approach as well as in suggested quantitative models. The most complete exposition of a particular estimation technique, COCOMO, is briefly explored next.

TABLE 14.6 EFFORT DISTRIBUTIONS FOR SOFTWARE DEVELOPMENT

Design Phase
Requirements Definition   40%
   Orientation (5%)
   Requirements Analysis (35%)
System Design   60%
   External Design (31%)
   Internal Design/Planning (29%)
     Design (27%)
     Implementation plan (2%)

Implementation Phase
Program Development   78%
   Detailed Design (15%)
     Orientation (3%)
     Design (12%)
   Coding (16.5%)
   Unit Test (30%)
   Subsystem/Integration Test (16.5%)
System Test and Demonstration   15%
Documentation for User   7%

Overall Project
Design Phase   20%
   Requirements Definition (8%)
     Orientation (1%)
     Requirements analysis (7%)
   System Design (12%)
     External design (6%)
     Internal design/planning (6%)
       design (5.5%)
       implementation plan (0.5%)
Implementation Phase   80%
   Program Development (62%)
     Detailed design (12%)
       orientation (2%)
       design (10%)
     Coding (13%)
     Unit test (24%)
     Subsystem/Integration Test (13%)
   System Test and Demonstration (12%)
   Documentation for User (6%)

14.5.3 Constructive Cost Model

B. W. Boehm [31], in a massive work, reviewed the literature in regard to software-engineering economics and related areas and set forth a COnstructive COst MOdel (COCOMO) as a basis for software cost estimation. The model is a hierarchy that consists of a basic, an intermediate, and a detailed level. It is considered by the industry to be the most complete and definitive work in this area. Only a short overview, however, is possible within the scope of this text.

At the basic level of COCOMO, the two key relationships are defined in terms of effort (person–months, each equal to 152 hrs) and development time (months) required through the general formula

$$PM = C(KDSI)^X \tag{14.1}$$

$$TDEV = D(PM)^Y \tag{14.2}$$

where PM is the effort, $KDSI$ is the number of thousands of delivered source instructions, $TDEV$ is the development time (not the full software life-cycle time), X and Y are empirical exponential factors, and both C and D are constants. Two additional relationships, derivable from Equations 14.1 and 14.2, are the productivity, P, and full-time equivalent staff, $FTES$, measures, given by

$$P = \frac{DSI}{PM} \tag{14.3}$$

$$FTES = \frac{PM}{TDEV} \tag{14.4}$$

Productivity is simply the number of delivered source instructions per person–month expended and $FTES$ is an average staffing level that may or may not correspond to the actual staffing level, which is not likely to be uniform.

For the basic COCOMO, Boehm presents considerable parametric data in relation to Equations 14.1 and 14.2. Here we examine these equations only with respect to what he calls the *mode* of the software development. The three modes considered are

1. organic—relatively small teams; extensive experience and familiarity; in-house stable environment
2. semidetached—intermediate and mixed levels of experience; mixture of experience with type of system under development; combination of organic and embedded modes
3. embedded—tightly constrained project with interacting and complex software, hardware, personnel, and operations; changes are costly and not readily accommodated

The cited relationships for these three modes are

Organic:	$$PM = 2.4(KDSI)^{1.05}$$	(14.5)
	$$TDEV = 2.5(PM)^{0.38}$$	(14.6)
Semidetached:	$$PM = 3.0(KDSI)^{1.12}$$	(14.7)
	$$TDEV = 2.5(PM)^{0.35}$$	(14.8)
Embedded:	$$PM = 3.6(KDSI)^{1.2}$$	(14.9)
	$$TDEV = 2.5(PM)^{0.32}$$	(14.10)

Thus it is a matter of direct calculation to find the effort, time, productivity, and average staffing levels, given an estimate of the delivered source instructions and the mode of the project. As an example, for a value of $DSI = 35,000$, the following computations can be made for the organic mode:

$$PM = 2.4(35)^{1.05} = 100.3 \text{ person–months}$$

$$TDEV = 2.5(100.3)^{0.38} = 14.4 \text{ months}$$

$$P = \frac{35,000}{14.4} = 349 \text{ DSI/PM}$$

$$FTES = \frac{100.3}{14.4} = 7 \text{ FTES}$$

With the exponent in the effort formula so close to unity, it is seen that the effort is almost linear with the delivered source instructions. Productivity is a decreasing function with DSI. For a linear relationship between effort and DSI, productivity would then be proportional to the reciprocal of the constant term in the effort equation, or approximately 417 DSI per person–month of effort. This is not very far from the 349 DSI/PM productivity value just calculated. If Equation 14.5 for effort is substituted into the one for time, Equation 14.6, the following approximate relationship is obtained:

$$TDEV \approx 3.5 \ (KDSI)^{.4}$$ (14.11)

For $KDSI = 35$, this equation yields $TDEV = 14.5$, which is less than 1% different from the above computation. Equations 14.5 and 14.11 are plotted in Figure 14.4 for DSI values up to 100,000.

A related parameter of interest for most software development tasks is the person-loading as a function of time. Experience shows [31] that this curve can be approximated quite well by a section of the Rayleigh distribution (see Chapter 5 for a discussion of this distribution). The relationship can be described as

$$FTES = PM \left(\frac{t}{T_m}\right) \exp \left(\frac{-t^2}{2T_m^2}\right)$$ (14.12)

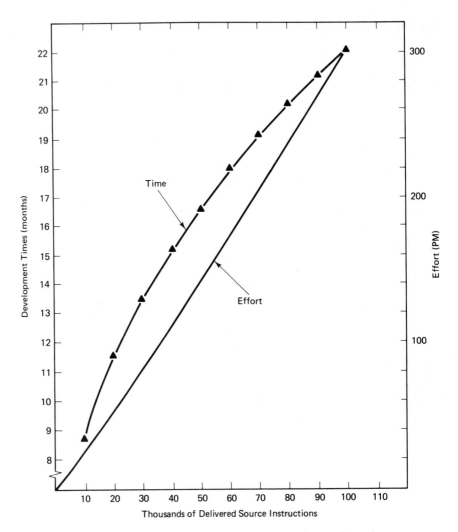

Figure 14.4 Time and effort as functions of delivered source instructions.

where t is the time variable and T_m is the time at which the development is at its maximum value of person-loading. Typically, T_m occurs when the project is between 50% and 70% completed. Also, the approximation is best between $0.3T_m$ and $1.7T_m$, becoming increasingly poorer at the beginning and at the right-hand tail of the distribution.

The aforementioned COCOMO relationships are not, of course, the only bases available for the estimation of effort and other related parameters. Boehm not only recognizes this but provides results of other investigators against which to compare COCOMO values. In regard to effort estimation, for example, investigators are cited and results are offered [31, p. 86] that yield constant (the value of C in Equation 14.1) estimates ranging from 0.7 to 28 and exponent estimates (the value of X in

Equation 14.1) from about 0.9 to 1.9. These results may be compared on a case-by-case basis with the previously cited COCOMO estimators. It is further noted that many differences in assumptions, conditions, and data analyzed were present in these investigations so that differences in estimators are to be expected. The point is that blind acceptance of any one relationship is not desirable without further examination of conditions and areas of applicability.

14.5.4 Cost Estimation

Given the effort estimation described in Section 14.5.3, it is a straightforward matter to develop costs, at least those specifically related to the expenditure of person–months of labor. The cost is thus simply

$$C = R \ (PM)$$

where C is the labor cost and R is the rate per person–month. At this point, we refer back to some of the cost-estimating relationships (CERs) considered in Chapter 11, and make the following observations:

- the (CERs) shown in Chapter 11 with respect to software development tasks are of the same type considered here, dealing with estimates of effort and consequent cost
- the CERs have generally taken the preferred form represented by Equations 11.4, 11.8, and 14.1, sometimes referred to as log-linear equations
- the CER methodology can and is being applied, on a continuing basis, to software estimation as well as a variety of other systems-engineering areas
- CER cost estimation for software development projects is only one component of a life-cycle cost model (LCCM) for a large-scale system, with the other cost elements for such a model as illustrated in Table 11.1
- the man–month and implied cost relationships (e.g., in COCOMO) must be used with care in that they are by no means necessarily the "full-up" costs of a software development project that includes extensive documentation such as the SDP, SSPM, SCMP, SQEP, and other data items referred to earlier in this chapter

Having made the special notes listed, we observe further that Boehm [31, p. 510] points us in the direction of other so-called software cost-estimation models that may be used to improve our ability to predict costs in this complex and extremely important area, still considered by many to be somewhat of a "black art." Such models include the

- 1965 SDC Model
- TRW Wolverton Model
- Putnam SLIM Model
- Doty Model (see Chapter 11)

- RCA Price S Model (Chapter 11)
- IBM-FSD Model
- 1977 Boeing Model
- 1979 GRC Model
- Bailey-Basili Meta-Model

The reader with a specific application requirement can examine Boehm's approach and results in relation to these other significant efforts.

14.6 CASE APPLICATION SUMMARY

The task at hand in this chapter has been software development and analysis, sometimes called software engineering. Using computer tools to support the latter can be referred to as computer-aided software engineering, a subset of computer-aided systems engineering. We have identified specific sets and categories of available tools to run on larger computers as well as on workstations and microcomputers, as follows:

Mainframes and Minis

Requirements Analysis
Software Design
Coding and Testing
Documentation
Configuration Management
Project Management

Workstations

Software Engineering
    Cross Development Tools
    General Development Tools
    Editors
    Languages
    Utilities

Micros (p-c)

Structured Analysis (SA)
Languages
General Support Tools
Utilities
Micro-to-Mainframe Links

In addition, applications of the basic tools of spreadsheets, DBMSs, word processors, and graphics are cited Section 14.4. Software measurement and estimation can be assisted by the spreadsheet and the use of programming languages with which the appropriate equations can be easily computed. It can also be expected that the mainframe, mini and workstation tools described will migrate over time to the microcomputer environment so that they can be used by the student and practicing engineer who have only modest resources available.

REFERENCES

1. Kull, D., "Profiles in Software," *Computer Decisions,* 17, no. 24, December 3, 1985.

2. "The DoD STARS Program," *IEEE Computer,* 16, no. 11, November 1983.

3. "Defense System Software Development," Department of Defense Standard 2167, June 4, 1985.

4. Caine, S. H., and E. K. Gordon. *PDL—A Tool for Software Design.* AFIPS Proceedings of the 1975 National Computer Conference, vol. 44, part I, Montvale, NJ, 1975.

5. Ross, D. T., "Structured Analysis (SA): A Language for Communicating Ideas", *IEEE Transactions on Software Engineering,* SE-3, no. 1, January 1977.

6. Ross, D.T., and K.E. Schoman, Jr. "Structured Analysis for Requirements Definition," *IEEE Transactions on Software Engineering,* SE-3, no. 1, January 1977.

7. Yourdon, E., "A Brief Look at Structured Programming and Top-Down Design," *Modern Data,* June 1974.

8. Ross, D. T., "Applications and Extensions of SADT," *IEEE Computer,* 18, no. 4, April 1985.

9. Yourdon, E., and L. L. Constantine, *Structured Design: Fundamentals of a Discipline of Computer Program and Systems Design.* Englewood Cliffs, NJ: Prentice-Hall, 1979.

10. DeMarco, T., *Structured Analysis and System Specification.* Englewood Cliffs, NJ: Prentice-Hall, 1979.

11. Gane, C. P., and T. Sarson, *Structured Systems Analysis: Tools and Techniques.* Englewood Cliffs, NJ: Prentice-Hall, p. 75, 1979.

12. Bohm, C., and G. Jacopini, "Flow Diagrams, Turing Machines and Languages With Only Two Formation Rules," *Communications of the ACM,* 9, no. 5, May 1966.

13. Yourdon, E., *Techniques of Program Structure and Design.* Englewood Cliffs, NJ: Prentice-Hall, 1975.

14. Yourdon, E., ed., *Classics in Software Engineering,* New York: Yourdon Press, 1979.

15. Houghton, R. C., Jr., "Features of Software Development Tools," NBS Special Publication 500-74, February 1981, National Bureau of Standards, U. S. Department of Commerce.

16. ———, "Software Development Tools: A Profile," *IEEE Computer,* 16, no. 5, May 1983.

17. Teichroew, D., and E. A. Hershey III, "PSL/PSA: A Computer-Aided Technique for Structured Documentation and Analysis of Information Processing Systems," *IEEE Transactions on Software Engineering,* SE-3, no. 1, January 1977.

18. Davis, A. M., "The Design of a Family of Application-Oriented Requirements Languages," *IEEE Computer,* May 1982.

19. Roman, G. C., "A Taxonomy of Current Issues in Requirements Engineering," *IEEE Computer,* 18, no. 4, April 1985.

20. Jenkins, A. M., "Surveying the Software Generator Market," *Datamation,* September 1, 1985.

21. Houghton, R. C., Jr., "Software Development Tools," NBS Special Publication 500-88, March 1982, National Bureau of Standards, U. S. Department of Commerce; see also NBS Special Publication 500-22, "Guide to Computer Program Directories," December 1977; Federal Software Testing Center (FSTC) "Software Tool Catalog," Report OIT/FSTC-84/021, March 1984; and FSTC "Software Tools Survey," Report OSD/FSTC-83/015, June 1983.

22. Lauber, R. J., "Development Support Systems," *IEEE Computer,* 15, no. 5, May 1982.

23. McDermid, J. A., ed., *Integrated Project Support Environments.* IEE Software Engineering Series, no. 1, Piscataway, NJ: IEEE Service Center, 1985.

24. Gillan, P., "Project Spotlight—Computer Aided Software Engineering: Automating DP," *ComputerWorld,* XVIII, no. 34, August 20, 1984.

25. Kolence, K. W., "Software Physics," *Datamation,* June 1975.

26. Halstead, M. H., *Elements of Software Science.* New York: Elsevier North-Holland, 1977.

27. Fitzsimmons, A., and T. Love, "A Review and Evaluation of Software Science," *Computing Surveys,* 10, no. 1, March 1978.

28. McCabe, T. J., "A Complexity Measure," *IEEE Transactions on Software Engineering,* December 1976.

29. "Special Issue on Software Reliability," *IEEE Transactions on Software Engineering,* part I-December 1985, part II-January 1986.

30. Brooks, F. P., Jr., *The Mythical Man-Month, Essays on Software Engineering.* New York: Addison-Wesley, 1975.

31. Boehm, B. W., *Software Engineering Economics.* Englewood Cliffs, NJ: Prentice-Hall, 1981.

*32. Fox, J. M., *Software and Its Development.* Englewood Cliffs, NJ: Prentice-Hall, 1982.

*33. *Directory of Computer Software—1986.* PB86-135357, U. S. Department of Commerce, National Technical Information Service (NTIS), 1986.

*Reference is relevant to text but is not specifically cited.

PROBLEMS

P14.1. For the area of SA, investigate Ross' ICOM box and Yourdon's data flow diagram and discuss, in three pages, their similarities and differences.

P14.2. Obtain data on one of the SA microcomputer tools listed in Appendix A. Write a three-page critique of this tool based upon explicit criteria that you develop for such a tool.

P14.3. Find an example of a PDL in the literature, and discuss its application in three pages.

P14.4. Select a requirements-analysis tool from Table 14.3, and discuss its significance, utility, and limitations, in three pages.

P14.5. Select a coding and test tool from Table 14.3 or another source, and discuss its significance, utility, and limitations, in three pages.

P14.6. Select a tool under the general category of IPSE, and discuss its significance, utility, and limitations, in three pages.

P14.7. Develop a list of 25 software-engineering tools you would wish to have at your own p-c workstation. Explain the purpose of each tool.

P14.8. Assume that you have an IBM 43XX computer and a p-c of your choice. Find a commercially available micro-to-mainframe link that will allow uploading and downloading between these two machines. Discuss the advantages and disadvantages of the link package that you have selected.

P14.9. Investigate Halstead's work (Ref. 26) and cite typical numeric values and computations for the six software science measures of program length, program volume, program level, language level, effort, and time. Interpret these results and measures.

P14.10. Obtain at least one significant source on software reliability modeling and write a five-page summary and critique of this source.

P14.11. Write programs in BASIC and Pascal that will compute effort (PM) and development time (TDEV), with the mode (organic, semidetached, embedded) as a selection on input by the user. Run these programs for KDSI from 10 to 100, in steps of 10, for all these modes.

P14.12. Based upon the results in Problem P14.11, plot productivity and full-time equivalent staff as a function of KDSI in the range cited, for all three modes. Discuss these results.

P14.13. Develop and plot equations for the TDEV as a direct function of KDSI for the semidetached and embedded modes. Compare and explain the results.

P14.14. Plot Equation 14.12 and its integral and interpret the results for each, for variable values that you select.

Light is the task when many share the toil
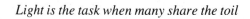

15

Homer

SYSTEMS ENGINEERING SUPPORT FUNCTIONS

A variety of tasks and activities can be viewed as support functions to the mainline systems engineering synthesis (design) and analysis processes. As discussed earlier, the core of systems engineering is the basic design of the system—the hardware, software, and personnel-related subsystems. Analysis activities are intended to optimize the design to the extent that this is possible. Given these cornerstones, a number of systems engineering functions are required as support, including several engineering-specialty areas cited in Chapter 2, for example

- safety
- security
- interface definition and control
- reliability-maintainability-availability analysis
- mission-requirements analysis
- integrated logistics support
- interoperability
- quality assurance
- training
- test and evaluation
- production-engineering analysis

- configuration management
- environmental compatibility

Of these, we will focus in this chapter on the following limited subset:

- reliability-maintainability-availability (RMA)
- integrated logistics support (ILS)
- quality assurance (QA)
- test and evaluation (T&E)
- configuration management (CM)

The reasons for the selections listed are twofold—their particular significance as systems engineering support functions and limitations of text space. Reference sources are cited for the reader who desires to examine other systems engineering support functions in greater detail.

It should be noted that once the fundamental top-level design approach of a system has been defined, essentially all of the listed support functions require both ingenuity of design and analysis as well as discipline and attention to detail. Design elements, for example, include such items as

- the full definition of a system-maintenance plan
- the construction of mechanisms, timetables, and stocking levels for logistics support
- test plans and procedures that fully verify performance and relate coherently to system requirements
- implementation of a quality-assurance plan that controls quality to a specified numeric level

These are crucial to the long-term operation, supportability, and cost of the system and should not be underestimated. Properly performing these systems engineering support functions is thus a thoroughly professional set of activities requiring important synthesis and analysis tasks that can enhance the basic system design and assure that the system is effective over its entire life cycle.

15.1 RELIABILITY-MAINTAINABILITY-AVAILABILITY

Reliability, maintainability, and availability can be broadly defined as follows:

- reliability—the likelihood of mission success, given that a system was available to operate at the beginning of the mission
- maintainability—the general ease of a system to be maintained, at all levels of maintenance
- availability—the likelihood that a system will operate successfully at the beginning of a mission

Reliability and availability are specific quantifiable top-level system parameters; maintainability may be viewed as characterized by a subordinate set of parameters that determine, in part, how reliable and available a system will be.

15.1.1 Reliability

A convenient starting point for a quantitative discussion of reliability is the so-called hazard function, which is defined as

$$h(t) = \frac{f(t)}{1 - F(t)} \tag{15.1}$$

where $h(t)$ is the hazard function, $f(t)$ is the probability density function of time to failure, and $F(t)$ is the cumulative distribution function (CDF) associated with $f(t)$ such that

$$F(t) = \int_0^t f(x) \, dx \tag{15.2}$$

The hazard function may further be interpreted as the conditional time-to-failure density function, given that failure has not occurred prior to time t.

If we assume that the hazard function is a constant, namely,

$$h(t) = \lambda, \qquad \lambda > 0 \tag{15.3}$$

then it can be shown (see problems at end of chapter) that the density function is exponential and given as

$$f(t) = \lambda \exp(-\lambda t) \tag{15.4}$$

in which case λ is the constant failure rate and its reciprocal is the mean-time-between-failure (MTBF).

If the hazard function is more complex, as for example

$$h(t) = \alpha \lambda t^{\alpha - 1} \qquad \alpha > 0, \lambda > 0 \tag{15.5}$$

then the density function (failure law) is Weibull (see Chapter 5) and is represented as

$$f(t) = \alpha \lambda t^{\alpha - 1} \exp(-\lambda t^{\alpha}) \tag{15.6}$$

This is a two-parameter distribution that clearly reduces to the exponential case when $\alpha = 1$. Whereas the mean value and variance for the exponential can be shown to be $1/\lambda$ and $1/\lambda^2$, respectively, these values for the Weibull turn out to be related to the Gamma function [1], which is tabulated in the more sophisticated mathematics or statistics reference sources.

Figure 15.1 depicts the notion of a changing failure rate where the hazard function is sketched as a function of time. For an initial time frame, the function starts out high and drops down, demonstrating the so-called burn-in or quality-control period. For components that are baked-out or burned-in, it may be expected that during this time period either defective components will be discovered, or that if they do

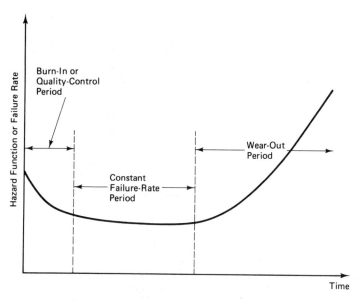

Figure 15.1 Sketch of hazard function or failure rate.

not fail or go out of spec, they will survive and be stabilized and enter the low-failure-rate time frame. In any case, quality-control defects are thought of as being discovered during the early-failure time period so that surviving components enter the next part of the curve at a lower failure rate. This constitutes the first portion of what is sometimes called the *bathtub* curve. In the middle portion of the curve the hazard function is viewed as more-or-less constant so that, as Figure 15.1 shows, the failure law can be adequately characterized as exponential. Indeed, the objective of many quality-control programs is to force the component line to the low and relatively constant portion of the curve. Finally, after the component has been in operation for some period of time, the hazard function or failure rate increases monotonically. This is indicative of a wear-out period and is often associated with electromechanical or mechanical components that clearly exhibit wear-out tendencies. Since the exponential failure law applies to the constant failure rate or hazard function period, it makes sense that other failure laws, such as the Weibull, can be applicable to the nonconstant portion of the bathtub curve. We will generally limit ourselves to considerations related to the exponential failure law.

If the failure law is exponential, then, from Equation 15.2, the failure probability is given by the CDF as

$$F(t) = 1 - \exp(-\lambda t) \tag{15.7}$$

Thus the probability of success, which is the probability of survival to time t, can be calculated as

$$R(t) = 1 - F(t) = \exp(-\lambda t) \tag{15.8}$$

To illustrate this simple equation, for an MTBF of 100 hours, the failure rate is

1/100 = 0.01 and the probability of survival for several selected time periods is listed here

Time (t)	$R\ (t)$
0	1.0000
10	0.9048
50	0.6065
100	0.3679
200	0.1353
500	0.0067
1000	0.000045

It is observed that the probability of success at the MTBF is 0.37 whereas there is a very small likelihood that a component will survive to ten times the MTBF.

The aforementioned reliability equation thus allows the systems engineer to calculate, or predict, the reliability of a piece of equipment, given some estimate of the MTBF or failure rate for that equipment. Later in this section we will comment upon the data that might be available for use in estimating failure rates. Here we concern ourselves with placing equipment in various series-parallel arrangements and the reliability calculations that are associated with these different configurations.

In Figure 15.2 we show seven series-parallel reliability block diagrams, all of which are based upon the exponential model. In configuration (a), two blocks are in series, meaning that both equipments must be operative in order for the overall function to be considered operative. In such a case, the reliabilities are multiplicative and the failure rates additive, as shown in the figure. The same holds true for n blocks in series, as indicated in part (b) of the figure. In part (c), two blocks are in parallel, representing on-line redundancy. The parallel block diagram means that only one block must be in working order for the entire configuration to be considered operative. Thus the case leading to unreliability is the one in which both equipments have failed. The resultant overall reliability is listed in part (c) of the figure. If the redundant equipment is placed in a standby mode [part (d) of the figure], the situation is different in that while it is in standby it is presumed to be operating at a reduced failure rate. It is fully turned on only when the mainline equipment has failed which is assumed to be automatic and instantaneous. In part (e), the on-line redundancy case is generalized to n equipments in parallel. This is seen as a simple extension of the two-equipment situation. In all redundant cases it is assumed that there is instantaneous knowledge of a failure and that the recovery mechanisms (e.g., switches) have not failed. Finally, two other configurations are shown in (f) and (g). The former has n equipments in series, with this string of equipments in parallel with a similar string. In (g), the n equipments are *cross-strapped,* which can also be viewed as a replication of case (c). Although both (f) and (g) have identically the same number of equipments, it is intuitively clear as well as mathematically provable that configuration (g) is uniformly more reliable than configuration (f). It is emphasized that in all of the diagrams of Figure 15.2, the representation is one of series and parallel reliability blocks rather than signal-flow block diagrams, as in Chapter 4.

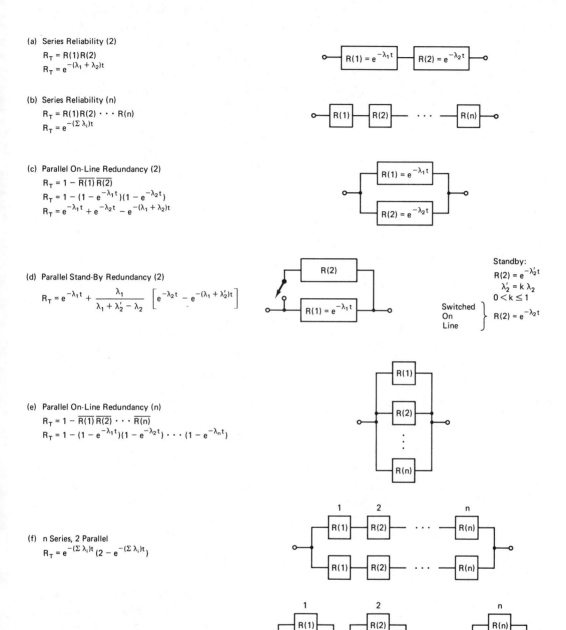

(a) Series Reliability (2)

$R_T = R(1)R(2)$

$R_T = e^{-(\lambda_1 + \lambda_2)t}$

(b) Series Reliability (n)

$R_T = R(1)R(2) \cdots R(n)$

$R_T = e^{-(\Sigma \lambda_i)t}$

(c) Parallel On-Line Redundancy (2)

$R_T = 1 - \overline{R(1)}\ \overline{R(2)}$

$R_T = 1 - (1 - e^{-\lambda_1 t})(1 - e^{-\lambda_2 t})$

$R_T = e^{-\lambda_1 t} + e^{-\lambda_2 t} - e^{-(\lambda_1 + \lambda_2)t}$

(d) Parallel Stand-By Redundancy (2)

$R_T = e^{-\lambda_1 t} + \dfrac{\lambda_1}{\lambda_1 + \lambda'_2 - \lambda_2} \left[e^{-\lambda_2 t} - e^{-(\lambda_1 + \lambda'_2)t} \right]$

Standby:

$R(2) = e^{-\lambda'_2 t}$

$\lambda'_2 = k \lambda_2$

$0 < k \le 1$

Switched On Line } $R(2) = e^{-\lambda_2 t}$

(e) Parallel On-Line Redundancy (n)

$R_T = 1 - \overline{R(1)}\ \overline{R(2)} \cdots \overline{R(n)}$

$R_T = 1 - (1 - e^{-\lambda_1 t})(1 - e^{-\lambda_2 t}) \cdots (1 - e^{-\lambda_n t})$

(f) n Series, 2 Parallel

$R_T = e^{-(\Sigma \lambda_i)t} (2 - e^{-(\Sigma \lambda_i)t})$

(g) n Cross-Strapped

$R_T = e^{-(\Sigma \lambda_i)t} \Pi(2 - e^{-\lambda_i t})$

Figure 15.2 Various reliability configurations.

With these and other combinations of series and parallel reliability block diagrams, it is possible to take a system and break it down into such combinations. Given such a breakdown, the equations in Figure 15.2 and other related equations can be used to represent the overall system reliability or significant portions thereof. In general terms, the effect of building redundancy into a system is to improve the system's reliability, as depicted in Figure 15.3. The price that is paid for this redundancy, of course, lies in the cost, weight, and volume associated with the redundant equipment.

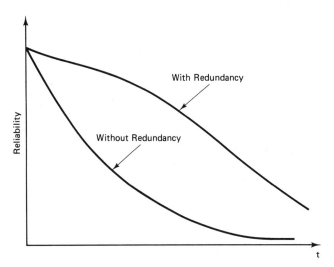

Figure 15.3 Reliability with and without redundancy.

15.1.2 Maintainability and Availability

In the discussion in Section 15.1.1, we noted that the mean value of the exponential distribution was the MTBF which, in turn, is the reciprocal of the failure rate. In a similar sense, one can envision a repair distribution that has a mean value, generally called and mean-time-to-repair (MTTR). Using a relatively simple expected-value representation, availability can then be calculated as

$$\text{Availability} = A = \frac{\text{MTBF}}{\text{MTBF} + \text{MTTR}} \tag{15.9}$$

This simple model can be thought of as an *up* time versus total time ratio. The numerator represents the up time, meaning that on the average, it is the time that the system is operating properly. When failures occur and are subject to (corrective) maintenance, the MTTR adds to the MTBF to produce the total time line, which is in the denominator. As a simple example, if the MTBF is 500 hr and it takes, on the average, 5 hr to repair a piece of equipment, then the availability is calculated as 500/505 = 0.9901.

Distinctions in the literature are made with respect to various kinds of availability [2, 3]. For example, the representation in Equation 15.9 has been called the *inherent availability* for which the MTTR refers to repair in a corrective-maintenance context. The notion is that each failure is followed by corrective maintenance. However, it is usual that a preventive maintenance plan will call for repair or replacement even in the absence of a failure. When the times of preventive maintenance are accounted for by adding them into the MTTR term, the availability is then sometimes called the *achieved availability*. Similarly, when logistics delay times and administrative delay times are also considered, they are added to the denominator and the resultant availability may be called the *operational availability*. These three availability forms are based upon the same two fundamental notions

1. an expected value approach to the problem
2. the addition to the denominator of time factors, which increase the time line over which the equipment is not in an operational state, thus decreasing the availability

The formulation also places in evidence the two ways in which availability may be increased

1. increase the inherent reliability of the equipment, thus increasing the MTBF
2. decrease the additive MTTR term in the denominator by a variety of techniques such as
   - the use of built-in test equipment
   - more efficient and thus shorter repair times
   - the use of automatically switched standby equipment so that corrective maintenance can be done when the equipment is still on the air

The latter technique is sometimes categorized as a fundamental increase in reliability through redundancy rather than an improvement in maintenance procedures. One of the areas in which knowledge-based expert systems are being examined is maintenance in an attempt to decrease the basic MTTR and thus increase overall availability.

15.1.3 System Effectiveness Revisited

In Chapter 12 we defined system effectiveness as related to three parameters

1. A = Availability (readiness reliability)
2. D = Dependability (conventional reliability)
3. C = Capability (performance)

Whereas the emphasis in Chapter 14 was on capability or performance, this chapter focuses on items 1 and 2.

An important systems engineering standard [4] cites both system/cost-effectiveness analysis and effectiveness-analysis modeling as key elements under the subject of system optimization. Under the latter, the following major points are made:

- use of system-effectiveness models when they contribute to the decision process
- variation of model input parameters to determine impacts on total system performance and life-cycle cost
- correlation of effectiveness-model parameters to performance characteristics allocated to system functions
- maintenance, updating, and modification of model-data files

Thus it is seen that RMA plays an important role, along with performance modeling and simulation, in system-effectiveness analyses, and that such analyses are viewed as significant in terms of required systems engineering activities.

15.1.4 Other Reliability-Maintainability-Availability-Related Areas

From the selected subjects discussed, we can infer that RMA is a rather important and complex area that cannot be fully covered in this text. However, comments are included here on a variety of related areas, namely

- failure rate data
- RMA allocation
- design reviews
- operating characteristic (OC) curves
- estimators

It is clear from examining the reliability equations that failure-rate data is required in order to use these equations in a practical application. Both constant failure-rate data as well as wearout data for various types of system components are available in the literature and at various government agencies that have focused upon this particular need [1]. The constant failure-rate data banks relate directly to the *lambda* of the exponential; wearout data can be related to the parameters of the Weibull distribution that accommodates a nonconstant failure rate. Some data banks also keep track of operating environments for the various components and use operating mode or *K factors* as modifiers of the failure rate data. In general, these factors become multipliers to the base failure data to increase the failure rates for more stressful environments.

The matter of allocation of RMA parameters relates to the top-down apportionment of RMA requirements and specifications, an example of which was shown in Figure 10.4. The latter example demonstrated how the MTBF or failure rate parameter might be allocated from a piece of equipment to three subordinate equipments. This process was seen to be an important part of top-level systems engineering since without such a process it is not possible to set requirements and specifications from the higher levels of design to the lower ones. In any case, the illustrative example shown in Chapter 10 for reliability allocation has direct analogs for both maintainability and availability allocation, where the latter deal with specific parameters (such as MTTR, maintenance hours per usage hour, and so on) that characterize these measures. The literature may be reviewed to obtain examples of allocation procedures with respect to maintainability and availability. In addition, various researchers have attempted to develop algorithms for "optimum" allocations. Such algorithms normally require the addition of other factors such as effort required in order to increase reliability from one value to another [1]. These are difficult but not impossible to apply to a real-world large-scale system with large numbers of functions and subsystems.

Design reviews are mentioned here since they represent a very important aspect of systems engineering. Formal milestones are set up in the systems engineering process (e.g., PDR-preliminary design review, CDR-critical design review) to specifically focus on all matters of design in which RMA plays a significant role. Although it is difficult to cite well-accepted analytic techniques related to the process of design review, good systems-design engineers will make enormous progress during design reviews toward improving the design of a system. Thus design reviews for RMA should be an integral part of any systems engineering program and the best design engineers should participate in these reviews.

Failure modes, effects, and criticality analyses (FMECA) were mentioned in Chapter 13 in regard to risk analysis. If not addressed in a formal risk-analysis program, they should be incorporated into an RMA program. These analyses deal with the following sequential steps:

1. identification of all significant and relatively likely failure modes
2. quantification of the likelihood of occurrence of these failure modes
3. determination of possible effects (consequences), should the first two failure modes occur
4. estimation of the criticality of the effects of 1 through 3, given that they occur

Often, various indices and ranking procedures are used for FMECA, similar to those delineated in the risk-analysis discussion in Chapter 13. The net result is that critical areas of the design are identified and rank ordered in terms of the combinations of tendency toward failure, consequences given failure, and criticality given the consequences. This allows fixes to be made to the high-priority items and thereby minimizes the risks due to system and subsystem failures.

The OC curve relates to the notion of sampling in order to determine whether or not components of a system are sufficiently reliable to satisfy the overall needs of the system or some well-defined level of quality [1]. The OC is a plot of the probability of acceptance as a function of the probability of failure, as illustrated in Figure 15.4. Thus, such a curve shows the likelihood that the given sampling plan will result in a decision to accept the component that is under investigation. If the OC curve is very steep, the plan will normally lead to a decision with decreasing error. A considerable amount of statistical theory, beyond the scope of this text, goes into the matter of establishing sampling plans and OCs. It can, however, be a very important aspect of systems engineering in relation to the issue of assuring that components have a prescribed level of performance and reliability.

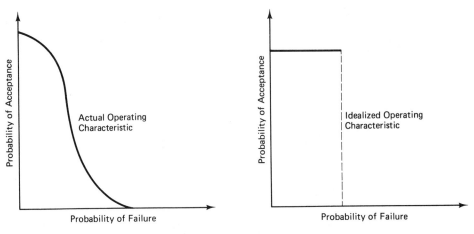

Figure 15.4 Operating characteristic curves.

Implicit in the process of sampling is the notion of estimating. We can think of this issue in terms of having a (possibly large) number of components of unknown reliability with the problem being that we desire to estimate the reliability of these components by sampling from the batch of components and measuring those that we sample. Sampling can be 100% or less of the total population. When we calculate parameters resulting from this sampling process, such as a mean value or a standard deviation, these are estimators of the so-called true values of these parameters. As estimators based upon sample statistics, they have certain characteristics such as bias, consistency, and variance error. These characteristics can be important in large-scale sampling plans and constitute a major subject within the field of statistics. The reader with a particular interest in this area should consult a text in statistics or one in reliability that emphasizes the mathematical foundations of reliability theory [1].

15.2 INTEGRATED LOGISTICS SUPPORT

Integrated logistics support (ILS) may be defined as *provisioning an operational system with all the necessary equipment, facilities, and personnel to find an optimum balance beween system effectiveness and life-cycle cost.* An overdesigned ILS system can help to sustain extremely high levels of effectiveness but at an unacceptably high cost. At the other extreme, for minimum cost investments, effectiveness may drop to low and unacceptable levels. Thus ILS considerations are extremely important in the overall field of systems engineering. They do not, in general, get the visibility that is afforded to design engineering but clearly come into play in a most serious manner when a system is fielded and in operation. Unfortunately, it is often very expensive to try to backtrack at that time to put into place a well-conceived ILS system.

ILS is specifically recognized in the DoD's systems engineering management standard [4] in the context of the relationship between ILS and systems engineering. The goal of the latter is cited as the achievement of a proper balance between operational, economic, and logistical factors. Logistics engineering is required as a mainstream engineering element that is focused on finding an optimal logistics-support concept for the deployment and operational phases of a system. Items cited as parts of logistics engineering include

- logistics-support analysis (LSA)
- maintenance-engineering analysis (MEA)
- repair-level analysis (RLA)
- logistics-support modeling (LSM)

LSA refers to the definition and evaluation of such items as spares provisioning, maintenance equipment, repair parts, personnel, documentation, and transportation. MEA involves considerations of complete maintenance requirements, the use of logistics data, specific quantities of maintenance equipment, personnel and spares, inputs to life-cycle cost and effectiveness analysis, and definition of measurement and calibration standards and requirements. RLA, as per this military standard, points the analyst to an Air Force requirement known as AFLCM/AFSCM 800-4. In broad terms, this requires a complete definition of how all equipments are to be repaired at the various repair levels (e.g., on-site, intermediate, depot) that are set up for maintenance of the system. LSM is a set of analytic activities involving

- the impact of logistics alternatives upon system and subsystem life-cycle cost
- availability analyses
- equipment and manpower loading
- provisioning of spare parts

Given the broad application areas of ILS, often considered as subsuming the subject of RMA, it is seen that logistics modeling can be far-reaching and complex. An example of existing logistics models is provided in Section 15.6 of this chapter; some elementary logistics computations are demonstrated next.

15.2.1 Spares Provisioning Example

An important issue for logistics support is to establish the level of spares to be provisioned in order to maintain a desired level of reliability for a given piece of equipment. In this regard we can use the Poisson distribution (see Chapter 5) to demonstrate a simple calculation. In the context of exponential failures and a constant failure-rate model we write the discrete Poisson distribution as

$$P(x) = \frac{(\lambda t)^x \exp{(-\lambda t)}}{x!} \tag{15.10}$$

where $P(x)$ is the probability of exactly x failures in time t and failures occur at the constant rate of λ failures per unit of time. For $x = 0$, of course, there are no failures and the Poisson becomes the simple exponential as shown in Equation 15.8.

If we wish to provide spares for each of the possible failures, and neglect spare replacement times, we can reason that we can use Equation 15.10 to calculate the likelihood that a spare is available, given the basic reliability of the equipment. If the reliability is $R = \exp(-\lambda t)$, then from Equation 15.10, the probability of having a spare available when needed is

$$P(s) = \sum_{x=1}^{s} \frac{[-(\ln R)^x](R)}{x!} \tag{15.11}$$

The relationship in Equation 15.11 is the cumulative Poisson distribution (CDF) as a primary function of the spares stock level, s, with the reliability R as a parameter. A plot of this cumulative distribution is provided here as Figure 15.5 [5]. As a simple example, if the failure rate were 0.01 and the time in question were 300 time units, then $\lambda t = 3$. For a spares level of $s = 4$, we enter the abscissa at 3 and move up to the curve for which $s = 4$. Reading to the left we find that there is a 0.8 likelihood that the spares level of 4 will be sufficient. If this level is to be raised to 0.99, for example, then the spares level must be increased to approximately 7. Thus the trade-off between spares level and reliability is evident from this simplified model.

15.2.2 Alternative Equipment Comparison

This example considers two alternative equipment configurations, which can represent different implementations of a system, subsystem, configuration item, or other assemblage of related components [6]. We will compare these two alternatives on the basis of overall annual cost. The same basic parameters will be considered for both configurations, (a) and (b).

The initial investment and cost of installation is I, designated as I_a and I_b for the (a) and (b) configurations, respectively. For a useful life of the systems, L, the ratio (I/L) is the annual amortized cost. Next, the annual cost of maintaining and servicing the configurations will be defined by the variable S. Finally, each system may operate in i performance states such that C_i is the cost of operating in the ith state and P_i is the fraction of time over the year that the configurations are expected to operate in the ith state. We then express the selection criterion in terms of deciding upon configuration (a) over configuration (b) if (a) is preferred on a cost basis, namely if

$$\frac{I_b}{L_b} + S_b + P_{bi}C_{bi} > \frac{I_a}{I_a} + S_a + P_{ai}C_{ai} \tag{15.12}$$

If we limit ourselves to just two performance states, namely, up and down (operative and inoperative), we can set the values of P as

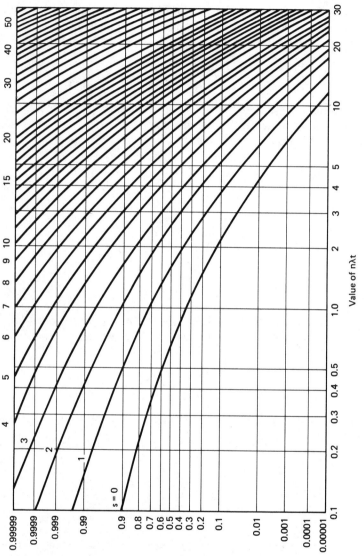

Figure 15.5 Cumulative Poisson probabilities. (From Ref. 5. Reprinted by permission of U.S. Navy.)

$$P_1 = \text{Prob (up state)} = \frac{T_f}{T_f + T_r} \tag{15.13}$$

$$P_2 = \text{Prob (down state)} = \frac{T_r}{T_f + T_r} \tag{15.14}$$

where T_f is the MTBF and T_r is the MTTR. We note that these equations are basically a reflection of the system's availability and unavailability (see Equation 15.9). Substitution of Equations 15.13 and 15.14 into 15.12 yields the decision criterion as

$$\frac{I_b}{L_b} + S_b + \frac{T_{bf}C_{b1}}{T_{bf} + T_{br}} + \frac{T_{br}C_{b2}}{T_{bf} + T_{br}} > \frac{I_a}{L_a} + S_a + \frac{T_{af}C_{a1}}{T_{af} + T_{ar}} + \frac{T_{ar}C_{a2}}{T_{af} + T_{ar}} \tag{15.15}$$

We now simplify the Equation 15.15 by assuming that both configurations provide essentially the same service in the up state, that is, $C_{a1} = C_{b1}$ and that for high reliability, $T_f \gg T_r$. This allows Equation 15.15 to be rewritten as

$$\frac{T_{br}C_{b2}}{T_{bf}} - \frac{T_{ar}C_{a2}}{T_{af}} - \Delta\left(\frac{I}{L}\right) - \Delta(S) > 0 \tag{15.16}$$

where

$$\Delta\frac{I}{L} = \frac{I_a}{L_a} - \frac{I_b}{L_b}$$

and

$$\Delta(S) = S_a - S_b$$

If we further set $\Delta(I/L) + \Delta(S) = K$, where K may be assumed for this example to be a constant and unrelated to the failure and repair times, then Equation 15.16 becomes

$$\frac{T_{br}C_{b2}}{T_{bf}} > \frac{T_{ar}C_{a2}}{T_{af}} + K \tag{15.17}$$

This may be converted (see problems at end of chapter) to the following form:

Choose configuration (a) over configuration (b) if

$$\frac{T_{af}}{T_{bf}} > \frac{T_{ar}}{T_{br}} \left[\frac{C_{a2}}{C_{b2} - \dfrac{KT_{bf}}{T_{br}}} \right] \tag{15.18}$$

We note that, for example, if (a) is a presumed improvement over (b), we show directly in Equations 15.12 through 15.18 the ratio of MTBFs and MTTRs as well as the cost of outages and the MTBF to MTTR ratio for (b). Equation 15.18 is plotted in Figure 15.6 with the MTTR ratio as the ordinate and MTBF ratio as the abscissa. The ratio of outage costs is the main parameter value. Configuration (a) is preferred in the region to the left of the parametric lines.

Systems Engineering Support Functions   Chap. 15

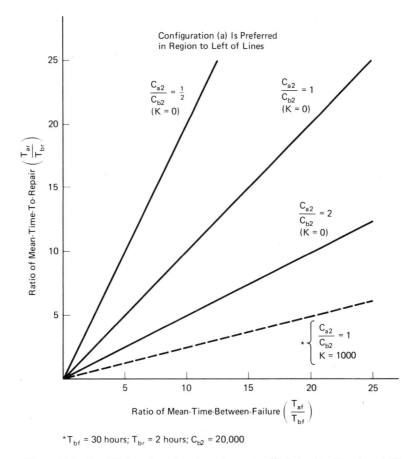

Figure 15.6 Simplified configuration-alternative-trade-off relationship (Equation 15.18).

This illustrative example and the final results in the form of equation 15.18 and Figure 15.6 demonstrates the following:

- a means of relating a variety of key logistics parameters, including
  useful life
  MTBF
  MTTR
  investment cost
  maintenance/service cost
  costs of up and down equipment conditions
- methods of simplifying a rather complex situation and set of relationships to one basic formula and curve

- the value of comparative (versus absolute) analysis
- some of the difficulties in attempting to quantify, in a simple manner, the factors that bear upon the issue of logistics support and related costs

It can thus be expected that the more serious quantitative logistics models that have been developed are computer based and contain a rather large number of parameters. In general, the relationships are not extremely sophisticated in terms of their inherent mathematics, but the multiplicity of parameters suggest data-driven models supported by some database management system. The illustrative examples shown in this section might be expected to constitute simple subroutines or procedures in an overall model. On the other hand, these examples by themselves are easily adapted to a spreadsheet or programming-language (e.g., BASIC) environment leading to both tabular and graphical parametric outputs.

15.2.3 Integrated Logistics Support Plan

An overview of the ILS area can be obtained by citing the various elements that generally go into what is called an *integrated logistics support plan* (ILSP). Such a plan actually contains a variety of interrelated subplans dealing with the following subject areas:

- logistics-support analysis
- test and support equipment
- supply support
- facilities
- transportation and handling
- publications and manuals
- personnel and training
- technical data
- computer resources
- reliability and maintainability
- system retirement

The list is indicative of the scope of activities normally considered within the context of integrated logistics support. As plans are developed and approved in each of the cited support areas, they carry forward and become the basis for preparing operations and maintenance (O&M) support manuals and documentation for the system in the field.

15.2.4 Department of Defense Integrated Logistics Support Standards

Since the DoD has been acquiring, operating, and maintaining large-scale systems for a long time, they have established many standards that relate to ILS. Among these standards are

1. DoD Directive 5000.39—Acquisition and Management of Integrated Logistics Support for Systems and Equipment, November 17, 1983
2. Mil. Std. 1388-1A—Logistics Support Analysis, April 1980
3. Mil. Std. 1388-2A—DoD Requirements for a Logistic Support Analysis Record, July 1984

Although there are other standards directed toward systems engineering subjects that relate to ILS (see Appendix C), the three documents listed will provide both an overview of the ILS area as well as requirements for specific types of logistics analyses and record keeping. Number 1 defines the overall requirements for ILS whereas the latter two standards focus on how to perform logistics-support analysis (LSA), a central part of ILS. Results of LSA are placed in records, known as LSARs. The reader with a further interest in the ILS area is urged to review the format and content of the three items listed and related military ILS standards.

15.3 QUALITY ASSURANCE

The issue of quality assurance (QA) for large-scale systems has been with us for some time. As new technologies have been developed (e.g., rockets, computers), this issue has become more visible, often resulting in reinforced efforts to assure or improve quality. Quality assurance, of course, applies to all elements of a program, including hardware, software and personnel. However, we will focus here on software QA since it is a newer field and is least understood.

Chapter 14 presented a discussion of software development and leaned heavily upon the requirements of Mil. Std. 2167. With respect to software quality assurance there is a companion, Mil. Std. 2168 [7] and we will emphasize the nature and requirements of that standard. Matters relating specifically to software quality factors or measures will be reserved for the end of Section 15.3.

15.3.1 Software Quality-Assurance Standard

Mil. Std. 2168 deals primarily with requirements for evaluating the quality of software for mission-critical computer systems. These requirements may apply as well to independent verification and validation (IV&V) activities if properly tailored to a particular program. The general requirement areas, which are independent of the phase of the software development cycle (as defined in Figure 14.1), are cited here in Table 15.1. One item to be noted is the requirement for the evaluation of software development tools, discussed in Chapter 14. Software-configuration management is another requirement and is discussed later in this chapter. The area of risk management was considered in Section 13.1.

TABLE 15.1 GENERAL REQUIREMENT AREAS FOR SOFTWARE-QUALITY EVALUATION

- Establishment and Implementation of a Software Quality-Evaluation Process
- Independence and Qualification of Evaluators
- Software Quality-Evaluation Planning
- Evaluation of Tools and Facilities
- Evaluation of Software-Configuration Management
- Evaluation of Software Development Library
- Evaluation of Documentation and Media Distribution
- Evaluation of Storage and Handling
- Evaluation of Nondeliverables
- Evaluation of Risk Management
- Quality of Subcontractor Products
- Commercially Available, Reusable, and Government-Furnished Software
- Quality-Evaluation Records
- Corrective-Action Process
- Quality-Evaluation Reports
- Evaluation of the Corrective-Action Process
- Assessment of Software-Quality Evaluation
- Certification
- Interface Between Development Contractor and Software IV&V Contractor
- Government Review at Contractor, Subcontractor, or Vendor Facilities
- Cost of Software-Quality Evaluation
- Deliverable Data
- Deviations and Waivers

Detailed requirements in the software QA standard are cited in regard to each phase of the software development cycle, including the

- requirements-analysis phase
- preliminary design phase
- detailed-design phase
- coding and unit-testing phase
- CSC integration and testing phase
- CSCI testing phase
- evaluations following the CSCI testing phase

The requirements are generally summarized, for each phase, in a matrix of evaluation criteria versus items to be evaluated. Illustrative evaluation criteria cutting across the various phases are listed in Table 15.2. These are related to but not consistently the same as the software-quality factors discussed later in Section 15.3.2. The items to be evaluated are, for example, the manuals, plans, and documents called for during the software development process (e.g., the SDP, SSPM, SCMP, SQEP—see Chapter 14). Thus, a rather complete evaluation process is set forth in a matrix format and in relation to all formal deliverables associated with the software development phases.

TABLE 15.2 EVALUATION CRITERIA FOR ASSESSMENT OF
SOFTWARE PRODUCTS

- Adequacy of Planned Tools, Facilities, Procedures, Methods, and Resources
- Adequate Test Coverage of Requirements
- Adherence to Required Format and Documentation Standards
- Appropriate Allocation of Sizing, Timing Resources
- Appropriate Content for Intended Audience
- Appropriate Degree of Completeness
- Appropriate Level of Detail
- Appropriate Requirement, Design, Coding Techniques
- Compliance with Contractual Requirements
- Consistency with Indicated Documents
- Feasibility
- Internal Consistency
- Technical Adequacy
- Traceability to Indicated Documents
- Understandability

15.3.2 Software-Quality Factors

The software QA standard [7] also defines, in an appendix, a set of software-quality
factors that may be used as requirements in the software-requirements specification.
These factors are listed here:

- correctness
- efficiency
- flexibility
- integrity
- interoperability
- maintainability
- portability
- reliability
- reusability
- testability
- usability

They are embedded in the evaluation process but are not explicitly or necessarily a
mandatory set of requirements. They may be compared with software properties for
verification and validation cited in the literature [8], as

- completeness—all parts of requirement spec fully included
- consistency—no internal design conflicts

Sec. 15.3   Quality Assurance

- correctness—input–output relationship is provably correct
- traceability—full antecedence in other specs
- feasibility—anticipated life-cycle costs less than life-cycle benefits
- equivalence—same behavior of two designs
- termination—sufficiently detailed design for full implementation

This set of quality factors may also be compared with those delineated in a study of the specification of software-quality attributes [9]. This study developed a hierarchy of factors, criteria, and metrics, as shown in Figure 15.7, with the factors sorted into three areas of acquisition concern, namely

1. performance
2. design
3. adaptation

The relationship between factors and criteria is shown in matrix form in Figure 15.8. Moving down the hierarchy, the metrics for each criterion are listed in Figure 15.9. Weighting and rating procedures or scorecards are suggested, building upward from the basic definitions of the metrics. Thus, a complete scheme for software-quality definition and evaluation is provided. This scheme is rather elaborate and oriented to the hierarchy of factors, criteria, and metrics whose full definitions can be found in the referenced study. Mil. Std. 2168 defines a top-level set of factors, as cited, but does not require formal ratings of these factors. Rating and weighting evaluation procedures have been discussed at some length in Chapter 13.

The final point relative to software-quality factors is simply that the definitions and evaluation notions outlined in this section are implicitly candidates for carrying out a software IV&V program, where V&V can be broadly stated as

- verification—confirmation that products of each software-development phase fulfill requirements imposed by an earlier phase
- validation—confirmation that all requirements are correct and when taken in combination will satisfy all system-level mission needs

The basic thrust of IV&V efforts is contained in these two definitions and the title, implying an activity independent of the mainline software development process. It is seen that software-quality assurance and IV&V thus bear a relationship to one another. Both are likewise in relatively early and formative stages of development and application, primarily since the computer and software areas are moving so quickly and are difficult to deal with in large-scale systems. Hopefully, the efforts discussed in this section will help to give us a better understanding of how to do so.

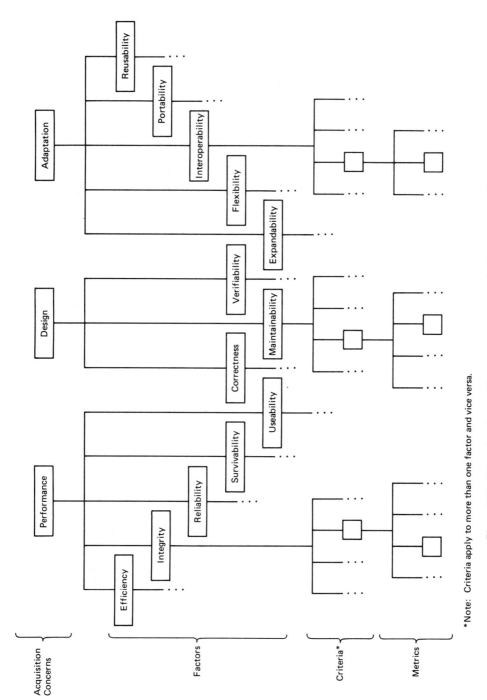

*Note: Criteria apply to more than one factor and vice versa.

Figure 15.7 Hierarchy of acquisition concerns, factors, criteria, and metrics [9].

Acquisition Concerns →	Performance					Design			Adaptation				
	Quality Factors												
Criteria	Efficiency	Integrity	Reliability	Survivability	Usability	Correctness	Maintainability	Verifiability	Expandability	Flexibility	Interoperability	Portability	Reusability
Accuracy			●										
Anomaly Management			●	●									
Application Independence													●
Augmentability									●				
Autonomy				●									
Distributedness				●									
Effectiveness (communication)	●												
Effectiveness (processing)	●												
Effectiveness (storage)	●												
Operability					●								
Reconfigurability				●									
System Accessibility		●											
Training					●								
Completeness						●							
Consistency						●	●						
Traceability						●							
Visibility							●	●					
Commonality											●		
Document Accessibility													●
Functional Overlap											●		
Functional Scope													●
Generality									●	●			●
Independence											●	●	●
Modularity				●		●	●		●	●	●	●	●
System Clarity													●
System Compatibility											●		
Self-Descriptiveness							●	●	●	●		●	●
Simplicity			●			●	●		●	●			●
Virtuality									●				

Figure 15.8 Relationships between quality factors and criteria [9].

Criteria	Metrics
Accuracy	Accuracy Checklist
Anomaly Management	Error-Tolerance Control
	Improper Input Data
	Computational Failures
	Hardware Faults
	Device Errors
	Communications Errors
	Node/Communication Failures
Application Independence	Database Management Implementation Independence
	Data Structure
	Architecture Standardization
	Microcode Independence
	Functional Independence
Augmentability	Data-Storage Expansion
	Computation Extensibility
	Channel Extensibility
	Design Extensibility
Autonomy	Interface Complexity
	Self-Sufficiency
Commonality	Communications Commonality
	Data Commonality
	Common Vocabulary
Completeness	Completeness Checklist
Consistency	Procedure Consistency
	Data Consistency
Distributedness	Design Structure
Document Accessibility	Access to Documentation
	Well-Structured Documentation
Effectiveness (Communication)	Communication-Effectiveness Measure
Effectiveness (Processing)	Processing-Effectiveness Measure
	Data-Usage-Effectiveness Measure
Effectiveness (Storage)	Storage-Effectiveness Measure
Functional Overlap	Functional-Overlap Checklist
Functional Scope	Function Specificity
	Function Commonality
	Function-Selective Usability
Generality	Unit Referencing
	Unit Implementation
Independence	Software Independence from System
	Machine Independence
Modularity	Modular Implementation
	Modular Design
Operability	Operability Checklist
	User Input Communicativeness
	User Output Communicativeness
Reconfigurability	Restructure Checklist

Figure 15.9 Criteria and related metrics [9], (continued on next page).

Self-Descriptiveness	Quantity of Comments
	Effectiveness of Comments
	Descriptiveness of Language
Simplicity	Design Structure
	Structured Language or Preprocessor
	Data-Flow and Control-Flow Complexity
	Coding Simplicity
	Specificity
	Halstead Level of Difficulty Measure
System Accessibility	Access Control
	Access Audit
System Clarity	Interface Complexity
	Program-Flow Complexity
	Application-Functional Complexity
	Communication Complexity
	Structure Clarity
System Compatibility	Communication Compatibility
	Data Compatibility
	Hardware Compatibility
	Software Compatibility
	Documentation for Other Systems
Traceability	Cross Reference
Training	Training Checklist
Virtuality	System/Data Independence
Visibility	Unit Testing
	Integration Testing
	CSCI Testing

Figure 15.9 *(cont.)*

15.4 TEST AND EVALUATION

Test and evaluation (T&E) is an extremely broad subject, in principle extending from the end of a system's development phase up through the decision to procure the system. In reality there is a considerable amount of test and evaluation during development, albeit generally at a subsystem or component level. The scope and pervasiveness of test and evaluation may be illustrated by the activities listed that may be properly considered a part of T&E

- test management
- test design
- development test
- operational test
- production test
- qualification test
- acceptance test
- environmental testing
- unit testing
- assembly-line testing

- sequential testing
- hypothesis testing
- system end-to-end testing

In spite of the scope and significance of T&E, it does not receive emphasis in the DoD's systems engineering management standard [4]. Explicit recognition is registered under the subject of system test planning and the following types of requirements:

- system tests are to be products of the systems engineering effort
- all engineering specialties are to be integrated to lead to an effective and economical total system test plan
- the test program and data must relate to
  TPM analysis
  program planning
  human performance
  personnel selection
  training
  man–machine interfaces (MMI)

A touchstone for the consideration of T&E requirements in the DoD, however, is DoDD 5000.3 [10] in which emphasis is given to the test and evaluation master plan (TEMP). An overview of the contents of a TEMP is provided in Table 15.3, and selected aspects of the TEMP are discussed next.

The TEMP is an overview document that describes all the top-level test considerations for the system. As noted in Table 15.3, development test and evaluation (DT&E), and operational test and evaluation (OT&E) are given particular emphasis in this document, which is updated each time a significant program milestone is reached. Such milestones represent continuing decisions to proceed with the system and the TEMP is an important part of that decisionmaking process. Development testing can be considered a general term for all significant testing during the development phase. Operational testing is the final verification that the system meets specification prior to the decision to acquire the system. During this latter type of testing, it is of special importance to create an environment for the testing that represents as closely as possible the true operating set of conditions that the system will see in the field.

Supporting the TEMP are numerous testing activities that are necessary as metal is bent and code written. Each of these lower level test plans and procedures are part of the system-development process and hopefully continue to add confidence that the design is appropriate and that the component selection is proper. Early testing is emphasized by the experienced program manager who wishes to verify, as soon as possible, that the basic system design is workable.

It is not possible here to examine even a small part of the T&E activity. Two special points, however, are made. The first is that there are numerous lower level tests that need to be made that have statistical analysis as a foundation. Such tests usually have to do with large numbers of components that can be sampled in order to

TABLE 15.3 ILLUSTRATIVE CONTENTS OF TEST AND
EVALUATION MASTER PLAN

Part I   Description
  • Mission Statement
  • System Description
  • Required Operational Characteristics
  • Required Technical Characteristics
  • Critical issues

Part II   Program Summary
  • Management Summary
  • Integrated Schedule
  • Test Documentation

Part III   Development Test and Evaluation (DT&E)
  • DT&E Objectives
  • DT&E to Date
  • Future DT&E
  • Critical DT&E Items

Part IV   Operational Test and Evaluation (OT&E)
  • OT&E Objectives
  • OT&E to Date
  • Future OT&E
  • Critical OT&E Items

Part V   Production Acceptance Test and Evaluation (PAT&E)
  • Planned PAT&E
  • Critical PAT&E

Part VI   Special Resource Summary
  • Test Articles
  • Special Support Requirements

determine their performance and reliability characteristics. Thus, the systems-test
engineer should be well versed in statistical testing, including both large and small
sample statistics. In the context of computer-aided techniques, it has been demonstrated in Chapter 7 that a variety of inexpensive statistical software packages are
accessible at the microcomputer level. These tools are available and as the p-c becomes commonplace on the desk of the systems engineer it can be expected that they
will be used in this capacity with increasing frequency and advantage. The second
point is that not all aspects and specifications of the system can be tested. This can
lead to considerable controversy in the "sign-off" of a system as well as uncertainty
as to whether or not the system will really perform as predicted. An overall test concept for dealing with such issues is a potential solution to such a problem and is discussed in Section 15.4.1.

15.4.1 Overall System Test Concept

In the final analysis, when system tests are made, results are compared with the system's approved specifications to verify that the system satisfies these specifications.
In broad terms, if the specifications are met, the system is accepted and becomes a

candidate for procurement. Conversely, if the specs are not satisfied, the system is rejected. Since there are numerous specifications, it may be that

- not all specifications are tested against
- of the specifications tested against, the system passes some and fails others

Thus, there are several gray areas in testing but the key principle remains—testing is carried out in relation to the currently approved system specifications.

The reasons that not all specifications are tested against are easy to understand. First, it may be impossible to do so directly, as in the case of verifying a satellite's in-orbit performance and reliability for ten years. Second, it may not be possible in terms of the operational scenario or threat, as is the situation for the SDI and testing against a threat of a large number of intercontinental ballistic missiles (ICBMs). Finally, it may not be feasible in terms of practical cost constraints, as with attempting to verify a point defense surface-missile system's kill probability against a set of targets. Statistically, it would be prohibitively costly to run enough targets through such tests. Thus it is necessary to formulate a system-test concept that addresses such issues. This type of concept involves the integration of real-world testing and test results with system modeling and simulation.

Under the concept of integrating real-world test results and system modeling, the following steps are appropriate:

1. development of a system model incorporating as many specification-related parameters as is feasible
2. for parameters in number 1, deciding on which ones to test directly and which ones to "verify" through modeling
3. designing and running real-world tests for the parameters that can be tested
4. using the test results from number 3 in the model to verify, through analysis, that the system meets specification

The aforementioned concepts and steps may be illustrated by reference to Figure 4.16. This figure shows a parameter dependency diagram (PDD) for the search mode of an air-defense fire-control system. That diagram, in conjunction with the supporting analytic structure (see Figures 4.17, 4.18, and 4.19), constitutes a model of the system. As shown in Figure 4.17, a total of 26 parameters make up such a model. Figure 15.10 shows a hypothetical sorting of parameters, as suggested in the previous discussion, which are

- cited in the specification
- to be real-world tested
- to be inferred by exercising the model
- known through other means (e.g., a constant)

From Figure 15.10 we observe that all parameters cited in the spec are covered either through real-world testing or inferential modeling (or are otherwise known).

Parameter	Cited In Spec?*	To Be Real-World Tested?*	Inferred Through Modeling?*	Known?*
Boltzmann's Constant	N	N	N	Y
Detection Probability	Y	N	Y	N
Detection Threshold	N	Y	N	N
Discriminator Gain	N	Y	N	N
False-Alarm Probability	Y	N	Y	N
I-F Amplifier Gain	N	Y	N	N
I-F Bandwidth	Y	Y	N	N
I-F Output Carrier Power	Y	Y	Y	N
I-F Output Noise Power	Y	Y	Y	N
I-F Output Noise Power Spectrum	Y	Y	N	N
I-F Output Signal Power	Y	Y	Y	N
Maximum Output Voltage	N	Y	Y	N
Output Noise Power	Y	Y	Y	N
Post Discriminator Filter Voltage Gain	N	Y	N	N
Post Discriminator Filter High-Frequency Cutoff	N	Y	N	N
Post Discriminator Filter Low-Frequency Cutoff	N	Y	N	N
Post Discriminator Noise Power Spectrum	N	Y	N	N
Range to Target	Y	Y	Y	N
Receiver Processing Power Gain	N	Y	N	N
Receiver Processing Power Losses	N	Y	N	N
Receiving Antenna Gain	Y	Y	N	N
System Noise Temperature	N	N	N	Y
Target Cross Section	Y	N	Y	N
Transmitted Power	Y	Y	N	N
Transmitting Antenna Gain	Y	Y	N	N
Wavelength	Y	Y	N	N

*Y = yes
 N = no

Figure 15.10 Sorting of parameters for test design.

Thus, through the mechanism of testing and modeling, an integrated approach can be developed. We note, in this illustrative case, the key role of the PDD technique as a modeling tool and how it provides a foundation for an overall system-test concept. Other modeling and simulation tools (see Chapter 12) are also candidates for application of this approach.

15.5 CONFIGURATION MANAGEMENT

Configuration management (CM) is a relatively straightforward set of activities that control and keep track of all significant changes in a system. CM, in a hardware context, has been going on for a long time and many CM systems are in place and operating very well. As major software systems have been developed, CM systems have not adapted adequately to software challenges. Indeed, it is because of the software issue that CM is included in this text.

A particularly cogent CM tutorial [11] reinforces the point that CM systems for software are of special importance and must be adapted to the kinds of problems posed by software. E. H. Bersoff, coauthor of the cited tutorial, has recognized the relevant issues and has set forth excellent descriptions [12] of the problem as well as suggested solutions.

In broad terms, the functional elements of a CM system can be thought of as fivefold

1. identification
2. control
3. auditing
4. status accounting
5. traceability

As a software (or hardware) system moves through its life cycle, various base lines are established. These base lines are normally identified with system-review milestones, but the basic concept is simply that a base line represents a point in a system's cycle when it is convenient to define a set of stable end products and move on to the next set of developmental activities. A set of such base lines might be identified as

1. functional/requirements
2. allocated
3. design
4. production
5. operational

Returning to the five functional elements of CM, the identification element is simply the proper description of all configuration items as well as all interfaces between items. This description is normally embodied in the tiers of specifications for the system. Documentation standards for these purposes are available, especially for the DoD and several are cited in Appendix C. The control element is applied to each base line and significant changes are most often controlled by a configuration control board (CCB). The boundaries as to what must go through a CCB must be defined in some fashion, but control is nonetheless exercised by some other responsible party. Thus, in principle, any authorized changes can only be made to a base line with the approval authority of some duly authorized body or persons. The audit element can be broken down into a functional configuration audit (FCA) and a physical configuration audit (PCA). The former validates the satisfactory completion of a configuration item, and the latter does much the same but normally operates just prior to the production of a system (i.e., at the end of the full-scale development phase). Audits can also be performed when there is reason to believe that the configurations are improperly identified or operating below required standards. The status accounting element, in a very real sense, supports elements 1 through 3. That is, it fully documents all configuration-management actions and results. Indeed, status accounting sometimes includes all documentation, though this is the exception rather than the rule. In any case, this function is the *scribe* or *recorder* of the full implementation of the other CM elements. Finally, the matter of traceability is considered although it is sometimes included in the status-accounting element. Traceability can be viewed as an overlay that correlates the configurations throughout all the base lines. This allows for the tracking of increases in detail and changes from the early functional and requirements base line through to the final operational base line. CM should also be applied to the operational fielded system. This area is often not properly tended to, resulting in extremely poor documentation, tracking, and accountability for systems that have been in operation for many years.

Another dimension of the software CM issue is to recognize that software can be thought of in terms of two forms: *nonexecutable* and *executable* [11]. The latter is the code itself that provides operating instructions to the computer. The former can be viewed as everything else in a software program that supports the efficient generation and application of code (e.g., data flow diagrams and data dictionaries).

One way to view the software CM area, then, is to map the three dimensions against one another, that is, the five functional elements of a CM system mapped against the five system base lines, mapped against the two forms of software. One can argue that if the five functional elements of a CM system address these other two dimensions clearly and efficiently, the major ingredients for an excellent software CM system are present.

In order to obtain some insight into how one official developer and acquirer of computer systems deals with CM, we cite some of the elements of an approach taken by a Data Directorate at NASA's Goddard Space Flight Center [13]. First, the objectives of CM in that application are set forth in terms of

- maximizing control at the proper levels of responsibility
- identifying items subject to configuration control
- providing traceability
- ensuring coordination of design changes with the established base lines
- providing uniform reporting and documentation
- incorporating management flexibility and visibility into design changes
- maximizing responsiveness and minimizing formality

As a matter of policy, several CCBs are established to implement the control element of the CM system. The five system base lines cited previously are accepted as the key ones subject to CM, as depicted in Figure 15.11. Software is not emphasized but a key function of the directorate is to produce and use software in fulfillment of its charter.

For a large-scale system, the paperwork involved in keeping track of all baseline configurations can be extensive. This area thus becomes an excellent candidate for the use of computer aids, particularly a database management system with a traceability capability or overlay. Various software developers have recognized this

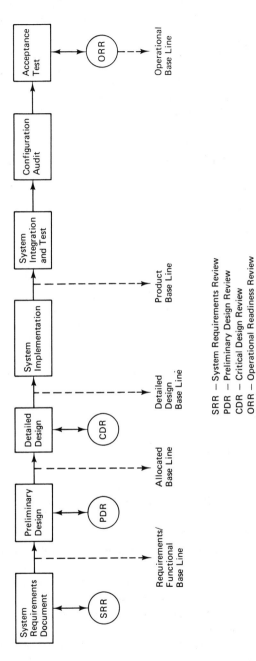

Figure 15.11 System base lines in relation to milestones.

SRR — System Requirements Review
PDR — Preliminary Design Review
CDR — Critical Design Review
ORR — Operational Readiness Review

need and have produced packages designed specifically for configuration management. Results in this regard are discussed in the next section, which deals with software tools applicable to systems engineering support functions in general.

15.6 SOFTWARE TOOLS

Software tools operating at the p-c level are generally not plentiful for the particular purpose of aiding in the execution of systems engineering support functions. However, there are some p-c-based tools available and we will first comment upon such tools.

In Chapter 7, a variety of mathematical and engineering tools were cited in a set of categories that included

- statistical
- engineering
- reliability
- simulation

For the RMA area discussed in this chapter, the reliability packages are clearly applicable. As mentioned in Chapter 7, reliability prediction programs are available that can satisfy many of the systems engineer's needs. For complex RMA problems that cannot be handled easily through analytic techniques, the user may be able to move to one of the simulation packages operating at the p-c level. In addition, if the RMA problem gets into the area of testing and statistical inference, the statistical packages may be directly applicable. A long list of some of the capabilities represented by these software packages was provided in Table 7.1. Another approach, if the software listed does not have the required features, is to program the problem and its equations directly with one of the programming languages (see Chapter 9).

The issues are somewhat the same for ILS since it is closely related to the RMA area. Limiting ourselves to the p-c level, we do not find a commercial focus to aid in ILS problem solving. Some educational institutions, however, have offered short courses that have addressed this area. For example, the University of California at Los Angeles (UCLA) has provided the following courses:

- ILS/LSA models and analysis for engineering and logistics managers
- Microprocessor programs and methods for integrated logistics support

The former is more oriented to modeling for ILS in general. The latter specifically goes into p-c software that may be used for ILS analyses. Both have a strong emphasis on life-cycle costing since there is a basic relationship between LCCMs and ILS. The University of Maryland University College, in conjunction with the UCLA Extension, offers these courses to working engineers as well as degree-seeking students. Thus there is a clear path to providing computer-aided tools at the micro level for support in the ILS arena, although specialized software tools are not generally available at your local p-c software retail store.

Much the same situation holds true for the areas of quality assurance, T&E, and configuration management. For T&E, however, one can expect that many of the statistical-analysis packages available on the commercial market are useful for problems in which formal statistical testing is required. One should not lose sight, in addition, of the overall utility of microbased spreadsheets and database management systems for application to these areas. As an example, it is clear that a DBMS has a major role to play in setting up a CM system, as referred to earlier in this chapter.

If one goes beyond what might be available commercially for the microcomputer, the situation changes dramatically in the sense that many models and programs can be found that apply to systems engineering support functions. Since the operating environment is the mini or mainframe, however, the software is not normally accessible by the student. Working engineers should be prepared to spend much more for these programs and their supporting documentation than for most commercial p-c software. We will focus on the ILS and CM areas to cite some relevant examples.

A list of selected models for the ILS function is provided in Table 15.4. This list includes the category of LCCMs that are related to logistics analysis, which were discussed in Chapter 11. Many of the models cited were sponsored by the federal government [14] and can usually be obtained by contacting the responsible government agency. In some cases, government printing office documents will be helpful in locating models that are sponsored by the government. Very specialized software is sometimes available through the authors of texts in this and related areas and the operating environments for this software is quite varied.

TABLE 15.4 SELECTED LOGISTICS-RELATED SUPPORT MODELS

ABLE	NRLA
ASTRA	ORLA
BDSM	SAMSON
COAMP	SCORE
GEMM	SKEM
HARDMAN	SPM
LCOM	TIGER
LORAM	Various LCCMs (see Chapter 11)

In the CM arena, we identify three specific packages that are sold to handle such applications, namely

1. Configuration Management Automation System (CMAS), BTG, Inc., Vienna, VA
2. Configuration Management Toolkit (CMT), Expertware, Mountain View, CA
3. Change and Configuration Control (CCC), Softool, Inc., Goleta, CA

These packages deal with the configuration-management elements discussed in this section in a direct and specific manner. Their design generally reflects an understanding of the government requirements for CM and specifically reflects the need to satisfy certain government standards, as cited in this chapter and reiterated in Appendix D. Again, these packages are not normally available to the student and are aimed at the practicing systems engineer who needs such a package for long-term application to large-scale systems engineering programs. As has been discussed in other chapters, we note finally that support tools are becoming increasingly available through third-party vendors for the workstation environment. As an example, Apollo Computer makes the CCC package from Softool available to the users of their workstations. These trends should continue. Also, as the p-c becomes more powerful more tools will probably become available for the microcomputer environment.

15.7 CASE APPLICATION SUMMARY

In summary, computer-aided applications to systems engineering support functions might be sorted into three categories. The first is the set of commercially available p-c software that can be used directly, such as the statistical, engineering, reliability, and simulation packages. The second is the set of commercially available p-c software that can be used indirectly, such as spreadsheets, DBMSs, and even the programming languages that can be utilized to program a special set of equations for a particular problem. This more-or-less exhausts the ways in which microcomputer packages can currently be used in this application area. The third possibility is to move up to the mini and mainframe environment, in which case there are other packages that can be found and accessed, such as those mentioned in the previous section. A variation on this theme is to access such a package and simplify it so that it can run in the more limited p-c environment. This latter approach is becoming increasingly popular as the p-c becomes more powerful. As a final means of addressing application areas in specific terms for systems engineering support functions, Table 15.5 lists some of the problems that can be handled by computer-aided techniques.

TABLE 15.5 CASE APPLICATION AREAS FOR SELECTED SYSTEMS
ENGINEERING SUPPORT FUNCTIONS

Reliability-Maintainability-Availability
- Use of reliability software package for reliability prediction
- Use of statistical software package for testing the life characteristics of a critical component
- Coding with a programming language to model the reliability of a system, including redundant configurations

Integrated Logistics Support
- Evaluation of spares-provisioning requirements using statistical software package
- Definition of ILS parameter model in a life-cycle cost matrix set up on a spreadsheet
- Use or adaptation of a mini or mainframe-based ILS model (see Table 15.4)

TABLE 15.5 (*cont.*)

Quality Assurance
  • Quality factor rating and weighting on decision support software (DSSW)
  • Quality factor rating and weighting on a spreadsheet or integrated package
  • Evaluation of software development products (e.g., SDP, SSPM, SCMP, SQEP) against evaluation criteria, as per Mil. Std. 2168

Test and Evaluation
  • DBMS application for tracking T&E program versus requirements and specifications
  • Word processor for test plans and procedures
  • Use of system model (e.g., simulation package) in conjunction with test data for integrated test planning

Configuration Management
  • Word processor for program and plans description
  • DBMS for use by configuration control board (CCB) to access current configuration-status data
  • DBMS and word processing for status accounting function

REFERENCES

1. Lloyd, D. K., and M. Lipow, *Reliability: Management, Methods and Mathematics*. Englewood Cliffs, NJ: Prentice-Hall, 1962.

2. *System Engineering Management Guide*. Defense Systems Management College, Fort Belvoir, VA, October 3, 1983.

3. Blanchard, B. S., and W. J. Fabrycky, *Systems Engineering and Analysis*. Englewood Cliffs, NJ: Prentice-Hall, 1981.

4. *Engineering Management*. Military Standard 499A, Department of Defense, May 1, 1974.

5. *Reliability Engineering Handbook*, vol. 2, NAVAIR 01-1A-32, Naval Air Systems Command, Washington, DC20361, U.S. Navy, July 1, 1977.

6. Brown, F. B., et al, "Techniques for the Evaluation of Surveillance Radar Systems," Defense Documentation Center No. AD 416320, October 31, 1961.

7. *Software Quality Evaluation*. Military Standard DoD-STD-2168 (draft), Department of Defense, April 26, 1985.

8. Yau, S. S., and J. J.-P. Tsai, "A Survey of Software Design Techniques," *IEEE Transactions on Software Engineering*, SE-12, no. 6, June 1986.

9. Bowen, T. P., G. Wigle, and J. T. Tsai, (Boeing Aerospace Company) *Specification of Software Quality Attributes*. RADC-TR-85-37, vol. I, Rome Air Development Center, Air Force Systems Command, Griffiss Air Force Base, AD-A153-988, February 1985.

10. *Test and Evaluation*. DoD Directive 5000.3, Department of Defense.

11. Bersoff, E. H., V. D. Henderson, and S. G. Segal, "Software Configuration Management: A Tutorial," *IEEE Computer Magazine*, January 1979.

12. Bersoff, E. H., V. D. Henderson, and S. G. Segal, *Software Configuration Management: An Investment in Product Integrity*. Englewood Cliffs, NJ: Prentice-Hall, 1980.

13. *Systems Management Policy*. MDOD-8YMP/0185, NASA/GSFC, Mission Operations and Data Systems Directorate, April, 1984.

14. Blanchard, B. S., *Logistics Engineering and Management*. Englewood Cliffs, NJ: Prentice-Hall, 1974.

*15. *Reliability and Maintainability*. DoD Directive 5000.40, Department of Defense, July 1980.

*16. *Configuration Management*. DoD Directive 5010.19, Department of Defense.

*17. *Configuration Control-Engineering Changes, Deviations and Waivers*. Military Standard 480A, Department of Defense, April 12, 1978.

*18. *Configuration Management Practices for Systems, Equipment, Munitions and Computer Programs*. Military Standard 483, Department of Defense, March 21, 1979.

*19. *Reliability Prediction of Electronic Equipment*. Military Handbook 217D, Department of Defense, January 1982.

*20. *Maintainability Program for Systems and Equipment*. Military Standard 470A, Department of Defense, 1983.

*21. *Maintainability Verification, Demonstration and Evaluation*. Military Standard 471A, Department of Defense, March 1973.

*22. *Definition of Effectiveness Terms for Reliability, Maintainability, Human Factors, and Safety*. Military Standard 721C, Department of Defense, 1980.

*23. *Reliability Modeling and Prediction*. Military Standard 756B, Department of Defense, November 1981.

*24. *Reliability Program for Systems and Equipment*. Military Standard 785B, Department of Defense, September 1980.

*25. *Reliability and Maintainability Management Handbook*. NAVAIR 01-1A-31, Naval Air Systems Command, U. S. Navy, July 1, 1977.

*26. Bazovsky, I., *Reliability Theory and Practice*. Englewood Cliffs, NJ: Prentice-Hall, 1961.

*27. Henley, E. J., and H. Kumamoto, *Reliability Engineering and Risk Assessment*. Englewood Cliffs, NJ: Prentice-Hall, 1981.

PROBLEMS

P15.1. Prove Equation 15.4, given a constant hazard function.

P15.2. Develop Equation 15.6 from the definition of the hazard function for the Weibull.

P15.3. Prove the expressions shown in (c) through (g) in Figure 15.2.

P15.4. Find a system's availability for the following parameters:
  a. exponential failure with a failure rate of 0.01 failures per hour
  b. MTTR of three hours
  c. average logistics delay time of two hours
  d. average administrative delay time of one hour

P15.5. For an equipment with an MTBF of 100 hr, how many spares are required to assure a 0.999 probability that a spare will be available over a time period of 1000 hours?

P15.6. Prove Equation 15.18.

*Reference is relevant to the text but is not specifically cited

P15.7. Write a computer program to calculate values of the ordinate and abscissa, as illustrated in Figure 15.6, using the following languages:
   a. BASIC
   b. Pascal

P15.8. Develop three more lines on the plot of Figure 15.6 by choosing your own parametric values.

P15.9. Select and obtain a copy of one of the ILS military standards cited in this chapter's text or list of references. Write a three-page summary of the main points of this standard and relate your summary to this chapter.

P15.10. Formulate a rating and weighting evaluation scheme and show an example for the evaluation criteria in Table 15.2 and four key software products. What specific software products would you select?

P15.11. Develop a four-level hierarchical rating and weighting scheme for a selected subset of the acquisition concerns, factors, criteria, and metrics in Figure 15.7. Show an example of how this scheme works.

P15.12. Structure an overall system-test concept similar to that discussed in the text but for software development and an illustrative data flow diagramming technique (see Chapter 4) in place of the PDD.

P15.13. Draw a functional flow diagram for the five elements of configuration management and explain the various blocks in the diagram.

P15.14. Write a one-page discussion of how each of the following categories of software might apply to the five systems engineering support functions discussed in this chapter:
   a. statistical
   b. engineering
   c. reliability
   d. simulation

I like the dreams of the future better than the history of the past

Jefferson

FORECASTING AND ARTIFICIAL INTELLIGENCE

This chapter addresses two subjects that in one way or another deal with directions for the future. These subjects are:

- forecasting
- artificial intelligence

With respect to forecasting, the issue is largely one of methodology. That is, there are many techniques involved in forecasting and an overview of these techniques is provided. This includes a brief examination of technology forecasting. All of these issues bear directly on systems engineering in that the systems engineer must be prepared to project demand upon or for a system and understand what technology might be available for future inclusion into the base-line design or for later versions of the system. This becomes even more relevant with the rapid improvements in technology and as system developments take longer.

In the case of artificial intelligence, the situation is somewhat different. Here we have a specific area of relatively new technology that is moving vigorously ahead. In general, it is not well known to the practicing engineer who has been out of touch with formal engineering training for some years. But beyond that it shows promise of being extremely important in the future, particularly in an area known as expert systems. The techniques that are part of such systems address diagnostic processes of various types (e.g., equipment maintenance) and even design procedures that may be involved in large-scale complex systems, to name just two areas. Thus, we take time in this chapter to consider some fundamentals regarding expert systems and cite some of the available software that can be accessed by the systems engineer.

16.1 FORECASTING

Forecasting represents a series of techniques that address projections into the future in applications such as

1. passengers or cargo that might be moved by a new transportation system
2. traffic between all nodes in a large-scale communications network
3. potential purchasers of a new data-communications device
4. the nature and capability of telecommunications technology for a system whose initial operation might be scheduled for the year 2000

Items 1 through 3 are related to more conventional questions concerning the demand on a system. Item 4 deals with technology forecasting. Both of these types of questions are considered in this section.

A particularly good overview of the general field of forecasting is reflected in the top-level taxonomy of forecasting methods listed below [1]:

- judgment methods
- counting methods
- time-series methods
- association or causal methods

In the judgment cases, a strong subjective element is present, although the results can be both qualitative and quantitative. Expert, executive, or informed opinion is usually a requirement, forming the basis for validation of the method. Research results may also be a major component in finding the expert opinion. The Delphi technique, discussed in Chapter 13, is an example of a judgment method. Counting methods are usually associated with survey techniques such as market or consumer surveys. Written or oral questionnaire instruments are used to collect data from a representative population. These data are then used to extrapolate to the characteristics of larger populations, often utilizing formal statistical survey mathematics. The federal government has used these techniques for years in dealing with such areas as the census, consumer price index, and evaluations of the effectiveness of social programs. Time-series methods are normally mathematical in orientation, forecasting future variable values from current and past values. Many of the microbased statistical software packages (see Chapter 7) incorporate time-series analysis techniques. Finally, the association or causal methods tend to also have strong mathematical foundations, utilizing such procedures as correlation, regression, econometrics, and input–output modeling.

Selection of a forecasting method appropriate to a given problem, given the aforementioned techniques, is a nontrivial task. Criteria have been formulated in relation to these methods [1]. The cited reference also qualitatively evaluates the methods in relation to these criteria to produce a particularly useful and compact guide to forecasting. The evaluation matrix contains 16 criteria, called *dimensions*, mapped against 20 methods, or a total of 320 individual cells. Strengths and weaknesses cited in the cells are color coded to assist the user in a more rapid evaluation.

16.1.1 Illustrative-Demand Forecasting

To illustrate one specific forecasting arena, we turn to the matter of demand forecasting and cases associated with transportation systems. The significance of forecasting in this area is readily apparent. Development lead times are extremely long and early estimation of the potential users of a system is crucial to the design and sizing of the system. In Figure 16.1 we show three types of demand-forecasting relationships applicable to

    a. passenger demand for air travel

    b. freight demand

    c. a transportation modal split

In the Figure 16.1(a) demand is projected as a function of fare, disposable income, and trip time, with the number of forecast passengers going up with decreasing fares and times, and down with decreasing income. We note the form of this equation, relating to a base value for each of the variables. The exponents are empirically derived, utilizing historical data and a variety of possible methods of statistical analysis (e.g., regression techniques). We also observe that in spite of the general notion that forecasting deals with a future time period, time (or future years) is not an explicit variable in the given demand relationship. This is typical of such approaches, allowing one to either associate a time period with the independent variables (in this case fare, time, and cost) or simply to estimate what would happen to demand if certain changes in the independent variables came about. Thus these types of relationships have the same essential character as the cost-estimating relationships (CERs) described in Chapter 11 and use basically the same tools for analysis of empirical data to estimate constants or base values.

Figure 16.1(b), also transportation related, deals with a gravity model that projects tons of freight flowing from one population center to another [2]. The exponents (a, b, and c) are likewise derived empirically. The underlying bases are that increasing population (or manufacturing employment) will drive demand up and increasing distance will drive it down. Here again, time is not an explicit variable.

Finally, for this illustration, Figure 16.1(c) shows a modal-split model that was calibrated in a study of the Northeast corridor [3]. The modal share is based upon an estimate of a fraction of traffic, S_m, for mode m. The equation contains travel times, travel costs, frequency of service, and a modal constant factor K_m that discriminates between mode selection by travelers when the modes have identical times, costs,

$$D = D_B \left(\frac{F}{F_B} \times \frac{I_B}{I} \right)^{-0.7} \left(\frac{T}{T_B} \right)^{-0.35}$$

D = Demand, number of passengers
F = Fare
I = Disposable income
T = Trip time
B = Base value for each respective variable

(a)

$$N_{ij} = \frac{K \, P_i^a \, P_j^b}{R_{ij}^c}$$

N_{ij} = Number of tons flowing from origin i to destination j
P_i = Population (or manufacturing employment) for region i
P_j = Population (or manufacturing employment) for region j
R_{ij} = Distance from population centers from i to j
K, a, b, c = Parameters to be estimated

(b)

$$\omega_m = K_m \, T_{ijm}^{-1.9135} \, C_{ijm}^{-0.8555} \left(1 - e^{-0.007Q_{ijm}} \right)^{0.5536}$$

$$S_m = \frac{\omega_m}{\Sigma \, \omega_m}$$

S_m = Modal share for mode m
T_{ijm} = Door-to-door travel time between cities i and j via mode m
C_{ijm} = Door-to-door travel cost between cities i and j via mode m
Q_{ijm} = Average daily service frequency between cities i and j via mode m
K_m = Modal constant (see text)

(c)

Figure 16.1 Typical demand-forecast relationship: (a) forecast demand, (b) freight demand [2], (c) modal split demand [3].

and frequencies. Illustrative values for this modal constant were determined to be in the following ranges:

Mode	Modal Constant (K_m)
air	0.641
high-speed ground	0.641 and 1.00
bus	0.127

We note these values for the sake of completeness and illustration; updated values would likely be necessary today.

16.1.2 Prediction Concepts

If we focus on the areas in which time is an explicit independent variable, we enter into the complex fields of time-series analysis, prediction theory, filtering, and similar subjects. One of our most advanced thinkers in the field of cybernetics, N. Wiener, helped to develop these concepts in a variety of books, one in particular deals with interpolation, extrapolation, and smoothing of stationary time series [4]. This fundamental approach, as well as others developed from it, looks at estimation or prediction from the point of view of constructing "filters" that perform the required computations on the input data. The latter can be viewed as a signal embedded in noise. The output is cited as a desired signal and the filter characterized by its transfer function (in the frequency domain), equivalent to its response to a unit impulse function in the time domain. Criteria for filter design include minimization of the mean square error or error probability and maximization of the signal-to-noise ratio. Classical theory in this area leans heavily upon the minimization of the mean square error which for the filtering problem can be expressed in terms of the

- impulse response of the filter
- autocorrelation function of the input signal
- autocorrelation function of the desired output signal
- cross-correlation function of the input and desired output signals

This leads directly to what is known as the Wiener–Hopf equation in order to develop an optimal form of the filter impulse response (or transfer function).

Using the above and derivative concepts and theory, a particularly useful filter design procedure was developed by R. E. Kalman [5]. The Kalman filter uses each new measurement, one at a time, to update estimates in a sequential filtering process. It is therefore especially good for real-time or on-line processing and estimation. It is usually better than a *single-shot* filter for the same total measurement data input.

The Kalman filter establishes a fit to a true set of measurements and estimates the current state of a system rather than the state at some epoch (or fixed time), that is, it has a moving epoch coincident with the time of measurement. It is also particularly good for error analyses since it provides updated estimates of the errors in the state vector that it produces in the calculation process. In broad terms, the filter provides a minimum variance estimate of unknowns (state vector), given an a priori estimate of these unknowns, noise characteristics and noisy measurements as a function of the unknowns.

As implied by this discussion as well as the taxonomy of methods, there are numerous technical-forecasting procedures based upon complex mathematics and statistics. These are beyond the scope of this text except by way of indicating that many software packages are available at the p-c level that contain the algorithms necessary to implement these methods. These packages are cited in Chapter 7 under the categories of statistics (primarily) and mathematics (secondarily) and are also addressed under a third category of forecasting software shown in Appendix A.

A particularly relevant forecasting subject area, in distinction to a methodology area, is that of technology forecasting. Its relevance with respect to systems engineering is straightforward; most of our large-scale and complex systems incorporate new technology at least in the important area of embedded computers. Section 16.1.3 discusses some of the features of this field.

16.1.3 Technology Forecasting

This subject area is relatively new, evolving in part from a broad societal need to better understand where our technology is or might be headed, how to take maximum advantage of it, and how to minimize possible deleterious effects and consequences. Indeed, an indicator of our interest is reflected by the establishment of the Congressional Office of Technology Assessment (OTA). The office's charter extends beyond technology forecasting alone; numerous studies performed on behalf of the OTA have included forecasting to establish a base line or direction for the future.

Technology forecasting to date has included both qualitative and quantitative methods. The qualitative approaches can be thought of as utilizing various judgment methods. The quantitative approaches most often are based upon association or causal methods. In the latter case, extrapolations are frequently used in the form of regression equations keyed to the most significant technical parameters of the technology in question. Relationships of this general type were discussed in Chapter 11 in relation to cost-estimating relationships. Clearly, parameters other than cost can be extrapolated by using the same basic methodology.

A large portion of technology forecasting is performed today based on judgment methods that involve expert opinion in a structured (e.g., Delphi) or unstructured manner. Many systems engineering technology forecasts are unstructured and focused on the specific technology items that might be part of the system being designed or its future upgrades. Growth potential, preplanned-product improvement (P^3I), improved price–performance, and similar motivations lead the systems engineer to serious considerations of future technology alternatives for our large-scale systems with lifetimes of 15 years and more. Long developmental lead times also force us into thinking of where the technology might be some 5 to 10 years from now rather than where it is today.

Some of the technology trends that are relevant to computer-aided system engineering are considered in Chapter 17. Here we show a few sample technology forecasting approaches.

The National Aeronautics and Space Administration (NASA) is continuously examining the nature of space systems today and where they might be going. Indeed, NASA is responsible for developing advanced space technology for future missions. In this regard, NASA has developed various versions and updates of what they call a *Space Systems Technology Model* [6]. Here we review some of the features and methodologies of this model.

The stated purpose of the model is to serve as a reference guide for

- planning technology programs and options
- identifying technologies required for planned and potential future missions

- assessing ongoing technology programs
- providing a technology reference source for mission planning

The three volume model is organized as shown in Table 16.1. As seen from this tabulation, technology trends, forecasts, and analysis represent major and significant portions of the model. The trends and forecasts almost constitute a functional design handbook so that systems engineers can determine what capabilities might be available for a future mission. A number of critical figures of merit are used to characterize these technological capabilities. The trends are based on historical data and the forecasts represent consensus expert opinion of persons active and knowledgeable in the various disciplines related to the specific space technologies. Thus, the technique used by NASA is largely a judgment method supplemented in some cases by quantitative extrapolation procedures and data.

TABLE 16.1 OVERVIEW OF NASA'S SPACE SYSTEMS TECHNOLOGY MODEL

  I. System/Program Descriptions and Technology Needs
     Mission Systems
       Chronology of mission systems/programs
       Systems by start date
       Detailed program information
       Summary of technology needs
     Instruments and Experiments
       Chronology of instruments/experiments
       Detailed instrument/experiment information
       Summary of key technology parameters
     Opportunity Systems
       Speculative systems and programs
 II. Technology Trends and Forecasts
     Systems Technology
     Discipline Technology
III. Analysis: Technology Needs and Opportunities
     Technology Analysis
     Landmark Mission Set

The model is rather broad in scope with the following technology categories, each of which is supported by a WBS.

- space-transportation-systems technology
- spacecraft-systems technology
- space-station-systems technology
- mission-support-systems technology
- propulsion

- structures and materials
- power systems
- thermal management
- navigation, guidance, and control
- communications
- processing
- manned systems
- environmental interaction
- payload
- ground support

The process of identifying relevant technologies is exemplified by the chart in Figure 16.2 [6] dealing with spacecraft-technology development. The user requirements are fed into the model and through an analysis procedure, system and subsystem requirements are mapped against missions. Requirements are selected and compared the state-of-the-art technology to establish technology gaps. Technology tasks then flow from the presumed need to fill these gaps. This leads to a technology program plan for future fiscal years. Essentially this same process is used in each of the technology categories in the previous list.

The specifics of the forecasts for each category of space technology may be illustrated by some summary data for both on-board spaceborne processing as well as for the ground systems that support spacecraft operations. Tables 16.2 and 16.3 show state-of-the-art (SOA) values (for 1984) together with values that were forecast to 1990 and beyond. Some of the attributes of the on-board processors in terms of performance and cost effectiveness are listed next and can be compared with the various general criteria for evaluation identified in earlier parts of this text

- Performance
    High data rate
    High throughput
    High density
    Low power
    Reliability
    Survivability
    Degree of autonomy
    Interface compatibility
- Cost Effectiveness
    adaptive evolutionary processors
    software transportability
    standard instruction sets
    higher order language programming
    efficient software and hardware testing
    higher device yields
    efficient software verification and validation
    performance trades across other subsystems

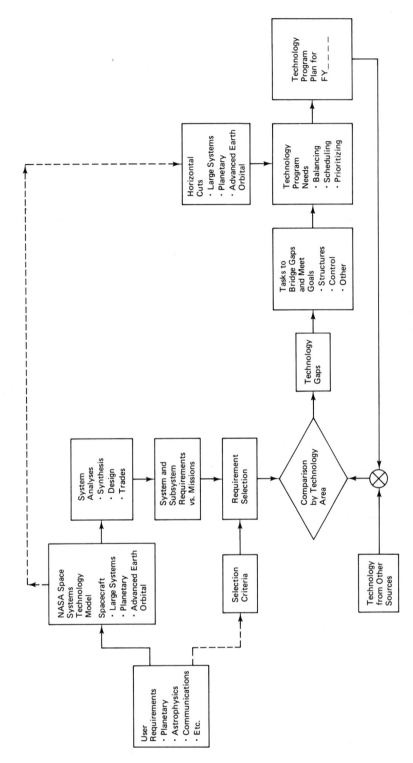

Figure 16.2 Spacecraft technology development.

TABLE 16.2 ON-BOARD SPACEBORNE TECHNOLOGY STATE-OF-THE-ART
AND FORECAST VALUES

Figure of Merit	State-of-the-Art Value	1991 Forecast Value
Spaceborne Computers		
Performance	3×10^6 ops	5×10^7 ops
Performance per Kilogram	3×10^4 ops/kg	1×10^5 ops/kg
Performance per Watt	1×10^4 ops/W	1×10^5 ops/W
Spaceborne Mass Storage Systems		
Capacity	5×10^9 bits	1×10^{11} bits
Transfer Rate	1×10^7 bps	3×10^8 bps
General Purpose Computing Throughput		
(without VHSIC)	2×10 MOPS	3×10^2 MOPS
Spaceborne Computer Performance		
Uniprocessor—Space	3 MOPS	10 MOPS
Uniprocessor—Airborne	10 MOPS	40 MOPS
Composite Performance of General		
Application Systems		
Uniprocessor	60 MOPS	400 MOPS
4-Unit Multiprocessor	110 MOPS	1000 MOPS
Array Processor	140 MOPS	2000 MOPS
Two Pipeline Uniprocessor	800 MOPS	5000 MOPS
64-Unit Array Processor	2000 MOPS	8000 MOPS
Computing Throughout (with VHSIC)		
Data Processing	0.9 MOPS	20 MOPS
Signal Processing	70 MOPS	1000 MOPS
Chip Complexity (Gate Count)	4000	10^5
Speed of Semiconductor Systems—		
Access Time		
Bipolar	10^{-2} s/Mbit	10^{-3} s/Mbit
MNOS	10^{-1} s/Mbit	5×10^{-2} s/Mbit
CCD	10 s/Mbit	1 s/Mbit
Mass Memory Technology		
CCD		
Access Time	100 micron/s	0.5 micron/s
Transfer Rate	1-5 Mbps	6 Mbps
Chip Capacity	1 Mbit	10^3 Mbit
Bubbles		
Access Time	1-3 ms	2 ms
Transfer Rate	0.1 Mbps	1.5 Mbps
Chip Capacity	0.13 Mbps	10^3 Mbit
Storage Cell Size	7×10^{-7} cm^2	—
Nonvolatile Memory Storage Capacity	100 Mbit/system	—
Spaceborne Mass Storage Systems		
Capacity	5×10^9 bits	8×10^{10} bits
Transfer Rate	10^9 bits/s	2×10^8 bits/s

TABLE 16.3 SPACECRAFT GROUND-SUPPORT TECHNOLOGY
STATE-OF-THE-ART AND FORECAST VALUES

Figure of Merit	State-of-the-Art Values	Forecast Values	Year
TDRSS Telemetry Data Rates			
Return Links SSA	12 Mbps	~1 Gbps	1993
KSA	300 Mbps		
MA	50 Kbps		
Forward Links SSA	300 Kbps		
KSA	25 Mbps		
MA	10 Kbps		
On-Line Archival Data Storage		10^{14} bits	1990
Off-Line Archival Data Storage		10^{15} bits	1990
Storage Densities		>500 Mbits/in^2	1990
Data Transfer Rates In/Out Archival		50–300 Mbits/s	1990
Processor Performance	4×10^8 ops/s	10^6 MOPS	1990
Ground Processing Costs	5×10^{-6} cent/bit	10^{-8} cent/bit	1990
Processor Throughput	2×10^6 bits/s	10^{10}–10^{12} bits/s	1990
Input Data Rates	1×10^8 bits/s	2×10^9 bits/s	1990

Thus the NASA model not only projects specific values of key parameters in a variety of technology areas, it identifies broad attributes of the technology categories and criteria for the evaluation of these technologies.

Another much more modest effort in the field of forecasting was referred to in Chapter 13 in regard to the use of the Delphi method and personal computers [7]. Forecasts and actual results were compared as of 1982 with the conclusion that the Delphi panel was almost completely accurate. Areas of consideration included the following:

- p-c networks
- a 32-bit p-c
- pirateless software
- p-c encryption
- dedicated p-c usage in new homes

The Delphi results represent just one sample case. This procedure should find its way more often into the systems engineering process for long lead-time systems and areas in which the larger companies can convene the appropriate panels of experts.

Another specific examples of technology forecasting is cited here in relation to forecasting the maximum speed of aeroplanes [8]. In this work a quantitative modeling approach was taken in which eight alternatives were examined and evaluated.

Data over the time period 1906 through 1965 were examined and growth curve, regression model, and stochastic model approaches were considered. As an example, the form of the growth model is seen by the following equation:

$$y(t) = \frac{L}{1 + A \exp\left[-b(t - t_o)\right]}$$

where L, A, and b are parameters, and t_o is a prespecified time. This simply identifies a particular growth curve model that is asymptotic to the numerator value as the time variable becomes large. Other growth models for technology forecasting have been used and can be investigated further by the reader with a particular interest in this field.

Finally, a review of some 25 years (1960–1984) of research of forecasting was carried out [9], examining such issues as

- decomposition
- extrapolation
- expert opinion
- causal-objective methods
- combinations of forecasts
- the assessment of uncertainty

Some of the more interesting conclusions were

- an endorsement of the efficacy of combining forecasts derived from different sources
- relative indifference to the more sophisticated mathematical approaches
- the relative sufficiency of crude estimating procedures where causal relationships were involved

Perhaps the bottom line is to note that in attempting to forecast into the future, the systems engineer should use the sophisticated mathematical techniques with care and temper the results with sound physical principles and engineering judgment.

16.1.4 Software Tools For Forecasting

Software packages applicable to forecasting tend to fall in the two method categories of time-series methods and association or causal techniques. However, there are a limited number of packages that have been designed to handle questionnaires and related surveys. Dominant areas of interest include taking moving averages, exponential smoothing, various types of filtering, time-series analysis, the Box–Jenkins technique, and correlation and regression analysis. Essentially all of these are statistical procedures and therefore the packages referred to in Section 7.1 are all relevant. These packages are reiterated in Appendix A together with several others that have been added. Software systems that deal with questionnaire or survey techniques are also cited but for large volumes of data the more sophisticated statistical packages normally have greater general-purpose capability. These packages tend to be extremely good value but the specific features should be investigated in relation to the intended application since forecasting methods generally represent a subset of a broader class of statistical routines.

16.2 ARTIFICIAL INTELLIGENCE

Artificial intelligence (AI) is a field of investigation in which interest in recent years has been increasing at a dramatic rate. The expanding literature on AI contains many definitions and there is no general consensus on a precise definition. Perhaps the simplest way of expressing AI is to say that *it involves all elements of investigation that simulate the features, attributes, and behavior of the human brain and related functions*. Some of these elements of investigation, have included

- robotics
- natural language processing
- vision
- learning
- expert systems
- fifth generation computers
- automated instruction systems
- symbolic inference
- intelligent man–machine interfacing
- database machines
- automated programming

Progress will probably continue in all of these areas in the years to come. In the long run, it is difficult to predict which areas will emerge as most important, at least in regard to the building of large-scale systems. In the short run, it appears that significant progress is being made in the expert systems area for applications that go beyond systems engineering, but which are also applicable to systems engineering. For this essential reason as well as limitations of text space, we will focus here mainly on the subset of AI known as expert systems.

16.2.1 Expert Systems

An overview of a general expert systems (ES), sometimes also referred to as a knowledge-based expert system (KBES), is shown in Figure 16.3. At the top left of the diagram are two blocks labeled *Expert* and *Knowledge Engineer*. An expert system must capture some body of knowledge or expertise so it is clear that an expert must be available and must input to the system. Whether or not a knowledge engineer is needed depends upon the design of the ES and the difficulty of entering the expertise into the system. In general, the knowledge engineer assists the expert in framing the knowledge and putting it into a format that can be accepted by the sys-

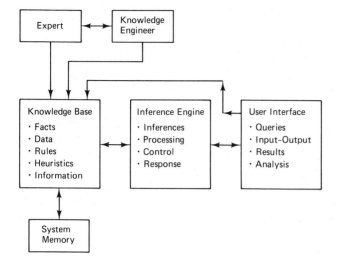

```
Expert  ◄──► Knowledge
              Engineer
```

Knowledge Base
· Facts
· Data
· Rules
· Heuristics
· Information

Inference Engine
· Inferences
· Processing
· Control
· Response

User Interface
· Queries
· Input-Output
· Results
· Analysis

System
Memory

Figure 16.3 Overview of general expert system.

tem. When this has been achieved, the knowledge becomes part of the ES and resides in the *knowledge base* of the system. This knowledge base, as illustrated in the figure, can contain facts, data, rules, heuristics, information, and other representations of knowledge. The notion of representing knowledge is an important part of designing an ES, and indeed there are several ways of doing so, including

- object-attribute-value structure
- rules
- logical relationships
- semantic networks
- frames

Often an object-attribute-value (O-A-V) approach is taken. If we were to apply this notion to a schedule for a large-scale system development, then the following might represent an O-A-V:

- object—schedule
- attribute—early, late, or on schedule
- value—three weeks late

Similar representations can be formulated for any triplet of object, attribute, and value. *If-then* rules, another way of describing knowledge, are cited in the example that follows.

The knowledge base is placed into the system memory and is accessed by the *inference engine*. This engine is a set of software that makes the appropriate inferences regarding the data in the knowledge base but is normally kept separate from the knowledge base. This allows new knowledge bases to be inserted as a plug-in module at some future time. The inference engine also performs the necessary processing and control in order to respond to the queries of the user. In effect, it constructs "meaning" out of the knowledge base. The user queries the system (or is queried by the system) with a problem through some type of input–output subsystem. When the results are provided by the system, the user is able to analyze them. If the system is so designed, it may be that the user can insert additional facts into the knowledge base, given the observation of a set of results. This allows users to improve the knowledge base.

The features of an ES may be better understood through an illustrative example. An important area of potential application is that of maintenance of a complex piece of equipment. In such a situation, the knowledge base consists of a representation of how the system is constructed and, in particular, what failure modes are likely and what symptoms would be observed as a function of these possible failures. These can be thought of as "rules" of behavior of the system. Rules can be conveniently expressed as if-then statements, such as

- IF the demodulator fails, THEN the system will lose target track
- IF the logic subsystem fails, THEN a stream of unintelligible alphanumeric characters will be output
- IF the power supply fails, THEN the screen will go blank

From these statements, the complexity of the knowledge base can be inferred in that for even a relatively small piece of equipment, a rather large knowledge base can be constructed. One way of characterizing an ES is by the number of rules that can be handled by the system. Thus it is necessary to discriminate between the primary and secondary rules of the knowledge base to capture its most important features.

The conclusions are reached by the inference engine, based upon a query or response by the user. In the maintenance example, the user would observe the malfunctioning behavior of the system and query the ES as to the most likely cause. We thus note that the queries are normally made from "*symptom to cause*" whereas the knowledge base, with if-then rules, is set up from *cause to symptom*. This "non-one-to-one" set of relationships is partly what makes the inference problem difficult. Clearly, as more and more symptoms are observed, the class of possible causes can be narrowed down to the most probable cause(s). It also has led to the design of ESs such that certainty factors and confidence ranges have been added to reflect the uncertainties in the inferences and therefore in the final conclusions.

It is not possible, in this short discussion, to explore in any detail the ways in which inference engines work. The latter example illustrates that a relationship can be established between the inference process and search processes, some of which include forward chaining, backward chaining, depth-first search, and breadth-first search, as well as the some of the methods described in Chapter 8. Thus we have another rationale for the inclusion of the latter in this text. In addition, it has been noted in previous chapters that a major issue for large-scale systems design is that of "searching" for solutions from among a combinatorially large number of possibilities. On these bases we have linkages between methods of drawing inferences, search theory, and large-scale systems synthesis. These matters will undoubtedly be afforded more attention in the literature in the years to come.

Two other fields of investigation that can be related to that of drawing inferences are logic and Bayesian analysis. Illustrative examples of each of these are given next to provide the reader with concrete applications tools.

16.2.2 Logic Terminology and Truth Tables

Formal procedures of logic have been used in ESs to embody the rules of the knowledge base and allow inferences to be made. Figure 16.4(a) shows an example of a set of eight logic functions. Some of these (e.g., AND, OR) were discussed in Chapter 5. Others have been added and this illustrative set of eight functions does not represent all possible logic functions. Representations of these eight functions are also shown in equation form in Figure 16.4(a). A bar line over a symbol indicates negation so that, for example, $\overline{X}$ is read as (not X). Figure 16.4(b) and (c) show zero–one and true–false representations, respectively, of all combinations of X and Y and the eight listed functions, $F1$ through $F8$. The zero–one representation is directly applicable to digital logic design and components are available that will implement these as well as other functions. The true–false representation in part (c)

Meaning	Name	Functional Representation
X and Y	AND	$F1 = XY$
X or Y	OR	$F2 = X + Y$
X or Y but not both	EXCLUSIVE OR	$F3 = X\overline{Y} + \overline{X}Y$
Not Or	NOR	$F4 = \overline{(X + Y)}$
Not And	NAND	$F5 = \overline{(XY)}$
X equals Y	EQUIVALENCE	$F6 = XY + \overline{X}\,\overline{Y}$
If Y then X	IMPLICATION	$F7 = X + \overline{Y}$
If X then Y	IMPLICATION	$F8 = \overline{X} + Y$

(a)

X	Y	F1	F2	F3	F4	F5	F6	F7	F8
0	0	0	0	0	1	1	1	1	1
0	1	0	1	1	0	1	0	0	1
1	0	0	1	1	0	1	0	1	0
1	1	1	1	0	0	0	1	1	1

(b)

X	Y	F1	F2	F3	F4	F5	F6	F7	F8
F	F	F	F	F	T	T	T	T	T
F	T	F	T	T	F	T	F	F	T
T	F	F	T	T	F	T	F	T	F
T	T	T	T	F	F	F	T	T	T

(c)

Figure 16.4 Representations of logic relationships: (a) selected logic functions, (b) zero-line binary representation, and (c) true–false representation (0 = false, 1 = true).

is also called a *truth table* and can be used to demonstrate how inferences might be made in an expert system. We will examine some of the characteristics of the if-then or implication functions ($F7$ and $F8$), since they clearly represent a method of maintaining knowledge in an expert system's knowledge base.

First we note intuitively that if both implication statements are true, then we would expect equivalence between X and Y, that is, that X equals Y. Stating that both implication statements hold is the same as writing

$$(\text{If } X \text{ then } Y) \text{ AND } (\text{If } Y \text{ then } X)$$

(If X then Y) is $F8$ and (if Y then X) is $F7$ and are shown here

X	Y	If X Then Y (F8)	If Y Then X (F7)
F	F	T	T
F	T	T	F
T	F	F	T
T	T	T	T

Taking ($F8$ and $F7$) should yield equivalence or $F6$. We note below that such is the case by comparing the truth table result for ($F8$ and $F7$) with that of $F6$ and accepting the notion that if two expressions have the same truth table they are said to be logically equivalent.

X	Y	F8 AND F7	F6
F	F	T	T
F	T	F	F
T	F	F	F
T	T	T	T

Another way of examining this issue is to formally use the functional representations in Figure 16.4. This is left as a problem at the end of the chapter.

A further example of the use of logic and truth tables lies in the examination of two well-known inferences, namely A and B here:

Inference A	Inference B
IF X THEN Y	IF NOT X THEN NOT Y
IF $\bar{Y}$ THEN $\bar{X}$	IF Y THEN X

The horizontal line for these inferences can be read as "implies." Inference A will be demonstrated next; inference B is left as an exercise for the reader.

Example 1

We are trying to demonstrate that the truth table for (IF X THEN Y) is the same as the truth table for (IF $\bar{Y}$ THEN $\bar{X}$). This is easily shown here

X	Y	$\bar{X}$	$\bar{Y}$	IF X THEN Y (F8)	IF $\bar{Y}$ THEN $\bar{X}$
T	T	F	F	T	T
F	T	T	F	T	T
T	F	F	T	F	F
F	F	T	T	T	T

The example provides sufficient background for a final word example that might be part of an expert system.

Example 2

In this case we begin by examining two statements ($S1$ and $S2$) for an automobile and ask whether or not the inference ($I1$) is valid, as

$S1 =$ IF the car battery is dead THEN the headlights are out
$S2 =$ (AND) the headlights are not out
_____
$I1 =$ the car battery is not dead

We will translate the above into X and Y terminology by setting

$Y =$ the car battery is dead

$X =$ the headlights are out

$\bar{X} =$ the headlights are not out

$\bar{Y} =$ the car battery is not dead

The question can then be expressed as

$$(\text{IF } Y \text{ THEN } X) \text{ AND } (\bar{X}) \overset{?}{=} \bar{Y}$$

The left side of this expression is

$$(X + \bar{Y})(\bar{X}) = X\bar{X} + \bar{Y}\bar{X} = \bar{Y}\bar{X}$$

which demonstrates that the $\bar{Y}$ is valid accepting also that $\bar{X}$ is valid, which is the premise in $S2$. Now consider the following statements and inference:

$S1 =$ IF the car battery is dead THEN the headlights are out
$S3 =$ the car battery is not dead
_____
$I2 =$ the headlights are not out

The question in this situation is

$$(\text{IF } Y \text{ THEN } X) \text{ AND } (\bar{Y}) \stackrel{?}{=} \bar{X}$$

The left side is then

$$(X + \bar{Y})(\bar{Y}) = X\bar{Y} + \bar{Y}$$

which does not imply $\bar{X}$. Thus the inference is not valid. The inference is simply that either the battery is not dead or it is not dead and (nonetheless) the headlights are out. The reader is asked to examine both of the latter "word" inferences by means of truth tables at the end of the chapter.

The applicability of Example 2 in terms of expert system constructs should be clear. The if-then rules become part of the knowledge base together with any other facts, statements, and so on. The user makes an observation or query that is then used in combination with the knowledge base to draw valid inferences. More observations will normally help to narrow down the set of valid inferences. The inference engine contains all the appropriate mechanisms necessary to reaching the proper conclusions. For a complex problem, the value of such a system can readily be seen.

16.2.3 A Bayesian Approach to Inference

Another possible approach to the matter of drawing inferences utilizes Bayes' Theorem, one form of which can be stated as

$$P(B_i \mid A) = \frac{P(AB_i)}{P(A)} = \frac{P(A \mid B_i)P(B_i)}{\Sigma P(A \mid B_i)P(B_i)} \tag{16.1}$$

where the B_i are exclusive and exhaustive events. With respect to expert systems, one module of Software A & E's (Arlington, Va.) KES expert system (see Tables 16.4 and 16.5 later in this chapter) appears to embody a Bayesian approach to drawing inferences [10].

To illustrate such an approach, we imagine a system with three major functions and corresponding subsystems that implement these functions. The overall system condition is said to be any one of the following:

1. *Condition One*   Excellent = fully ready to perform its mission
2. *Condition Two*   Good = not completely ready, but more likely than not to perform its mission
3. *Condition Three*   Poor = in need of repair and not likely to perform its mission

We further take as a knowledge base the data listed in Figure 16.5. These data represent:

- the prior probabilities of the system being in the above conditions (i.e., excellent, good, poor)

Functional Subsystem/Output	System Condition		
	Excellent (Probability = 0.2)	Good (Probability = 0.5)	Poor (Probability = 0.3)
A.1 Output Within Spec	0.9	0.8	0.6
A.2 Output Out of Spec	0.1	0.2	0.4
B.1 Output Within Spec	0.8	0.7	0.5
B.2 Output Out of Spec	0.2	0.3	0.5
C.1 Output Within Spec	0.7	0.6	0.4
C.2 Output Out of Spec	0.3	0.4	0.6

Figure 16.5 Illustrative knowledge-base probabilities for Bayesian inference example.

- the probabilities of the functional subsystem outputs being in or out of specification for each of the three conditions listed

We now make a measurement M that contains three parts

1. the output of functional subsystem A
2. the output of functional subsystem B
3. the output of functional subsystem C

An example of such a measurement is $M = $ (A.2, B.1, C.1), which means that

- the output of A is out of spec
- the output of B is within spec
- the output of C is within spec

Given this measurement and the knowledge base, we now inquire as to the (a posteriori) likelihood that the system condition is excellent, good, or poor. For excellent conditions, this may be expressed, from Bayes' Theorem, as

$$P(Ex|M) = \frac{P(Ex)P(M|Ex)}{P(Ex)P(M|Ex) + P(Good)P(M|Good) + P(Poor)P(M|Poor)}$$

(16.2)

Similar expressions hold for the *good* and *poor* conditions.

The prior probabilities $P(Ex)$, $P(Good)$, and $P(Poor)$ are contained in the columns of the knowledge base and have the numerical values of 0.2, 0.5, and 0.3, respectively. The conditional probabilities in the numerator are likewise calculated from the knowledge base, assuming independence, as follows:

$$P(M \mid Ex) = (0.1)(0.8)(0.7) = 0.056$$

$$P(M \mid Good) = (0.2)(0.7)(0.6) = 0.084$$

$$P(M \mid Poor) = (0.4)(0.5)(0.4) = 0.08$$

Substitution of the appropriate values into Equation 16.2 yields

$$P(Ex|M) = \frac{(0.2)(0.056)}{(0.2)(0.056) + (0.5)(0.084) + (0.3)(0.08)} = \frac{0.0112}{0.0772} = 0.145$$

Similar calculations result in the following probability estimates:

$$P(Good|M) = \frac{(0.5)(0.084)}{0.0772} = 0.544$$

$$P(Poor|M) = \frac{(0.3)(0.08)}{0.0772} = 0.311$$

These three values add to unity, as they should. They also represent the probabilities that the sytem is in each of the three conditions after the triplet measurement, M, is made. Thus an inference is made as to the likely condition of the system, given a knowledge base of prior probabilities and a measurement of the system. Note also that an assumption is required as to the exclusivity of the sets of conditions in order to fully apply this Bayesian approach. In this particular example, given the illustrative measurement, the system would be most likely to be in a *good* condition, supported by a probability estimate of 0.544.

Applications of the expert system described can readily be envisioned, as for example with

- remote monitoring and measurement of the status of a not easily accessible satellite tracking station, or
- "go-no go" aircraft cockpit systems and displays with a choice as to which system is more likely than another to perform its intended mission

From the discussion in this section we can see that expert systems can be extremely complex. Notwithstanding this fact, such systems have been constructed and are available at reasonable prices for the p-c. An overview of some of the expert system packages that can be obtained and used in a microcomputer environment is provided in the following section.

16.2.4 Software Packages for Expert Systems

A list of expert systems that are available for the p-c is contained in Table 16.4 (see also Appendix A). Selected features for each of these packages are also provided in the table. A more general set of characteristics of these commercially available expert systems may be expressed in terms that include the following:

- the language in which it is written (e.g., Pascal, Lisp, Prolog)
- hooks to other languages
- hooks to databases
- example versus rule based

TABLE 16.4 MICROBASED EXPERT SYSTEMS
AND SELECTED FEATURES

ES/P ADVISOR
- Attribute-value pair expressions
- Backward chaining inferences
- Knowledge-base entry with standard word processor
- Sample systems and tutorial

Expert-Ease
- Example based
- Induces own rules
- Table of attributes and variables
- No computer skills required

EXSYS
- Mathematical computation capability
- Embedded certainty factors
- Able to call external programs
- If-then rule based

GURU
- If-then rules
- Certainty factors
- Why and how logic tracing
- Built-in integrated package functions

Insight 1 & 2
- Backward reasoning from likely conclusion
- Has confidence factors
- If-then-else rules
- Inexpensive

KDS
- Example-case history based
- Programming capability not required
- Able to use confidence factors
- rapid run-time

KES
- Hypothesis, if-then, and Bayes modules
- Mathematical computation capability
- Able to incorporate outside program data
- Able to diagnose problems with inspection module

M.1
- Rule-based shell
- Goal seeking
- Powerful and expensive
- Training course available and necessary

MICRO-PS
- Subset of KES
- Limited capability
- Good teaching device
- Very inexpensive

PERSONAL CONSULTANT PLUS
- Able to capture graphical data
- Rule-based shell

TABLE 16.4 *(cont.)*

- O-A-V triplet
- Has confidence factors

SeRIES-PC
  - If-then rules
  - Backward chaining and depth-first search
  - Some use of A-V pairs
  - Built-in knowledge-base editor

TIMM-PC
  - A-V pairs
  - Example based
  - Contains certainty factors
  - Has decision-tree methodology

- type of rule system
- number of rules
- certainty factors
- mathematics capability
- inference method (e.g., forward versus backward chaining)
- interpreted versus compiled
- menu driven
- simple versus multiple objects
- simple versus multiple values
- speed of execution
- price

The reader is referred to the literature (References 11 through 16) for discussions of these and other characteristics of expert systems. It is noted that many expert systems are called *shell* systems, largely since they provide the overall structure for the system but a knowledge base needs to be added for the system to operate properly. Also, the investigator of commercially available p-c packages, such as those listed here, will find very large differences in price, ranging from about $100 to $10,000. Of course, features are generally more extensive for the more expensive systems, but the potential user would do well to examine the less costly systems in some detail in relation to specific needs or as a vehicle for prototyping or learning how to construct a larger system.

These characteristics apply as well to expert systems hosted on the larger mini and mainframe computers. For the practicing engineer who may have access to the resources necessary to operate in a more powerful computer environment, Table 16.5 shows some of the expert systems that can be obtained through purchase or license arrangements [15, 16]. Selected features of these packages are also summarized for each of the cited packages. In general, these software systems are considerably more powerful and expensive than those developed for the p-c environment.

TABLE 16.5 OTHER COMMERCIAL EXPERT
SYSTEM PACKAGES

ART
  • O-A-V triplet and facts
  • Forward and backward chaining
  • Has confidence factors
  • Powerful and flexible
EXPERT
  • If-then rules
  • Has confidence factors
  • A-V pairs
  • Statistical analysis capability
KEE
  • Frame and rule-based reasoning
  • Graphics capability
  • Backward or forward chaining
  • Training course available
KES
  • Hypothesis, if-then, and Bayes modules
  • Mathematical computation capability
  • Able to incorporate outside program data
  • Able to diagnose problems with inspection module
LOOPS
  • Object/frame oriented
  • Rule package
  • Knowledge engineering environment
  • Inexpensive as unsupported tool
OPS 5
  • General programming environment
  • If-then rules
  • O-A-V triplets
  • Forward chaining
S.1
  • O-A-V triplets
  • If-then rules
  • Backward chaining
  • Confidence factors

Some of the expert or AI systems, or so-called AI languages, that are specifically available in the workstation environment (e.g., Apollo Computer, Sun Microsystems) are listed in Table 16.6 [17, 18]. Table 16.6 lists several versions of AI "languages" that will be discussed briefly later in this section. It is seen from the tables cited that expert and related systems are becoming widespread as commercial products at all computer size levels.

In addition to the aforementioned bonafide expert systems, there are several software packages for the p-c that are advertised as related to artificial intelligence or expert systems [12, 19, 20]. Although these packages would not generally qualify as true AI systems, they are noted here in Table 16.7 (and Appendix A) for the poten-

TABLE 16.6 SELECTED ARTIFICIAL INTELLIGENCE SOFTWARE AVAILABLE IN WORKSTATION ENVIRONMENTS [17, 18]

CSI LISP
D-PROLOG
DUCK
Franz Lisp
KES
LeLisp
Mprolog
NISP/DUCK
POPLOG
Portable Common Lisp
PROLOG/P
Quintus Prolog Release 1.0
Rulemaster
SILL

tial user who wishes to investigate their features for possible application in a systems engineering context.

16.2.5 Other Artificial Intelligence Activities

The subject of AI is extensive and cannot be covered in any real detail in the context of this book. However, we briefly comment on a few additional areas within AI, emphasizing expert systems, as follows:

- Lisp and Prolog
- a special approach to advanced ESs
- DBMSs
- natural language processing
- decision support
- defense applications
- well-known historical systems

Table 16.6 identified workstation software for AI applications that contained specific packages pertaining to Lisp and Prolog. These are both programming languages, are available at the microcomputer level, and have qualities that are particularly applicable to the AI area. Lisp (LISt Processing language) was developed by J. McCarthy at MIT in 1958. It embodies symbolic expressions and processes lists. It uses dynamic data structures with pointers in a manner similar to that previously described for Pascal (see Chapter 9). It also relies on recursion relationships for

TABLE 16.7 MICROCOMPUTER SOFTWARE CITED AS
RELATED TO ARTIFICIAL INTELLIGENCE

Clout 2
- Natural language interface with R:base 4000/5000 DBMSs
- Facilitates DBMS queries
- Uses augmented transition networks and heuristics
- Dictionary base

Javelin
- Natural language datasheets
- Can use diagram, chart, graph, table, worksheet
- Identifies possible errors
- Good for modeling

Logic-Line Series 1
- Connects text references to search requests
- Sophisticated database manager
- Mathematical computation for search
- Pattern matching

NaturaLink
- Contains acceptable words and phrases for query
- Window assisted
- Able to front-end several other packages
- Has tool kits for other program use

Paradox
- Machine-reasoning concepts for database queries
- Relational DBMS
- Simplified, efficient data manipulation
- Allows access to other software packages

Q & A
- Natural language interface
- Facilitates flat-file DBMS queries
- Built-in vocabulary
- Word processing feature

Savvy PC
- Database manager
- Pattern-recognition algorithm
- Natural-language capability
- Built-in updatable lexicon

much of its processing power and efficiency. Both data and programs can be incorporated as lists. Relatively, there are not many functions in Lisp, but user-defined, high-level functions can be used extensively. Lisp has been implemented in many forms, sometimes called dialects. There are attempts at standardization, with Common Lisp apparently having an edge in this regard. An example of two well-known Lisp operators are the CAR and the CDR. They operate on a list or set of items, say $X = (A,B,C,D)$ such that CAR returns the first element of a list and CDR returns the list obtained when the first element is deleted. Symbolically, this can be expressed as:

$$(CAR\ X) = A \qquad \text{and} \qquad (CDR\ X) = B, C, D$$

There are, of course, many other operations in Lisp and the interested reader should consult the extensive literature on this language [21].

Prolog was developed in 1972 by A. Colmerauer and P. Roussel at the University of Marseilles. The language is based upon the use of predicate calculus and therefore represents a true logical language. As such it relies heavily on symbolic computation, as does Lisp. It is extremely popular outside the U.S. but it is readily available in several forms in the U.S. as Table 16.6 shows. Typical inputs would include rules and facts regarding objects and relationships between them. In this sense, it is a simple matter to input a set of if-then relationships as discussed earlier. The language structure would then provide the logic necessary to draw inferences from the inputs. We can see how Prolog applies directly to the expert system construction and therefore why it has been a popular tool in the AI field. Its popularity outside the U.S. is reflected by its extensive use as part of Japan's Fifth Generation Computer initiative.

The direction of further development of expert systems might well be influenced by an approach taken by D. Lenat in his construction of a system called EURISKO [22]. Unlike the ESs that have a fixed knowledge base, EURISKO incorporates heuristics that lead to a type of learning. The program, in effect, modifies its strategies by repeated applications and testing of results. Strategies that suceeed are adopted and ones that fail are rejected. From a search theory point of view, it provides a method for converging to desired solutions within a large search space. Further developments along the lines set forth by Lenat may prove to be extremely fruitful.

Another trend that may develop relates to the possible influence of AI research and results in terms of database design for DBMSs [23]. New AI structures and formalisms may lead to "intelligent" databases that could be extensions of the knowledge data base that is part of an expert system. At the same time, improvements in the use of natural languages, as briefly discussed next, could help the user interact with future DBMSs in a more efficient manner.

The movement toward natural language systems is simply an attempt to bring the user's language closer to the machine's language by translating the parts of speech in a sentence into a set of commands that the computer will understand [12]. In such a process, a *parser* identifies and designates the various parts (noun phrase, verb phrase, and so on) of each sentence. An output can be, for example, a parse tree or a sentence diagram. A semantic analysis of the sentence parts is performed that gives meaning to the parser results with the aid of a dictionary. These results are then used to develop code that is understandable by the computer. As indicated, several of the packages in Table 16.7 have a natural language capability, which is considered rudimentary at the micro level.

A trend that has already been started is the attempt to embed an expert system capability in a larger commercial software product under the category of decision support software (DSSW—see Chapter 13) or financial modeling [24]. In general, these efforts are focused on larger systems and larger host machines (mini, mainframe). The costs reflect that environment, running from a low end of $4000 up to $100,000 and more.

AI applications in the defense world are quite extensive, considering the relative infancy of this field. Military-oriented subjects in which such applications are present include the following [25]:

- command and control
- communications
- intelligence
- tactical systems
- information systems
- robots and other unmanned systems
- transportation
- decision aids

The military will continue to look to AI and expert systems especially in areas in which decisions need to be made quickly and with a quality that approximates what might be achieved by the human element in a system. One example is that of a rapid-situation assessment [26] in which new data inputs could be quickly correlated with a knowledge base to provide the decision maker with an array of alternatives in relation to the specifics of the current situation, on a near real-time basis. Another example is the strategic defense initiative program in which time is of the essence, especially during the boost phase. The battle managers require extremely rapid situation assessment and, as a consequence of the short time lines, various tactical decisions might ultimately be relegated to an expert system without benefit of human intervention and decision making.

The student or practicing engineer with a deeper interest in AI and expert systems in particular may wish to track through the earlier history of the development of these systems in order to better understand where we are today and what trends might evolve. Well-known systems and their principal areas of application are listed here

- Dendral—chemistry-export system with algorithms developed by Nobel Laureate J. Lederberg
- Eliza—natural-language processing
- Hearsay—oriented toward natural-language processing
- Internist/Caduceus—internal-medicine diagnostics
- Macsyma—massive general purpose mathematical analysis and computation
- Mycin—medical diagnostics (first large ES of its type)
- Prospector—geological exploration
- Puff—rapid prototyping for medical diagnostics (related to Mycin)
- Shrdlu—natural-language processing

Most serious references devoted to AI for example, Reference 15, will provide additional information on these as well as other historical and existing systems.

16.2.6 CASE Application Areas

It is fair to say that there has not been an extensive attempt to apply AI or expert systems thus far to general systems engineering tasks and activities. We can foresee, however, serious applications of expert systems in a variety of areas in which a body of knowledge or expertise is available and necessary to address the solution of a systems engineering problem. Applications in the field of maintenance have already been alluded to. Clearly, an important part of maintenance is knowledge based and the diagnostic procedures carried out by skilled maintenance personnel can be embodied in an expert system.

It is envisioned that even small expert systems will find their way into program planning and control. As an example, the if-then rules of an ES can be used in a direct manner to incorporate the constraints (e.g., facilities, timing, other resources) and options perceived by a program manager. The logical inferences that flow from these constraints and options could help the manager find an effective course of future action. The same concept applies as well to the use of expert opinion in terms of capturing the knowledge of leading technical experts that might be called in to consult on a program.

Conceptually, an expert system's search and inference capability could prove to be the most significant of all the application areas. Here we return to the issue of design or synthesis. A morphological approach to design (see Chapter 2) is systematic and all inclusive, leading to the articulation of all possible design configurations. Each such configuration is characterized by a set of attributes that are not necessarily commensurable. Thus the "space" of feasible design solutions is extensive and multidimensional. If efficient search algorithms and if-then constraints are utilized, it will be possible to narrow down these alternatives more quickly, converging to answers that might not be forthcoming through more conventional unstructured engineering judgment coupled with standard analytic tools. Other related notions with respect to a potential future CASE environment are considered in Chapter 17.

REFERENCES

1. Georgoff, D. M., and R. G. Murdick, "Manager's Guide to Forecasting," *Harvard Business Review,* January-February 1986.
2. "Transport of Solid Commodities Via Freight Pipeline, Demand Analysis Methodology," vol. IV, DOT-TST-76T-38, Department of Transportation, July 1976.
3. "High Speed Ground Transportation Alternatives Study," PB 220 079, Department of Transportation, January 1973.
4. Wiener, N., *Extrapolation, Interpolation and Smoothing of Stationary Time Series.* NY: The Technology Press of MIT and John Wiley, 1960.
5. Kalman, R. E., "A New Approach to Linear Filtering and Prediction Problems," *Journal of Basic Engineering,* March 1960.
6. "NASA Space Systems Technology Model," Fifth Issue, National Aeronautics and Space Administration, Office of Aeronautics and Space Technology, Code RS, January 1984.

7. Gray, P., and J. M. Nilles, "Evaluating a Delphi Forecast on Personal Computers," *IEEE Transaction on Systems, Man, and Cybernetics,* SMC-13, no. 2, March/April 1983.

8. Murthy, D. N. P., and K. C. Staib, "Forecasting Maximum Speed of Aeroplanes—A Case Study in Technology Forecasting," *IEEE Transactions on Systems, Man, and Cybernetics,* SMC-14, no. 2, March/April 1984.

9. Armstrong, J. S., "The Ombudsman: Research on Forecasting: A Quarter-Century Review, 1960-1984," *Interfaces 16, An International Journal of The Institute of Management Sciences and the Operations Research Society of America,* January-February 1986.

10. *Knowledge Engineering System: Knowledge Base Author's Reference Manual.* Arlington, VA: Software Architecture & Engineering, June 1984.

11. Williamson, M., *Artificial Intelligence for Microcomputers.* Englewood Cliffs, NJ: Brady Communications Co., 1986.

12. Fersko-Weiss, H., "The Intelligent Computer-Part II, Natural Language" and "Expert Systems," *Personal Computing,* 9, no. 11, November 1985.

13. Gold, J., "Do-It-Yourself Expert Systems," *Computer Decisions,* 18, no. 2, January 14, 1986.

14. Helliwell, J., "Guru: Brave New Expert System?" *PC Magazine,* 5, no. 10, May 27, 1986.

15. Harman, P., and D. King, *Expert Systems: Artificial Intelligence in Business.* New York: John Wiley, 1985.

16. Seaman, J., "AI Moves Into the Mainstream," *Computer Decisions,* 17, no. 24, December 3, 1985.

17. *Catalog of Applications for Domain Systems.* Chelmsford, MA: Apollo Computer, 1985.

18. *Catalyst: A Catalog of Third-Party Referral Software and Hardware.* Mountain View, CA: Sun Microsystems, 1985.

19. Gold, J., "The Search for AI Micro Software," *Computer Decisions,* 17, no. 25, December 17, 1985.

20. Rothfeder, J., "Is There Intelligent Life in the PC?" *PC Magazine,* 5, no. 1, January 14, 1986.

21. Winston, P. H., and B. K. P. Horn, *Lisp.* Reading, MA: Addison-Wesley, 1981.

22. Schrage, M., "Artificial Intelligence: Teaching Computers Power of Creative Stupidity," *Washington Post* newspaper, December 1, 1985.

23. Bic, L., and J. P. Gilbert, "Learning From AI: New Trends in Database Technology," *IEEE Computer Magazine,* 19, no. 3, March 1986.

24. Kull, D., "Decision Support With 20/20 Foresight," *Computer Decisions,* 18, no. 10, May 6, 1986.

25. Andriole, S. J., ed., *Applications in Artificial Intelligence.* Princeton, NJ: Petrocelli Books, 1985.

26. Ben-Bassat, M., and A. Freedy, "Knowledge Requirements and Management in Expert Decision Support Systems for (Military) Situation Assessment," *IEEE Transactions on Systems, Man, and Cybernetics,* SMC-12, no. 4, July/August 1982.

*Reference is relevant to the text but is not specifically cited.

*27. Kull, D., "Expert Help in the Data Center," *Computer Decisions,* 18, no. 11, May 20, 1986.

*28. Feigenbaum, E. A., and P. McCorduck, *The Fifth Generation, Artificial Intelligence and Japan's Computer Challenge to the World.* New York: New American Library, 1983.

*29. Simon, H. A., *The Sciences of the Artificial.* Cambridge, MA: MIT Press, 1969.

*30. Rich, E., *Artificial Intelligence.* New York: McGraw-Hill, 1983.

*31. Brooks, R. A., *Programming in Common Lisp.* New York: John Wiley, 1985.

*32. Burnham, W. D., and A. R. Hall, *Prolog Programming and Applications.* New York: Halsted Press, 1985.

*33. Bonnet, A., *Artificial Intelligence: Promise and Performance.* Englewood Cliffs, NJ: Prentice-Hall, 1985.

*34. "Technology Assessment and Forecast," S/N 003-004-00542-4, 1977; S/N 003-004-00580-7, U. S. Government Printing Office, Wash., DC, 1981.

*35. Issue on Artificial Intelligence, *SIGNAL Magazine,* Armed Forces Communications and Electronics Assoc., (AFCEA), Burke, VA, 40, no. 10, June 1986.

PROBLEMS

P16.1. Select one of the time series or association/causal forecasting methods and discuss in specific terms how a software package of your choice is able to implement the given method.

P16.2. Obtain Reference 1 and write a two-page critique of the forecasting guide. Explain how one might set up a formal weighting and rating evaluation scheme, given the prescribed methods and dimensions of the guide.

P16.3. Select one of the three demand relationships in Figure 16.1 and explain, in a three-page discussion, how the empirical parameters may be estimated.

P16.4. Write a five-page discussion of the utility of and essential methodology employed by the Kalman filter.

P16.5. Obtain a reference on technology forecasting and compare the methods suggested with the general methods of forecasting delineated in Table 16.1.

P16.6. Demonstrate the AND, OR, EXCLUSIVE OR, NOR, and NAND logic relationships by means of Venn diagrams (see Chapter 5).

P16.7. Use the functional representations in Figure 16.4 to demonstrate that the expression [if X then Y) AND (if Y then X)] yields equivalence between X and Y.

P16.8. Demonstrate that (IF NOT X THEN NOT Y) implies (IF Y THEN X).

P16.9. Show the validity and lack of validity of the inferences in Example 2 in the text by using truth-table constructions.

P16.10. Discuss the pros and cons of taking a Bayesian approach to the construction of an expert system.

P16.11. Obtain additional information on one of the expert-system software packages cited in the text and discuss, in three page or more, its advantages and disadvantages in terms of an application that you select.

*Reference is relevant to the text but is not specifically cited.

P16.12. Discuss, in three to five pages, a search or inference procedure of an expert system and how it might be used to deal with fundamental systems engineering design issues.

P16.13. Investigate and then write a simple program in Lisp. Write a program in BASIC or Pascal that performs the same function. Compare the features of Lisp and the other language you select in regard to this program.

P16.14. This is the last problem in the book and the student who has reached this point deserves a reward in the form of at least the option of a minor respite. If this problem is assigned by your instructor, then you the student have the option to submit a reasonable problem or question to your instructor with the expectation that you will receive a well-considered response.

Progress is the activity of today and the assurance of tomorrow

Emerson

TOWARD
A FUTURE CASE
ENVIRONMENT

In previous chapters we have examined a variety of introductory concepts and structures for systems engineering, specific tasks and activities of systems engineering, mathematical and logic-based techniques that support systems engineering, and computer tools that are applicable to systems engineering. In this final chapter we briefly explore:

- a summary of the microbased software categories defined in this book and map them against the tasks and activities of systems engineering
- the software categories that are most mature and applicable to the needs of the systems engineer
- the gaps that are apparent in regard to systems engineering tasks and activities that could be further assisted by improvements in commercially available software
- some further trends and directions
- a hypothetical future CASE environment

In the hypothetical CASE environment the systems engineer will more effectively be able to carry out the tasks and activities of systems engineering. These tasks and activities were discussed in Chapters 1 and 2 and explored in greater detail in Part III. The strong force exerted by the federal government in defining the nature of systems engineering was also noted, and liberal reference has been made to the various standards, directives, and approaches set forth by the government. This then represents a base line of requirements and, as such, changes relatively slowly with time. A major thesis of this book is that responses to the base line of requirements can be materially improved through the use of computer-aided techniques. Here we examine some ways in which these techniques may evolve over time. First, however, we start with where we are today.

17.1 CURRENT SOFTWARE AND GAPS

We have presented and discussed in various parts of this text 29 categories of software packages currently available at the personal computer level. These categories are summarized in Table 1.3 and represent a good portion of today's software base line upon which future software developments will be built. In effect, this software constitutes a current "library" that can be tapped for immediate application today. Future software will certainly be even more capable, moving forward with certain improvements in the speed of computation and storage capacities of our hardware.

From a systems engineering point of view, we have demonstrated throughout this text some of the ways in which these 29 categories of software can be applied. Many of these applications were quite direct and mature. For others, software is not plentiful, at least in the context of relatively inexpensive commercially available packages. This situation can be explored further by mapping the categories of software against the tasks and activities of systems engineering to look for areas of coverage as well as gaps. In effect, this is a morphological approach to examining areas in which new developments are needed in order to more fully bring this software to the systems engineering environment. Figure 17.1 shows such a mapping and the following text discusses current applications as well as gaps in the matrix that represent areas for future development of software that would be more responsive to the needs of the systems engineer.

If we look at the 29 categories of software, we see some that are broadly applicable to a wide range of systems engineering areas (such as spreadsheets, DBMSs, multitask managers, and integrated packages) and others that are more specialized and therefore less broadly applicable (such as reliability and expert systems packages). Some packages may be thought of as very highly developed in terms of sophistication and capability such that it is difficult to imagine what might be done to improve them other than making them more user friendly. Perhaps word processors and some of the statistical packages fall into such a category. Others seem to be awaiting the improvement in price and performance of the p-c such as the micro-based simulations and the expert systems. For these cases, early implementations were carried out on larger host machines and they migrated to the p-c environment after a while. Still others appear to be at the high-cost end (e.g., workbenches, structured analysis tools) while a number of these tools are real bargains at under $100. Thus we see a rather large range of capabilities and prices that in the aggregate represent an enormous set of resources. Over time, they will certainly find increasing acceptance in the world of systems engineering.

Figure 17.1 Software categories mapped against systems engineering tasks and activities.

Software Categories (X = Significant current or near-term future application)	Definition of Needs/Goals/Objective	Requirements Analysis	Requirements Allocation	Functional Analysis	Functional Allocation	Specification Development	System and Subsystem Design	System and Subsystem Analysis	Trade-Off/Alternatives Evaluation	Software Development and Analysis	Interface Definition	Schedule Development	Life-Cycle Costing	Technical Performance Measurement	Program and Decision Analysis	Risk Analysis	ILS	P³I	RMA	Integration	Test and Evaluation	Configuration Management	Quality Assurance	Training	Documentation	Production	Installation	Operation	Operations Evaluation/Modification
A. Spreadsheets		X	X			X		X	X	X			X	X	X	X													X
B. DBMS		X	X			X				X							X								X	X			X
C. Word Processors	X					X						X	X												X	X			X
D. Graphics				X	X			X				X													X	X			
E. Integrated Packages		X	X			X						X													X	X			
F. Multitask Applications Managers	X			X	X																				X	X			
G. Project-Management Aids							X					X								X					X	X			
H. Alternatives and Preference Evaluators							X	X	X															X					
I. Decision-Support Systems							X	X	X																				
J. Toolchests/Desk Managers							X								X														
K. Idea Processors	X	X		X	X																				X				
L. Management Evaluators/Aids	X												X							X									
M. Statistical			X				X	X	X				X		X	X				X				X					X
N. Mathematical		X	X				X	X	X																				X
O. Engineering							X	X	X																				
P. Reliability		X	X					X	X										X										
Q. Mathematical Programming								X	X																				
R. Simulation							X	X	X											X	X	X							X
S. Curves and Plotting				X	X																				X	X	X	X	
T. CAD/CAM/Graphics							X																						
U. Workbenches				X	X															X	X				X		X	X	
V. Structured-Analysis Tools		X	X			X	X														X						X		
W. Languages			X							X										X	X								
X. General-Support Tools																													
Y. Utilities																													
Z. Micro-to-Mainframe	X	X				X		X	X				X			X	X			X	X	X	X	X	X	X			
AA. Forecasting	X	X																											X
BB. Expert Systems		X	X																										X
CC. AI-Oriented Systems		X	X			X		X	X						X														X

Another way to examine the matrix of Figure 17.1 is to look at the various columns that represent key tasks and activities of systems engineering, in relation to the various software categories. A summary of the results of such a process, highlighting the main points, is provided here:

1. definition of needs, goals, and objectives
   - word processors clearly useful
   - idea processors may find greater application

2. requirements analysis
   - dearth of p-c packages that can handle this area
   - specialized and powerful mainframe packages likely to migrate into the p-c environment
   - requirements traceability and the like can be handled by overlays on existing DBMSs, and integrated packages will provide additional capability for summarizing and plotting of results
   - workbenches and structured-analysis tools begin to provide capability to analyze and display requirements and their interrelationships
   - AI-oriented tools provide advanced DBMS capabilities

3. requirements allocation
   - needs rational algorithms for definition of allocation procedures, which may be found in statistical, and either mathematical or engineering packages or both
   - new algorithms, simple or complex, can be programmed using one of several available programming languages

4. functional analysis
   - diagramming techniques useful here; these are embedded in workbenches and structured analysis tools
   - graphics and CAD/CAM packages may be adapted
   - correlation of requirements and functions easily built in to DBMSs
   - multitask applications managers provide overlay capability to bring results of aforementioned applications together

5. functional allocation
   - same as number 3, requirements allocation

6. specification development
   - word processors have clear and direct application
   - DBMSs maintain traceability between specifications, functions, and requirements
   - diagramming tools in workbenches, structured-analysis tools, graphics packages, and some project management aids provide specification-tree capabilities

7. system and subsystem design
   - primary design tools include diagramming in workbenches, graphics, CAD/CAM, alternative and preference evaluators, structured-analysis tools, and engineering packages
   - HEMATs can be set up on spreadsheets and integrated packages
   - idea processors may be helpful in looking for new approaches
   - expert systems' search capabilities ultimately adaptable to narrow down design alternatives
   - forecasting tools may help to project trends and system demand for design purposes

8. system and subsystem analysis
   - all of design tools in number 7 applicable
   - simulation packages play a specialized and potentially major role
   - spreadsheets and integrated packages help to sort out relatively simple comparisons and constitute simple models
9. trade-off/alternatives evaluation
   - all of analysis tools in number 8 applicable
   - alternatives and preference evaluators particularly useful
10. software development and analysis
    - workbenches, languages, and structured-analysis tools directly applicable
    - general-support tools and utilities provide overall assistance
    - spreadsheets, statistical and mathematical tools help to develop software estimation models (e.g., person–months and time for development)
11. interface definition
    - diagramming capabilities of workbenches allow interfaces to be shown in diagram format
    - spreadsheets or DBMSs, set up in matrix form, show point to point interfaces
12. schedule development
    - principal use of project management aids
13. life-cycle costing
    - spreadsheets and integrated packages ideal for setting up LCCMs
    - statistical and mathematical tools assist in developing CERs
    - languages usable to define new LCCM equations and relationships
    - project management aids correlate schedule and cost
14. technical performance measurement
    - all systems-analysis tools apply
    - project management aids place TPMs in a schedule context
15. program and decision analysis
    - alternatives and preference evaluators as well as decision-support systems apply directly
    - project management aids handle various types of top-level program analyses
16. risk analysis
    - alternatives and preference evaluators have capability to incorporate risk analysis
    - analytical character of risk analysis addressable by statistical and some mathematical packages
17. integrated logistics support
    - spreadsheets and integrated tools allow simple model building
    - programming languages available for new algorithm computations
    - word processors assist with documentation requirements

18. preplanned-product improvement
    • project management aids help structure plans for such improvements
19. Reliability-Maintainability-Availability
    • reliability tools apply directly
    • statistical tools useful for more sophisticated analyses
20. integration
    • DBMSs help to track and trace equipment hierarchies
    • spreadsheets and integrated packages can define interfaces
21. test and evaluation
    • same as for number 19, integration
    • workbench and simulation tools help to model test variables and parameters
    • word processors assist with test plans and procedures
22. configuration management
    • overlays on DBMSs can provide CM capability
23. quality assurance
    • software QA may be based upon alternatives and preference evaluators and matrix structures available in spreadsheets and integrated packages
24. training
    • idea processors may help develop training curricula
    • word processors assist in preparation of training materials
25. documentation
    • direct application of word processors
26. production
    • CAD/CAM tools help to prepare production diagrams and prints
    • information and data about system stored in DBMSs and other tools can be accessed
27. installation, operation, and operations evaluation/modification
    • combinations of various tools that have appropriate information and data regarding system characteristics

These points reinforce the position that currently available microbased software has numerous systems engineering applications. However, products on the open market cannot handle some of the tasks and activities of systems engineering. This does not necessarily imply that no such software exists. As an example, a generalized life-cycle cost model is not addressed directly in what we would normally consider the commercial market for p-c software. However, such models do exist and have been developed by industrial firms in response to specific needs, often in relation to a particular requirement of a given program. These kinds of models, in general, have not migrated to the commercial market and are considered a gap in this discussion.

A summary of a dozen of these apparent gaps is provided in Table 17.1. These gaps will probably receive attention by the industry since needs exist that are not being fulfilled.

TABLE 17.1 P-C SOFTWARE GAPS IN SYSTEMS ENGINEERING APPLICATIONS

1. Requirements languages and traceability matrices (e.g., as per PSL/PSA)
2. Less-expensive workbenches and structured-analysis tools
3. Less expensive but complete simulation packages
4. Tools specifically designed for risk analysis
5. Tools specifically designed for systems engineering support functions (see Chapter 15)
   - RMA
   - integrated logistics support
   - quality assurance (especially software)
   - configuration management
   - test and evaluation
   - others
6. Tools specifically designed for other systems engineering specialty areas (see Chapter 2):
   - safety
   - security
   - environmental compatibility
   - electromagnetic compatibility
   - others
7. Expert systems as design tools
8. Specialized algorithm and heuristic optimum-seeking design tools
9. LCCM tools
10. Complete and inexpensive language compilers (e.g., Ada language)
11. Better natural-language capabilities and interfaces
12. Capability of capturing diagramming information directly in computer files (e.g., DBMSs)

We noted the significance of requirements languages to capture the system requirements as early as possible in a computer-based format. These languages are available, but a gap exists in the micro environment. In the case of workbenches and structured-analysis tools, the situation is different. Packages are available today for the p-c but they need to come down in price. Precisely the same is true for simulation packages, as discussed in Chapter 12. Despite the relatively large number of decision-analysis tools in the area of risk analysis, a specific response is needed. However, adaptation of the existing decision support software is certainly workable.

There are few packages that specifically address the systems engineering support functions (Chapter 15) or the engineering-specialty areas (Chapter 2). The tools that do exist are largely for the more powerful computers such as those cited in regard to configuration management. Those that exist at the p-c level (such as IBM's SPQR/20 for software reliability and quality assessments) tend either to be expensive for micro software or have been constructed by a large aerospace firm and therefore are not generally available on the open market.

Expert systems are available across the board with respect to price and the gap lies in their currently unfulfilled potential as design tools. This, however, may be a matter of application since the expert systems are shells that do not, in general, contain the specific application. With respect to the matter of system design or synthesis, a gap exists with respect to specialized algorithm and heuristic optimum-seeking design tools. Rectifying this will take considerable imagination, since it deals with the creative endeavor of developing generalized-design approaches. LCCM tools, as with the systems engineering support functions, have been developed by industry but have not generally migrated to the open commercial marketplace.

In the case of languages, a gap exists in the area of low-cost complete compilers for computer program development and prototyping of larger programs. Natural-language capabilities and interfaces are improving but are still relatively immature.

The matter of capturing the real information contained in diagrams of various types was discussed in Chapter 4. Notwithstanding the value of a diagram in its own right, software needs to be developed that will automatically convert the diagram information into computer data that can then be manipulated (e.g., for requirements and specification analysis and traceability).

17.2 SELECTED TRENDS AND DIRECTIONS

The fields of computer technology and software are developing so quickly and in so many directions that a full text would be required to properly analyze and predict trends. The issue here, however, is more specific in that our concern is only with those aspects of potential change that relate to systems engineering. More particularly, we have focused on the personal computer, which further limits the overall scope of investigation. In regard to the future of systems engineering, the micro, and a computer-aided systems engineering environment, the three dozen items listed in Table 17.2 appear to be of special importance. Some are more hardware oriented; others relate more to software. Yet others are conjectural in the sense that they follow from the brief gap discussion above. That is, it is assumed that if a gap exists it is reasonable to conclude that it will be filled in the near future. In this context some of the trends and directions are suggested rather than observed through the examination of hard empirical data. Several others are indeed supportable by real-world data either from advanced development groups, R & D programs, or observations of market products.

TABLE 17.2 TRENDS THAT MAY IMPACT ON COMPUTER-AIDED SYSTEMS ENGINEERING

---

1. Advances Toward Supercomputer Power
2. Broader Infusion of Desktop Publishing
3. Better and More Extensive Programming Language Compilers/Software (e.g., Ada language)
4. Better Natural-Language Capabilities and Interfaces
5. Better Techniques for Software RMA and QA
6. Bulk Electronic Data Transfer Between Offices
7. Data Security Improvements
8. Fourth Generation Languages (4GLs) Available at Micro Levels
9. Further Infusion of Electronic Mail
10. Government-Supplied Information on Diskettes
11. Higher Resolution Graphics

TABLE 17.2 (*cont.*)

12. Higher Quality Flat Displays
13. Improved Image-Processing Machines and Software
14. Improved Price–Performance of Micros; Further Migration Toward Current (32-bit) Workstations and Beyond
15. Improvements in Cost and Performance of Micro-to-Mainframe Links
16. Increased Availability of Specialized Systems Engineering Packages (e.g., LCCMs and CM)
17. Increased Capabilities of Laptops and Portables
18. Increased Use of Higher Order Languages (HOLs)
19. Increased Acceptance of Videoconferencing
20. Insistence Upon Automated Requirements and Structured Analysis
21. Larger, More Complex Systems With Embedded Computers
22. Modest Increases in Telecommuting
23. More Compact, Less Expensive Mass-Storage Devices
24. More Capable and Compatible Networking
25. More Flexible and Capable Overlay Software (e.g., multitask managers)
26. More Powerful and User-Friendly Software-Development Tools, Including Automated-Program Generators
27. Rapid Prototyping of Software
28. Reduced Cost for High-End P-C Packages (e.g., simulations, workbenches)
29. Slow Entry of Voice/Speech Recognition Systems
30. Smaller, More Powerful Chips
31. Software Designed for Use in Group Interactions and Meetings
32. Software Designed to Capture and Process Diagramming Information
33. Stronger Buyer Tendencies Toward Standardization (e.g., Ada, Unix)
34. Transcription Devices
35. Neural-Net and Other Advanced Computer Architectures
36. Widespread Use of Expert Systems Including System-Design Applications

17.3 A HYPOTHETICAL FUTURE CASE ENVIRONMENT

We postulate here that the centerpiece of a future CASE environment will be a computer workstation. Such a workstation would have at least the same capability as today's workstations, as typified by the Apollo and Sun classes of workstations, and IBM's advanced RT with reduced instruction set computing (RISC) technology. The situation may be expressed as shown in Figure 17.2. The CASE workstation is shown in the center of the figure and is designated as a WS/p-c, representing a workstation personal computer. This has at least two essential implications

1. the WS/p-c will represent an upward migration in performance from today's personal computers so that it eventually has at least the capability associated with today's workstation technology
2. the WS/p-c will be properly classified as (the top end of) a personal computer, accessible to the practicing engineer as a personal purchase for both work and home use

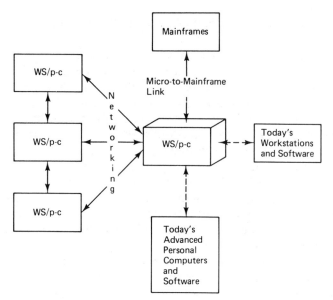

Figure 17.2 Postulated future workstation environment.

The second implication is particularly significant because of the emphasis in this text on the personal computer. Systems engineers, in contrast to software or computer scientists, began to use the computer more extensively when the micro became accessible both at work and at home.

Figure 17.2 also shows a downward-postulated compatibility with the advanced personal computers and vast amounts of software which will continuously be generated over time. This implies that advanced versions of the computer packages at the p-c level, as discussed in this book, will be usable at the WS/p-c. There is also a postulated sideways compatibility that has at least two dimensions

1. compatibility with a broad class of commercially available WS/p-c hardware so that these units can be networked together
2. compatibility with advanced versions of today's workstation software in which considerable investments have already been made

Finally, we have upward compatibility in the sense that the WS/p-c is presumed to have the micro-to-mainframes links that will assure access to even greater computing power as they are connected to future versions of minis and mainframes.

Thus, the systems engineers in this environment will have a powerful capability on their desks; they will be able to fully access advanced and compatible hardware and software. The software will contain, as a minimum, future versions of the 29 categories of software discussed, and will constitute a large library to draw upon.

On the requirements side, it may eventually be expected that the customers will place their goals, objectives, needs, and requirements documentation in a machine-readable form. This will allow the developers to avoid further keying of information and begin the systems engineering process with data already captured on the computer. Compatibility of data formats, and so on, will likely remain an issue for this type of environment but conversion capabilities can be made available to solve these types of problems.

Finally, it is recognized that a significant amount of systems engineering is performed in team situations. These may be design teams, consultation groups, Delphi panels, or review committees. Whatever the form, it would certainly be useful to be able to provide computer assistance to the deliberations of such teams. A process of computer support can be envisioned in several different forms, ranging from a simple transcription of information into a word processor to the use of an expert system. An example of the latter would be in a situation involving a panel of experts and a knowledge engineer who is able to take the expert opinion and place it in a format compatible with an expert system. These may be if-then statements of expertise which, in the aggregate, would constitute a knowledge base. In principle, the expert system would then be able to draw inferences after the consultation sessions that would reveal the implications of the various expert judgments expressed.

Although this is perhaps somewhat futuristic, it is clear that we are moving through a time of technological expansion, one that promises even greater computer-aided capabilities in terms of carrying out the tasks and activities of systems engineering. It has been brought about by the accessibility of the computer, especially the personal computer, to the working systems engineer. With some good sense, and a heritage of rather extraordinary accomplishments, we are likely to experience many years of high productivity, rapid change, technological and economic growth, and lots of excitement.

Part V Appendices

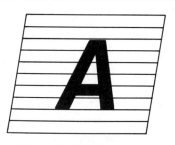

A SAMPLE OF MICROCOMPUTER-BASED SOFTWARE

This appendix contains a list of various software categories and packages for the microcomputer that can be employed in relation to computer-aided systems engineering (CASE). Altogether, 29 categories of such software are set forth in various chapters of this text, with a complete listing of these categories in this appendix as well as in Table 1.3 and Figure 17.1.

The specific, commercially available software packages are listed in Table A.1 in alphabetic order under each of the 29 categories, together with coded and incomplete names of the vendors who have developed and market the packages. These lists do not represent a judgment as to the best software available; they constitute a broad representation of the types of software for various machines, in different price ranges, and so on. Many, if not all of the packages, have appeared in product reviews in one or more of the standard computer magazines, journals, and catalogs. The list can be thought of as simply a sampling of the commercial software that is available for the p-c and the interested reader is urged to dig more deeply into the features that may be contained in these and similar software packages in relation to specific needs. Also, the 29 categories do not represent all categories of available software; the reader with a highly specialized interest (e.g., accounting packages or educational packages) may wish to consult other sources.

Some of the categories in Table A.1 bear a relationship to one another and some of the specific software packages have more than one inherent function. As a consequence, the categories overlap and several packages can be thought of as fitting into more than one category. The following points are to be noted in this connection:

- the integrated packages category (E), normally contain at least two of the capabilities represented by the (A) Spreadsheets, (B) Database Management Systems, (C) Word Processors, and (D) Graphics packages
- many of the categories and specific packages may be considered Decision-Support Systems (I), particularly the Alternatives and Preference Evaluators (H)
- many packages contain a Database Management System (B) as an underlying feature that is used to carry out more specialized functions
- the General-Support Tools (X) are sometimes also considered Utilities (Y) and vice versa
- many packages listed under Forecasting (AA) are also directly usable under Statistical (M), and vice versa
- the AI-Oriented (CC) packages are advertised as containing one or more features of an AI or Expert System (BB) but are not considered to be true or full AI systems

Finally, Table A.2 is an alphabetic list of the vendors shown in Table A.1. Full names and addresses are provided in distinction to the short coded names in Table A.1. Direct inquiries can thus be made of these vendors, recognizing that a number of vendors might have changed in regard to location, and even name.

TABLE A.1 MICROCOMPUTER SOFTWARE CATEGORIES AND SELECTED COMMERCIAL PACKAGES, AND PRODUCERS/VENDORS

A. Spreadsheets
    1. Cambridge Spreadsheet Analyst   Cambridge SW
    2. FlashCalc   Paladin
    3. MagiCalc   Artsci
    4. Magic Worksheet   Structured Systems
    5. Mycalc   The SW Toolworks
    6. Multiplan   Microsoft
    7. Number Works   Paperwork SW
    8. Perfect Calc   Thorn
    9. PFS: Plan   SW Publishing
   10. PlannerCalc   IBM
   11. PractiCalc   PractiCorp
   12. SuperCalc   Computer Associates
B. Database Management Systems
    1. dBase II, III, III Plus   Ashton Tate
    2. Cornerstone   Infocom
    3. DataEase   SW Solutions
    4. Enrich   Migent
    5. Formula IV   Dynamic
    6. Overvue   Provue

7. KnowledgeMan	MDBS
8. Oracle/PC	Oracle
9. Paradox	Ansa
10. PC/Focus	Info Builders
11. PFS:File, Report, Professional File	SW Publishing
12. PowerBase	PowerBase
13. R-Base 5000, System V	Microrim
14. Revelation	Cosmos
15. Zim	Zanthe

C. Word Processors

1. DisplayWrite	IBM
2. FinalWord	Mark/Unicorn
3. Microsoft Word	Microsoft
4. MultiMate	Multimate
5. Perfect Writer	Thorn
6. PC-Write	Quicksoft
7. PFS:Write	SW Publishing
8. Textra	Ann Arbor SW
9. Volkswriter	Lifetree SW
10. WordPerfect	Satellite SW
11. WordStar	MicroPro
12. ZenWord	Knowledge Eng

D. Graphics

1. Artwork and Brushwork	West End
2. Freelance	Graphic Comm
3. Imigit Plus	Chorus Data
4. Lumena	Time Arts
5. Newsroom	Springboard SW
6. Paintpad Video Art Software	Vectrix
7. PC Paint	Mouse
8. PC Storyboard	IBM
9. The Grafix Partner	Brighthill-Roberts
10. Videogram	Softel

E. Integrated Packages

1. AppleWorks	Apple
2. Corporate MBA	Lemain
3. Enable	The SW Group
4. Excel	Microsoft
5. Framework	Ashton Tate
6. Integrated T/Maker	T/Maker
7. Jazz	Lotus
8. Lotus 1-2-3	Lotus
9. PractiCalc	PractiCorp
10. Samna+	Samna
11. SuperCalc	Computer Associates
12. Symphony	Lotus
13. Three Easy Pieces	Haba
14. Twin	Mosaic
15. VP-Planner	Paperback SW

F. Multitask Applications Managers

1. APX Core Executive	Application Exec

    2. DESQ-view   Quarterdeck
    3. E-Z-DOS-IT   Hammer
    4. GEM Desktop   DRI
    5. Topview   IBM
    6. WindowMaster   Structured
    7. Windows   Microsoft

G. Project Management Aids
    1. CPM/PERT   Elite
    2. EX-PERT/80   Decision Support
    3. GSC/DSS   GSC
    4. MicroPERT 0   Sheppard
    5. Milestone   Digital Marketing
    6. The Harvard Project Manager   Harvard
    7. Time Line   Breakthrough SW
    8. Trigger   Thoughtware

H. Alternatives and Preference Evaluators
    1. Arborist   Scientific Press
    2. Expert Choice   Decision Support
    3. Expert Ease   J. Perrone, Human Edge
    4. Lightyear   Lightyear
    5. Supertree   Strategic Decisions

I. Decision-Support Systems
    1. Clout   Microrim
    2. Decision   Once Begun
    3. Decision Aide   Kepner-Tregoe
    4. Decision-Analyst   Executive SW
    5. Encore   Ferox
    6. Helix   Odesta
    7. IFPS/Personal   Execucom
    8. MAC-PAC/PC   Arthur Anderson
    9. pcExpress   Management Decision
    10. Reflex   Analytica
    11. The Confidence Factor   Simple SW
    12. The Deciding Factor   SW Publishing
    13. The Decision Maker   Alamo
    14. see Alternatives & Preference Evaluators (H)

J. Toolchests/Desk Managers
    1. Get Organized   Electronic Arts
    2. Homebase   Amber
    3. My Desk   Third Floor
    4. Office   Applied MicroSystems
    5. PC-Desk III   SW Studios
    6. Polywindows Desk   Polytron
    7. Pop-Up Deskset Plus   Bellsoft
    8. Sidekick   Borland
    9. Spotlight   Lotus
    10. The Desk Organizer   Warner

K. Idea Processors
    1. daVinci   Applied MicroSystems
    2. Factcruncher   InfoStructures
    3. MaxThink   Maxthink

4. PC-Outline	SoftWorks
5. Ready!	Living Videotext
6. ThinkTank	Living Videotext
7. Thor	Fastware
L. Management Evaluators/Aids	
1. Communication Edge	Human Edge
2. Management Edge	Human Edge
3. Negotiation Edge	Human Edge
4. Sales Edge	Human Edge
5. Thoughtware	Thoughtware
6. Wholistic Performance Appraisal	James Campbell
M. Statistical	
1. Abstat	Anderson Bell
2. SAS	SAS
3. SL-Micro	Dynacomp
4. SPSS/PC	SPSS
5. StatPac, StatPac Gold	Walonick
6. StatWorks	Heyden
7. The Statistics Series	HSD
N. Mathematical	
1. Basic Scientific Routines	Dynacomp
2. Fast Fourier Transform Master	Dynacomp
3. Math Master Series	Dynacomp
4. Math Pack	Dynacomp
5. MATRIX 100	Stanford
6. muMath	Microsoft
7. TK!SOLVER	SW Arts
O. Engineering	
1. Active Circuit Analysis Program (ACAP)	Dynacomp
2. Beam Deflection	Dynacomp
3. Heat Loss	Dynacomp
4. Micro-Logic/Cap II	Dynacomp
5. TK!SOLVERPACKS	SW Arts
6. Transfer Function Analyzer	Dynacomp
7. WatNet	Disco/Morton
P. Reliability	
1. PSI Reliability Prediction Program	Elliam
2. RAP217 PREDICTION	Prompt
Q. Mathematical Programming	
1. AMPS & MPP II	Micro Vision
2. LINDO, GINO, VINO	Lindo
3. Linear Programmar	Dynacomp
4. LP 83/MIP 83	Sunset
5. MASCOPE IV	MAS
6. MP Integrated Package	SCI
7. What's Best	General Opt
R. Simulation	
1. ACES	Modulo 2
2. CAPS	Hutchinson
3. micro-DYNAMO	Addison-Wesley
4. GPSS	Minuteman

5. ISIM	Crosbie
6. MicroNet	Pritsker
7. PC Simscript II.5	CACI
8. Siman, Blocks & Playback	Systems Modeling
9. SLAM II	Pritsker
S. Curves and Plotting	
1. Curve Fitter/Scientific Plotter	Microware
2. Data Smoother/Xtrapolator	Dynacomp
3. LSF	Prentice-Hall
4. Regression Analysis	Dynacomp, Apple Core
5. Statgraphics	STSC
6. BPS Business Graphics	BPS
T. CAD/CAM/Graphics	
1. AutoCad	Autodesk
2. Cadkey	Micro Control
3. CadPlan	Personal CAD
4. Cadvance	CalComp
5. Diagram Master	Decision Resources
6. Execuvision	VCN
7. PC Draw	Micrografx
8. VersaCAD	T & W Systems
9. 3 Design	Tritek
U. Workbenches	
1. Analyst Workbench	Visible
2. CASE 2000	Nastec
3. Excelerator	InTech
4. DesignAid	Nastec
5. Information Engineering Workbench	KnowledgeWare
V. Structured-Analysis Tools	
1. Action Diagrammer	Database Design '
2. Dezign-PC	Zeducorp
3. PCSA	StructSoft
4. ProKit Analyst	McDD
5. ProMod	Promod
6. Structured Architect	ISDOS
7. see Workbenches (U)	
W. Languages	
1. Ada*	Telesoft
2. Forth	Forth
3. Microsoft Fortran	Microsoft
4. IQLisp	Integral Quality
5. Lattice C Compiler	Lifeboat
6. MBasic, QuickBASIC	Microsoft
7. Turbo Prolog	Borland
8. Modula-2 86	Logitech
9. Nevada Cobol	Ellis
10. PL/1-86	DRI
11. Turbo Pascal	Borland
X. General-Support Tools	
1. Appgen	SW Express
2. Keyworks	Alpha SW

3. Newkey	FAB SW
4. Pick	Pick
5. Prokey	RoseSoft
6. RE/Call	Yes SW
7. SmartKey	SW Research
8. SuperKey	Borland
9. Supersort II	MicroPro
10. Team-Up	Unlimited

Y. Utilities

1. Diskette Manager	Lassen
2. Disk Magic	Norell
3. Diskmap/PC	Unicorn
4. Disk Mechanic	MLI
5. Double DOS	SoftLogic
6. DU	NY Amateur
7. Power!	Computing!
8. TallScreen	Qualitas
9. The Norton Utilities	P. Norton
10. SYMDEB	Microsoft

Z. Micro-to-Mainframe Links

1. ADR PC	ADR
2. Executive Peachpak II	MSA
3. Filenet Series, Forte PJ	Forte
4. IDEAcomm 3278	IDEAssociates
5. Irma	Digital Comm
6. Multipro	Cap Gemini
7. OmniLink	On-Line SW
8. PC 204	CCA

AA. Forecasting

1. Automatic Time Series Analysis	Transaction
2. INSTAT-R: Multiple Regression	Statistical
3. Interactive Multiple Prediction (IMP)	Math SW
4. Multilinear Regression	Dynacomp
5. Regression Analysis	Dynacomp, Apple Core
6. Regression I, II	Dynacomp
7. Smartforecasts II	Smart SW
8. Solomon	Personnel
9. Survtab	Stat Comp
10. TeloFacts	Dilithium
11. see Statistical (M)	

BB. Expert Systems

1. ES/P ADVISOR	Expert Systems
2. Expert-Ease	Human Edge
3. EXSYS	Exsys
4. Guru	MDBS
5. Insight 1 and 2	Level Five
6. KDS	KDS
7. KES	SW A&E
8. M.1	Teknowledge
9. MICRO-EXPERT	McGraw-Hill
10. MICRO-PS	Ashton-Tate

11. PERSONAL CONSULTANT PLUS	TI
12. RULEMASTER	Radian
13. SeRIES-PC	SRI
14. TIMM-PC	GRC
CC. AI-Oriented	
1. Clout 2	Microrim
2. Javelin	Javelin
3. Logic-Line Series 1	Clarity
4. NaturaLink	TI
5. Paradox	Ansa
6. Q & A	Symantec
7. Savvy PC	Savvy

*Ada is a registered trademark of the U.S. Government (Ada Joint Program Office)

TABLE A.2 ALPHABETIC LIST OF P-C SOFTWARE PRODUCERS/VENDORS

Addison-Wesley	Addison-Wesley Publishing Company, Inc., Jacob Way, Reading, MA 01867
ADR	Applied Data Research, Route 206 and Orchard Road, CN-8, Princeton, NJ 08540
Alamo	Alamo Learning Systems, Suite 500, 1850 Mount Diablo Blvd., Walnut Creek, CA 94596
Alpha SW	Alpha Software Corp., 30 B St., Burlington, MA 01803
Amber	Amber Systems, Inc., 1171 Saratoga-Sunnyvale Rd., San Jose, CA 95129
Analytica	Analytica Corp., 3155 Kearny St., Fremont CA 94538
Anderson Bell	Anderson Bell Co., P. O. Box 191, Canon City, CO 81212
Ann Arbor SW	Ann Arbor Software, 407 N. Main, Ann Arbor, MI 48104
Ansa	Ansa Software, 1301 Shoreway Rd, #221, Belmont, CA 94002
Apple	Apple Computer Co., Inc., 20525 Mariani Ave., Cupertino, CA 95014
Apple Core	San Francisco Apple Core, 1515 Sloat Blvd., Suite 2, San Francisco, CA 94132
Application Exec	Application Executive Corp., 600 Broadway, Suite 4C, New York, NY 10012
Applied MicroSystems	Applied MicroSystems Inc., P. O. Box 832, Roswell, GA 30077
Arthur Anderson	Arthur Anderson & Co., Suite 1500, 816 Congress Ave., Austin TX 78701

Artsci	Artsci, Inc., 5547 Satsuma Ave., N. Hollywood, CA 91601
Ashton-Tate	Ashton Tate Inc., 10150 W. Jefferson Blvd., Culver City, CA 90230
Autodesk	Autodesk, Inc., 2320 Marinship Way, Sausalito, CA 94965
Bellsoft	Bellsoft, 2820 Northup Way, Bellevue, WA 98004
Borland	Borland International, 4113 Scotts Valley Dr., Scotts Valley, CA 95066
BPS	Business & Professional Software, Inc., 143 Binney St., Cambridge, MA 02142
Breakthrough SW	Breakthrough Software Corp., 505 San Marin Dr., Novato, CA 94947
Brighthill-Roberts, Ltd.,	Suite 421, University Bldg., 120 E. Washington St., Syracuse, NY 13202
CACI	C.A.C.I., 3344 North Torrey Pines Court, La Jolla, CA 92037
CalComp	CalComp, Personal Systems Unit, 200 Hacienda Ave., Campbell, CA 95008
Cambridge SW	Cambridge Software Collaborative, University Place, #200, Cambridge, MA 02138
Cap Gemini	Cap Gemini Software Products, Dallas, TX
CCA	Computer Corporation of America, Four Cambridge Center, Cambridge, MA 02142
Chorus Data	Chorus Data Systems, Inc., P. O. Box 370, 6 Continental Blvd., Merrimack, NH 03054
Clarity	Clarity, Chesterland, OH
Computer Associates	Computer Associates International, Micro Products Div., 2195 Fortune Drive, San Jose, CA 95131
Computing!	Computing!, 2519 Greenwich St., San Francisco, CA 94123
Cosmos	Cosmos, Inc., 19530 Pacific Coast Hgwy., South, Suite 102, Seattle, WA 98188
Crosbie	Roy E. Crosbie, Dept. of Computer Science, California State University, Chico, CA 95927
Cullinet	Cullinet Software, Inc., 400 Blue Hill Drive, Westwood, MA 02090
Database Design	Database Design, Inc., 2020 Hogback Road, Ann Arbor, MI 48104
Decision Resources	Decision Resources, 25 Sylvan Rd. S., Westport, CT 06880

TABLE A.2 (*cont.*)

Decision Support	Decision Support Software, Inc., 1300 Vincent Place, McLean, VA 22101
Digital Comm	Digital Communications Associates Inc., 100 Alderman Drive, Alpharetta, GA 30201
Digital Marketing	Digital Marketing Corp., 2363 Boulevard Circle, Walnut Creek, CA 94595
Dilithium	Dilithium Press Ltd., P. O. Box 606, Beaverton, OR 97075
Disco/Morton	Disco-Tech/Morton Technologies, 600 B St., P. O. Box 1659, Santa Rosa, CA 95402
DRI	Digital Research, Inc., 60 Garden Court, Box DRI, Monterey, CA 93942
Dynacomp	Dynacomp, Inc., 1064 Gravel Rd., Webster, NY 14580
Dynamic	Dynamic Microprocessor Associates, Inc., 545 Fifth Ave., New York, NY 10017
Electronic Arts	Electronic Arts, 2755 Campus Drive, San Mateo, CA 94403
Elite	Elite Software Development, Inc., P. O. Box 1194, Bryan, TX 77806
Elliam	Elliam, 24000 Bessemer St., Woodland Hills, CA 91367
Ellis	Ellis Computing, Inc., 3917 Noriega St., San Francisco, CA 94122
Execucom	Execucom Systems Corp., 3410 Far West Blvd., Austin TX 78731
Executive SW	Executive Software, Inc., Two North State St., Dover, DE 19901; Bay St., Shanty Bay, Ontario, Canada L0L 2L0
Expert Systems	Expert Systems International, 1150 First Ave., King of Prussia, PA 19406
Exsys	Exsys, Inc., P. O. Box 75158, Contr. Sta. 14, Albuquerque, NM 97194
FAB SW	FAB Software, P. O. Box 336, Wayland, MA 01778
Fastware	Fastware Inc., 200 Freeway Drive East, East Orange, NJ 07018
Ferox	Ferox Microsystems, Inc., Suite 611, 1701 North Fort Meyer Drive, Arlington, VA 22209
Forte	Forte Communications, 2205 Fortune Drive, San Jose, CA 95181
Forth	Forth, Inc., 2309 Pacific Coast Hgway., Hermosa Beach, CA 90254
General Opt	General Optimization Inc., 2251 N. Geneva Terr., Chicago, IL 60614

Graphic Comm	Graphic Communications Inc., 200 Fifth Ave., Waltham, MA 02254
GRC	General Research Corp., P. O. Box 6770, Santa Barbara, CA 93160
GSC	General Software Corp., 8401 Corporate Drive, #502, Landover, MD 20785
Haba	Haba Systems, 6711 Valjean Ave., Van Nuys, CA 91406
Hammer	Hammer Computer Systems, Inc., 900 Larkspur Landing Circle, Suite 250, Larkspur, CA 94939
Harvard	Harvard Software, Inc., 521 Great Rd., Littleton, MA 01460
Heyden	Heyden & Son Inc., 247 S. 41st St., Philadelphia, PA 19104
Human Edge	Human Edge Software Corp., 2445 Faber Place, Palo Alto, CA 94303
Hutchinson	Hutchinson Associates Ltd., 3404 Colette Court, Mequon, WI 53092
HSD	Human Systems Dynamics, 9010 Reseda Blvd., Suite 222, Northridge, CA 91324
IBM	IBM, 1000 N. W. 51st St., Boca Raton, FL 33432; Old Orchard Road, Armonk, NY 10504; P. O. Box 2328, Menlo Park, CA 94025
IDEAssociates	IDEAssociates Inc., 35 Dunham Road., Billerica, MA 01821
Info Builders	Information Builders Inc., 1250 Broadway, New York, NY 10001
Infocomp	Infocomp Inc., 55 Wheeler St., Cambridge, MA 02138
InfoStructures	InfoStructures Inc., P. O. Box 32617, Tucson, AZ 85751
InTech	Index Technology Corp., Five Cambridge Center, Cambridge, MA 02142
Integral Quality	Integral Quality, Inc., P. O. Box 31970, Seattle, WA 98103
ISDOS	ISDOS Inc., 325 East Eisenhower Parkway, P. O. Box 4179, Ann Arbor, MI 48106
James Campbell	James C. Campbell & Associates, Inc., 2822 Columbine Pl., Nashville, TN 37204
Javelin	Javelin Software Corp., One Kendall Square, Bldg. 200, Cambridge, MA 02139
J. Perrone	Jeffrey Perrone & Associates, 3685 17th St., San Francisco, CA 94114

TABLE A.2 (*cont.*)

KDS	KDS Corp., 934 Hunter Rd., Wilmette, IL 60091
Kepner-Tregoe	Kepner-Tregoe, Inc., Research Road, P. O. Box 704, Princeton, NJ 08540
Knowledge Eng	Knowledge Engineering Inc., 1439 Circle Ridge, Austin TX 78746
KnowledgeWare	KnowledgeWare, Ann Arbor, MI
Lassen	Lassen Software, Inc., P. O. Box 1190, Chico, CA 95927
Lemain	Lemain, Inc., P. O. Box 3010, Agoura, CA 91301
Level Five	Level Five Research, 4980 South Hgwy. A1A, Melbourne Beach, FL 32951
Lifeboat	Lifeboat Associates, 1651 Third Ave., New York, NY 10028
Lifetree SW	Lifetree Software Inc., 411 Pacific St., Monterey, CA 93940
Lightyear	Lightyear Inc., 2465 E. Bayshore Rd., Suite 301, Palo Alto, CA 94303
Lindo	Lindo Systems Inc., P. O. Box 148231, Chicago, IL 60614
Living Videotext	Living Videotext Inc., 2432 Charleston Rd., Mountain View, CA 94043
Logitech	Logitech, Inc., 805 Veterans Blvd., Redwood City, CA 94063
Lotus	Lotus Development Corp., 55 Cambridge Pkwy., Cambridge, MA 02142
Management Decision	Management Decision Systems, Inc., 630 Fifth Ave., Suite 1661, New York, NY 10111
Mark/Unicorn	Mark of the Unicorn, 222 Third St., Cambridge, MA 02142
MAS	Management Analytic Support, Inc., Route 4, Box 4127, La Plata, MD 20646
Math SW	Mathematical Software Co., P. O. Box 12349, El Cajon, CA 92022
Maxthink	Maxthink, Inc., 230 Crocker Ave., Piedmont, CA 94610
McDD	McDonnell Douglas Information Systems Group, P. O. Box 516, St. Louis, MO 63166
McGraw-Hill	McGraw-Hill Book Company, Inc., 1221 Avenue of the Americas, New York, NY 10020
MDBS	Micro Data Base Systems, Inc., P. O. Box 248, Lafayette, IN 47902
Micro Control	Micro Control Systems, Inc., 143 Tunnel Rd., Vernon, CT 06066

Micrografx	Micrografx, Inc., 1701 N. Greenville, Suite 305, Richardson, TX 75801
MicroPro	MicroPro International Corp., 33 San Pablo Ave., San Rafael, CA 94903
Microrim	Microrim, Inc., 3380 146th Place SE. Bellevue, WA 98007
Microsoft	Microsoft Corp., 10700 Northrup Way, Box 97200, Bellevue, WA 98009
Micro Vision	Micro Vision, 145 Wicks Commack, NY 11725
Microware	Microware, Inc., P. O. Box 139, State College, PA 16804
Migent	Migent Software, 8300 N. Hayden Rd., Suite 203, Scottsdale, AZ 85258; P. O. Box 6062, Ineline Village, NC 89450
MLI	MLI Microsystems, Box 825, Framingham, MA 01701
Minuteman	Minuteman Software, P. O. Box 171, Stow, MA 01775
Modulo 2	Modula 2 Company, P. O. Box 58781, Tukwila, WA 98188
Mosaic	Mosaic Software, Inc., 1972 Massachusetts Ave., Cambridge, MA 02140
Mouse	Mouse Systems, 2336H Walsh Ave., Santa Clara, CA 95051
MSA	Management Science America Inc., 3445 Peachtree Rd. NE, Atlanta, GA 30326
Multimate	Multimate International Corp., 52 Oakland Ave., East Hartford, CT 06108
Nastec	Nastec Corp., 24681 Northwestern Highway, Southfield, MI 48075
Norell	Norell Data Systems, 3400 Wilshire Blvd., P. O. Box 70127, Los Angeles, CA 90010
NY Amateur	New York Amateur Computer Club, Inc., P. O. Box 106, Church Street Station, New York, NY 10008
Odesta	Odesta Corp., 3186 Doolittle Drive, Northbrook, IL 60062
Once Begun	Once Begun Computations, Searsport, ME 04974
On-Line SW	On-Line Software International, Inc., Two Executive Drive Fort Lee, NJ 07024
Oracle	Oracle Corp., 2710 Sand Hill Rd., Menlo Park, CA 94025
Paladin	Paladin Software, 2895 Zanker Rd., San Jose, CA 95134
Paperback SW	Paperback Software International, 2612 Eighth St., Berkeley, CA 94710

Personal CAD	Personal CAD Systems, Inc., 981 University, Los Gatos, CA 95030
Personnel	Personnel Software, Inc., 317 Barton Court, Danville, CA 94526
Pick	Pick Systems, 1691 Browning, Irvine, CA 92714
P. Norton	Peter Norton Utilities, 2210 Wilshire Blvd., Santa Monica, CA 90403
Polytron	Polytron, P. O. Box 787, Hillsboro, OR 97123
Powerbase	PowerBase Systems Inc., 12 W. 37th St., New York, NY 10018
PractiCorp	PractiCorp International, The Silk Mill, 44 Oak St., Newton Upper Falls, MA 02164
Prentice-Hall	Prentice-Hall, Inc., Rte. 9W, Englewood Cliffs, NJ 07632
Pritsker	Pritsker & Associates, P. O. Box 2413 A, West Lafayette, IN 47906
Programming Logic	Programming Logic Systems, Inc., 31 Crescent Drive, Milford, CT 06460
Promod	Promod, Inc., 22981 Alcalde Drive, Laguna Hills, CA 92653
Prompt	Prompt> The Software Company, 393 Englert Court, San Jose, CA 95133
Provue	Provue Development Corp., 222 22nd St., Huntington Beach, CA 92648
Qualitas	Qualitas Inc., 8314 Thoreau Drive, Bethesda, MD 20817
Quarterdeck	Quarterdeck Office Systems, 150 Pico Blvd., Santa Monica, CA 90405
Quicksoft	Quicksoft, 219 First Ave. North, #224, Seattle, WA 98109
Radian	Radian Corp., 8501 MoPac Blvd., P. O. Box 9948, Austin, TX 78766
RoseSoft	RoseSoft Inc., 4710 University Way N. E. #601, Seattle, WA 98105
Samna	Samna Corp., 1700 N.E. Expressway, Atlanta, GA 30345
Satellite SW	Satellite Software International, 288 W. Center St., Orem, UT 84057
Savvy	The Savvy Corp., 122 Tulane SE, Albuquerque, NM 87106
SCI	SCI Computing, 1167 Wilmette Ave., Wilmette, IL 60091
Scientific Press	The Scientific Press, 540 University Ave., Palo Alto, CA 94301

SAS	SAS Institute Inc., Box 8000, SAS Circle, Cary, NC 27511
Sheppard	Sheppard Software Co., 4750 Clough Creek Rd., Reading, CA 96002
Simple SW	Simple Software Inc., 2 Pinewood, Irvine, CA 92714
Smart SW	Smart Software Inc., 392 Concord Ave., Belmont, MA 02178
Softel	Softel Inc., 34 1/2 St. Marks' Place, New York, NY 10003
SoftLogic	SoftLogic Solutions, 530 Chestnut St., Manchester, NH 03101
SoftWorks	SoftWorks Development, 750 Stierlin Rd.,#142, Mountain View, CA 94043
Sorcim/IUS	Sorcim/IUS Corp., 2195 Fortune Drive, San Jose, CA 95131
Springboard SW	Springboard Software, 7808 Creek Ridge Circle, Minneapolis, MN 55435
SPSS	SPSS Inc., 444 N. Michigan Ave., Chicago, IL 60611
SRI	SRI International, Advanced Computer Science Dept., 333 Ravenswood Ave., Menlo Park, CA 94025
Stanford	Stanford Business Software, Inc., 4151 Middlefield Rd., Suite 215, Palo Alto, CA 94303
Statistical	Statistical Consulting Services, 517 East Lodge Drive, Tempe, AZ, 85283
Stat Comp	Statistical Computing Consultants, 10037 Chestnut Wood Lane, Burke, VA 22015
Strategic Decisions	Strategic Decisions Group, 3000 Sand Hill Road, Menlo Park, CA 94025
StructSoft	StructSoft, Inc., 24 Homer St., Parsippany, NJ 07054
Structured	Structured Systems Group, 111 Broadway, Suite 203, Oakland, CA 94607
STSC	STSC, Inc., 2115 East Jefferson St., Rockville, MD 20852
Sunset	Sunset Software, Suite 153, 1613 Chelsea Rd., San Marino, CA 91108
SW A&E	Software Architecture and Engineering, Inc., 1500 Wilson Blvd., Suite 800, Arlington, VA 22209
SW Arts	Software Arts., Inc., 27 Mica Lane, Wellesley, MA 02181 (or Lotus)
SW Express	Software Express, 2925 Briarpark Drive, 7th Floor, Houston, TX 77042

TABLE A.2 (*cont.*)

SW Publishing	Software Publishing Corp., 1901 Landings Drive, Mountain View, CA 94043
SW Research	Software Research Technologies, Inc., 3757 Wilshire Blvd., #211, Los Angeles, CA 90010
SW Solutions	Software Solutions Inc., 12 Cambridge Drive, Trumbull, CT 06611
SW Studios	Software Studios, 8516 Sugarbush, #104, Annandale, VA 22003
Symantec	Symantec Corp., 10201 Torre Ave., Cupertino, CA 95014
Systems Modeling	Systems Modeling Corp., Calder Square, P. O. Box 10074, State College, PA 16805
Teknowledge	Teknowledge, Inc., 525 University Ave., Palo Alto, CA 94301
TeleSoft	TeleSoft, 10639 Roselle St., San Diego, CA 92121
The SW Group	The Software Group, Northway Ten Executive Park, Ballston Lake, NY 12019
The SW Toolworks	The Software Toolworks, 15233 Ventura Blvd., Suite 1118, Sherman Oaks, CA 91403
Third Floor	Third Floor Systems, 1630 Oakland Road, Suite A114, San Jose, CA 95131
Thorn	Thorn EMI Computer Software, 1881 Langley, Ave., Irvine, CA 92714
Thoughtware	Thoughtware, 2699 S. Bayshore Drive, Suite 1000A, Coconut Grove, FL 33133
TI	Texas Instruments, P. O. Box 809063, Dallas, TX 75380
Time Arts	Time Arts Inc., 3436 Mendocino Ave., Santa Rosa, CA 95401
T/Maker	T/Maker Co., 2115 Landing Drive, Mountain View, CA 94043
Transaction	Transaction Systems, Inc., 8164 South Elwood, Tulsa, OK 74132
Tritek	Tritek Vision Systems, 4710 University Way N.E., Suite 1512, Box C-56789, Seattle, WA 98105
T & W Systems	T & W Systems, Inc., 7372 Prince Drive, #106, Huntington Beach, CA 92647
Unicorn	Unicorn Systems Co., 3807 Wilshire Blvd., Los Angeles, CA 90010
Unlimited	Unlimited Processing Inc., 8382 Baymeadows Rd., #8, Jacksonville, FL 32216

TABLE A.2 (*cont.*)

VCN	VCN, 238 Main St., Cambridge, MA 02142
Vectrix	Vectrix Corp., 2606 Branchwood Drive, Greensboro, NC 27408
Visible	Visible Systems, 336 Baker Ave., Concord, MA 01742
Walonick	Walonick Associates, 6500 Nicollet Ave. So., Minneapolis, MN 55423
Warner	Warner Software, 666 Fifth Ave., New York, NY 10103
West End	West End Film, Inc., 2121 Newport Place, N. W., Washington, D.C. 20037
Yes SW	Yes Software Inc., 390-10991 Shellbridge Way, Richmond, B. C. Ontario, Canada V6X 3C6
Zanthe	Zanthe Information Inc., 1785 Woodward Drive, Ottawa, Ontario, Canada K2C 0R1
Zeducorp	Zeducorp, P. O. Box 68, Stirling, NJ 07980

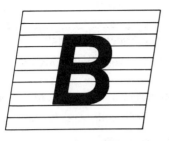

LARGE-SCALE DATABASE MANAGEMENT SYSTEMS

This appendix contains a list of database management systems (DBMSs) oriented toward minicomputers and mainframe computers. A selected sample of DBMSs for the microcomputer is provided in Appendix A.

In spite of the personal-computer orientation of this text, we focus here on the DBMS for the larger machine environments largely since these systems have a broad domain of possible applications. They can be used for numerous "stand-alone" systems engineering application areas, ranging from keeping vendor lists to personnel assignments to equipment hierarchies (with associated data files). With appropriate overlays they can and have been used in such systems engineering areas as requirements analysis and traceability, configuration management, specification trees, test and evaluation, technical performance measurement, program management, and many others. Put simply, when the p-c DBMSs are too limited (principally by size) to do the required task, then the systems engineer must move to a larger package and environment. Hence, we provide a very limited window into the kinds of packages that are available. In general, they are much more elaborate and expensive than the microbased packages. They may require specialized personnel (e.g., database administrators) to keep them running and up to date. Some of the packages are advertised also as fourth generation language (4GL) DBMSs, referring largely to the advanced nature of the command structure. Although the general trend appears to be toward relational DBMSs, others (hierarchical, network) are available and in definitive use. Most of these DBMSs are relatively mature, but improvements are being made as vendors attempt to improve their market positions. The list is by no means all inclusive but might be considered representative of what is available. A serious potential user has much investigation and analysis to perform in order to venture into this world of DBMSs.

As with the previous appendix, this one has two tables. Table B.1 is an alphabetic list of large-scale DBMSs, by package name and coded producer or supplier name. Table B.2 contains an alphabetic list of the full names and addresses of the suppliers of the various packages.

TABLE B.1 SELECTED LARGE-SCALE DATABASE MANAGEMENT SYSTEMS (DBMSs)

System Name	Producer/Supplier
Accent R	NIS
Adabas*	SW AG
Basis	Battelle
CA-Universe	Comp Assoc Intl
Data Base Plus	Tominy
Datacom/DB	ADR
DB2*	IBM
DL/1	IBM
DM-IV	Honeywell
DMS-II	Burroughs
DMS 1100	Sperry
DRS	Advanced Data Mgmt
Focus*	Info Builders
Freestyle	On-Line SW
IDMS/R*	Cullinet
IMS*	IBM
Info	Henco
Ingres*	Relational
Inquire*	Infodata
Magnum	McDonnell Douglas
MIMS	Geisco
Minisys	Systemhouse
Model 204	CCA
Nomad2	D & B
Natural	SW AG
Oracle*	Oracle
PRO-IV	PCS
Ramis-II*	Mathematica
Rapport	Logica
RTFile	Contel
Seed	Seed SW
Sir/DBMS	SIR
SQL/DS*	IBM
System 1022, 1032	SW House
System 2000	SAS
Total Information System	Cincom
TurboIMAGE	HP
Ultra	Cincom
VAX DBMS	DEC

*One of the more popular packages.

TABLE B.2 SUPPLIERS OF LARGE-SCALE DATABASE MANAGEMENT
SYSTEMS

ADR	Applied Data Research, Route 206 and Orchard Rd., CN-8, Princeton, NJ 08540
Advanced Data Mgmt	Advanced Data Management, Inc., 15-17 Main St., Kinston, NJ 08528
Battelle	Battelle Software Products Center, 505 King Ave., Columbus, OH 43201
Burroughs	Burroughs Corp., Burroughs Place, Detroit, MI 48232
CCA	Computer Corporation of America, Four Cambridge Center, Cambridge, MA 02142
Cincom	Cincom Systems, Inc., 2300 Montana Ave., Cincinnati, OH 45211
Comp Assoc Intl	Computer Associates International, Inc., 125 Jericho Turnpike, Jericho, NY 11753
Contel	Contel Information Systems, Inc., 4330 East-West Hgwy., Bethesda, MD 20814
Cullinet	Cullinet Software, Inc., 400 Blue Hill Drive, Westwood, MA 02090
D & B	D & B Computing Services, 187 Danbury Rd., Wilton, CT 06897
DEC	Digital Equipment Corp., 146 Main St., Maynard, MA 01754
Geisco	General Electric Information Services Co., 401 North Washington St., Rockville, MD 20850
Henco	Henco Software Inc., 100 Fifth Ave., Waltham, MA 02154
Honeywell	Honeywell Information Systems, 200 Smith St., Waltham, MA 02154
HP	Hewlett-Packard Company, Computer Systems Division, 19420 Homestead Rd., Cupertino, CA 95014
IBM	IBM, Old Orchard Road, Armonk, NY 10504 (DB2); 1133 Westchester Ave., White Plains, NY 10604 (IMS, DL/1)
Info Builders	Information Builders, Inc., 1250 Broadway, New York, NY 10001; 1655 N. Fort Myer Drive, Suite 200, Arlington, VA 22209
Infodata	Infodata Systems Inc., 5205 Leesburg Pike, Falls Church, VA 22041
Logica	Logica Database Products Inc., 666 Third Ave., New York, NY 10017
Mathematica	Mathematica (subs. of Martin Marietta), P. O. Box 2392, Princeton, NJ 08540

TABLE B.2 (*cont.*)

McDD	McDonnell Douglas Information Systems Group, P. O. Box 516, St. Louis, MO 63166
NIS	National Information Systems, Inc., 20370 Town Center Lane, Suite 130, Cupertino, CA 95014
On-Line SW	On-Line Software International, Inc., Two Executive Drive, Fort Lee, NJ 07024
Oracle	Oracle Corp., 2710 Sand Hill Rd., Menlo Park, CA 94025
PCS	Pro Computer Sciences, Inc., 23181 Verdugo Drive, Suite 103A, P. O. Box 30770, Laguna Hills, CA 92654
Relational	Relational Technology Inc., 1080 Marina Village Parkway, Alameda, CA 94501
SAS	SAS Institute Inc., Box 8000, SAS Circle, Cary, NC 27511
Seed SW	Seed Software Corp., 2121 Eisenhower Ave., Alexandria, VA 22314
SIR	SIR, Inc., 5215 Old Orchard Rd., Suite 800, Skokie, IL 60077
Sperry	Sperry Corporation, Computer Systems, P. O. Box 500, Blue Bell, PA 19422
SW AG	Software AG, 11800 Sunrise Valley Drive, Reston, VA 22091
SW House	Software House, 1105 Massachusetts Ave., Cambridge, MA 02138
Systemhouse	Systemhouse Inc., 1655 N. Fort Myer Drive, 5th Floor, Arlington, VA 22209
Tominy	Tominy, Inc., 4221 Malsbary Road, Cincinnati, OH 45242

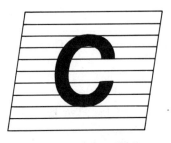

SELECTED DEPARTMENT OF DEFENSE SOURCE DOCUMENTS

The main body of this text has cited many Department of Defense (DoD) source documents in relation to the various tasks and activities of systems engineering. The DoD, responsible for the design, development, construction, and operation of numerous large-scale systems, has produced the largest and most coherent set of standards and related source materials in relation to systems engineering. In addition, the DoD has sponsored considerable research and analysis of a broad range of systems engineering subject areas. Table C.1 contains a selected list of source documents, with emphasis on military standards, that the systems engineering student and practicing engineer might find useful. Further bibliographic listings can be obtained from the Superintendent of Documents, U.S. Government Printing Office, Washington, D. C., 20402.

TABLE C.1 SELECTED DEPARTMENT OF DEFENSE SOURCE DOCUMENTS

AMCP 706-133	Maintainability Engineering Theory and Practice, January 1976
AMCP 706-134	Maintainability Guide for Design, July 1972
DoDD 4245.3	Design to Cost, 1983
DoDD 5000.1	Major System Acquisitions, March 19, 1980
DoDD 5000.2	Major System Acquisition Procedure
DoDD 5000.28	Design To Cost, May 23, 1975
DoDD 5000.3	Test and Evaluation, December 26, 1979
DoDD 5000.31	Interim List of DoD Approved HOL Programming Languages, November 24, 1976
DoDD 5000.39	Acquisition and Management of Integrated Logistic Support for Systems and Equipment, November 17, 1983
DoDD 5000.40	Reliability and Maintainability, July 1980
DoDD 5010.19	Configuration Management
DoD Guide LCC-1	Life Cycle Costing
DoD Guide LCC-2	Life Cycle Costing Casebook
DoD Guide LCC-3	Life Cycle Cost for Systems
DoD-Std-1679A	Software Development-October 22, 1983
DoD-Std-2167	Defense System Software Development, June 4, 1985
DoD-Std-2168	Software Quality Evaluation, April 26, 1985
Mil-Hdbk-217D	Reliability Prediction of Electronic Equipment, January 1982
Mil-Hdbk-259	Life Cycle Cost in Navy Acquisitions, April 1, 1983
Mil-Hdbk-470A	Maintainability Program Requirements for Systems and Equipment, 1983
Mil-Hdbk-471A	Maintainability Verification, Demonstration, Evaluation, March 27, 1973
Mil-Hdbk-472	Maintainability Prediction, May 24, 1966
Mil-Hdbk-63038-1	Technical Manuals Writing Handbook, May 1, 1977
Mil-Hdbk-63038-2	Technical Writing Style Guide, May 1, 1977
Mil-Q-9858A	Quality Program Requirements, August 7, 1981
Mil-S-52779A	Software Quality Assurance Program Requirements, August 1, 1979
Mil-Std-105D	Sampling Procedures and Tables for Inspection by Attributes, November 1, 1978
Mil-Std-454G	Standard General Requirements for Electronic Equipment, March 15, 1980

TABLE C.1 (*cont.*)

Mil-Std-480A	Configuration Control—Engineering Changes, Deviations and Waivers, April 12, 1978
Mil-Std-481A	Configuration Control—Engineering Changes, Deviations and Waivers (Short Form), October 18, 1972
Mil-Std-482A	Configuration Status Accounting Data Elements and Related Features, April 1, 1974
Mil-Std-483	Configuration Management Practices for Systems, Equipment, Munitions and Computer Programs, March 21, 1979
Mil-Std-490	Specification Practices, May 18, 1972
Mil-Std-499A	Engineering Management, May 1, 1974
Mil-Std-680	Contractor Standardization Plans and Management, April 28, 1977
Mil-Std-690B	Failure Rate Sampling Plans and Procedures, August 1, 1974
Mil-Std-721C	Definitions of Effectiveness Terms for Reliability, Maintainability, Human Factors, and Safety, June 12, 1981
Mil-Std-765B	Reliability Modeling and Prediction, November 1981
Mil-Std-781C	Reliability Design Qualification and Production Acceptance Tests: Exponential Distribution, March 20, 1981
Mil-Std-785B	Reliability Program for Systems and Equipment Development and Production, September 15, 1980
Mil-Std-790C	Reliability Assurance Program for Electronic Parts Specifications, November 2, 1979
Mil-Std-881A	Work Breakdown Structure for Defense Material Items, April 25, 1975
Mil-Std-883B	Test Methods and Procedures for Micro Electronics, March 1, 1976
Mil-Std-1369	Integrated Logistics Support Program Requirements, April 29, 1977
Mil-Std-1379A	Training Operations and Training Data, July 20, 1977
Mil-Std-1388-1A	Logistics Support Analysis, April 11, 1983
Mil-Std-1388-2A	DoD Requirements for a Logistic Support Analysis Record, July 20, 1984
Mil-Std-1389A	Design Requirements for Standard Electronic Modules, December 21, 1978
Mil-Std-1390B	Level of Repair, December 1, 1976
Mil-Std-1472C	Human Engineering Design Criteria for Military Systems, Equipment and Facilities, May 2, 1981
Mil-Std-1519	Test Requirements Document, August 1, 1977

TABLE C.1 (*cont.*)

Mil-Std-1521A	Technical Reviews and Audits for Systems, Equipment and Computer Programs, December 21, 1981
Mil-Std-1535	Supplier Quality Assurance Program Requirements, February 1, 1974
Mil-Std-1543	Reliability Program Requirements for Space and Missile Systems, July 22, 1977
Mil-Std-1561A	Provisioning Procedures, March 17, 1981
Mil-Std-1629A	Procedures for Performing a Failure Mode, Effects and Criticality Analysis, November 24, 1980
Mil-Std-1679	Weapon System Software Development, December 1, 1978
Mil-Std-2068	Reliability Development Tests, March 21, 1977
Mil-Std-9858A	Quality Program Requirements, December 16, 1963
OMB Circular A-76	Cost Comparison Handbook
OMB Circular A-109	Major System Comparison, April 5, 1976

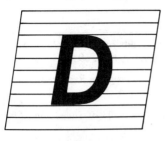

STUDENT OVERVIEW
GUIDELINES

This appendix contains a summary table of areas of study for each of the individual chapters. The areas that have been cited are the principal ones that have been discussed. In general, areas suggested for further investigation in the text are not cited here. This summary is mainly provided for students who want an overview of the text in relation to the subject matter at large or for purposes of studying the subject matter in detail. The practicing engineer, of course, can use this summary for the same purposes. In a similar vein, an instructor may find this summary useful as a source for preparation of lectures or examinations.

TABLE D.1 SUMMARY OF AREAS FOR STUDY BY CHAPTER

Chapter I   Systems Engineering and the Computer
  • Typical Systems Engineering Tasks and Activities
  • Mathematical and Logic-Based Techniques
  • Categories of Microcomputer-Based Packages
  • Characteristics of a Good Systems Engineer
  • Elements of CASE

Chapter II   Formal Systems Engineering Structures
  • Elements of a Systems Approach
  • Elements of Systems Engineering
  • The Morphological Approach
  • Heuristic Matrix (HEMAT) Applications
  • Elements of Systems Engineering Management
  • Systems Acquisition Process

Chapter III   Basic Computer Tools
  Features, Attributes, and Applications of
  • Spreadsheets
  • Database Management Systems (DBMSs)
  • Word Processors
  • Graphics Packages
  • Integrated Packages

Chapter IV   Diagramming Techniques
  Features, Attributes, and Applications of
  • Simple Process-Flow Charts
  • Signal-Flow Block Diagrams
  • Hierarchical Decomposition Diagrams (HDDs)
  • Data Flow Diagrams (DFDs)
  • Functional Flow Diagrams and Descriptions (F^2D^2)
  • HIPO Diagrams
  • Warnier–Orr Diagrams (WODs)
  • Michael Jackson Diagrams
  • Action Diagrams (ADs)
  • Sequence and Timing Diagrams (STDs)
  • Parameter Dependency Diagrams (PDDS)
  • Logic Flowcharts
  • Nassi–Shneiderman Charts (NSCs)
  • Decision-Network Diagrams (DNDs)

Chapter V   Overview of Probability Concepts
  • Discrete Events, Probabilities and Key Relationships
  • Reliability Relationships
  • Discrete Distributions
      Uniform
      Binomial
      Poisson
  • Continuous-Density and Cumulative-Distribution Functions
      Uniform
      Normal (Gaussian)
      Exponential
        Reliability Applications
      Weibull
      Rayleigh

TABLE D.1 (*cont.*)

- Averages
    - Mean Values
    - Higher Order Moments
    - Variance and Standard Deviation
    - Means and Variances for Various Distributions
- Sums of Random Variables
    - Means
    - Variances
    - Distributions
    - Error-Analysis Application
- Functions of Random Variables
    - General Equation
    - Standard Normal Application
- Joint Distributions
    - Rectangular and Polar-Coordinate Systems
    - Dependence and Independence
    - Joint Normal and Rayleigh
    - Covariance and Correlation Coefficient
    - Correlation Applications
- Summary of Important Relationships

Chapter VI   Specialized Computer Tools
Features, Attributes, and Applications of
- Multitask Applications Managers
- Project Management Aids
- Alternatives and Preference Evaluators
- Decision Support Systems
- Toolchests/Desk Managers
- Idea Processors
- Management Evaluators/Aids

Chapter VII   Mathematical and Engineering Tools
Features, Attributes, and Applications of
- Statistical Packages
- Mathematical Packages
- Engineering Packages
- Reliability Packages
- Mathematical Programming Packages
- Simulation Packages
- Curves and Plotting Packages
- CAD/CAM/Graphics Packages
- Workbenches

Chapter VIII   Information and Search Theory
- Self-Information
- Entropy
- Redundancy
- Coding
    - Binary
    - Alphabetic
    - Huffman
    - Code Length
    - Efficiency

- Capacity
    Binary Channel
- Fundamental Theorem
- Diagnostics Using Information Theory
- Search Using Information Theory
- Optimum Search Algorithm
- Searching for an Optimum
    Calculus
    Lagrange Multipliers
- Computer Search Procedures

Chapter IX   Programming Languages
- Three Fundamental Logical Constructs
- BASIC
    Assignments, Operators, Functions, and PRINT
    NEW, String Variables, and Printing Messages
    REM, READ..DATA
    INPUT, PRINT to Skip Line
    INPUT Message, IF..THEN, GOTO
    FOR..TO, STEP..NEXT, Comma, and Reassignment
    Subscripted Variables and Steping Summation
    Nested Loops, Subroutines, Restore, Random-Number Generation, User-Defined Functions

- Pascal
    Declarations and Definitions of Constants and Variables
    Assignment
    String Variables and WRITELN Messages
    Annotations and READLN
    IF..THEN..ELSE
    Output Formats and REPEAT..UNTIL
    ARRAYS and FOR..DO
    WHILE..DO
    Program Structures and Procedures
    Functions
    Files
    Records
    Pointers and Linked Lists

Chapter X   Top-Level Systems Engineering
- Requirements Analysis, Allocation, and Traceability
- Requirements Tools and Languages
- Specifications
- Design Alternatives
- Measures of Merit (MOMs) and Measures of Effectiveness (MOEs)
- Evaluation Criteria
- Overall Evaluation Frameworks
- Commensurability
- Heuristic Matrix (HEMAT) Application
- Cost Effectiveness

Chapter XI   Scheduling and Costing
- Scheduling
    Gantt Charts
    PERT Charts
    Decision and Stochastic Networks

TABLE D.1 (*cont.*)

- Costing
    Life-Cycle Cost Structure
    Quantity-Unit Cost Products
    Bottoms-Up Unit-Cost Estimation
    Parametric Cost Estimating Relationships (CERs)
        linear CERs
        nonlinear CERs
    Software Cost Estimation
    Discounting and Cost Escalation
    Uncertainty in Costing
    Life-Cycle Cost Models (LCCM)
    Coupling Schedule and Cost
    Econometrics

Chapter XII   Performance Modeling and Simulation
- Conceptual, Configurational, Flow, and Computational Models
- Output Analysis
- Effectiveness
- Technical Performance Measurement (TPM)
- Technical Performance Models
- Monte Carlo Simulation
- Random-Number Generation
- Weighting and Rating Procedures
- Setting Up a TPM Program
- Modeling–Simulation Packages
    GPSS
    SIMSCRIPT
    GASP
    SLAM
    DYNAMO
    Others

Chapter XIII   Risk and Decision Analysis
- Risk Analysis
    Technical Performance Risk Methods
    Risk-Abatement Activities
    Societal Risk
    Fault-Tree and Event-Tree Methods
- Decision Analysis
    Decision Networks and Trees
        Q-GERT Software Package
    Decision Theory and Signal Detection
    System-Concept Evaluations
    Subjective Estimation
        Delphi technique
        other approaches
    Decision Support Software (DSSW)

Chapter XIV   Software Development and Analysis
- Software Development Process
- Requirement Areas
- Standards
- Structured Analysis

TABLE D.1 (*cont.*)

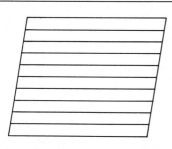

INDEX

Systems engineer, 7
Systems engineering:
  definition of, 17
  management, 25–37
  management plan (SEMP), 25
  process, 26, 31–34
  support functions, 396–431
  tasks and activities, 3, 20, 468–71
Systems evaluation framework, 55
System type and cost example, 60–61

**Tasks and activities of systems engineering,
  3, 20, 468–71**
T&E (*see* Test and evaluation)
Technical performance analysis, 295
Technical performance measures, examples,
  295–96
Technical performance measurement (TPM),
  29–31, 235–36, 295–315, 333
Technical performance models, examples,
  297–313
Technical performance risk, 332–36
Technical program planning and control, 25–27
Technical reviews, 31
Techniques, of systems engineering, logic-
  based and mathematical, 3
Technology forecasting, 440–46
Teichroew, D., 374
TEMP (*see* Test and evaluation master plan)
Test:
  concept of, 422–24
  planning of, 29
  tools, 375
Test and evaluation (T&E), 420–24
  master plan (TEMP), 421–22
Testing, of system, 16
Time dimension, of systems engineering, 18–19
Toffler, A., 1
Toolchests, 145–46
Tools:
  software, 48–66, 139–66, 315–23, 356–59,
    371–83, 428–31, 445–46, 454–58,
    467–73, 477–93
  system design, 20
Top-down synthesis, 17
Top-level systems engineering, 229–58
TPM (*see* Technical performance measurement)
Traceability, requirements, 235–36
Training, 36–37

Transinformation, 177–80
Transportation, examples, 241–44, 246–48,
  250–53, 308–13, 351–52
Trends, 473–76
Trip time, 310–12
Truth table, 449–51
Tversky, A., 354

Uniform distribution, 109–10, 112
Upgrade, of system, 16
Urban dynamics, 3, 321
Utilities, 382
Utility theory, 3

Value theory, 3
Variance, 120–21
  for selected distributions, 122–23
  of sum of random variables, 123, 126
Vendor reviews, 31
Verification and validation (V&V), 416
Villanueva, R., 318
V&V (*see* Verification and validation)

Warnier, J., 78
Warnier–Orr diagrams, 78–80
Weapon System Effectiveness Industry Advi-
  sory Committee (NSEIAC), 294
Weibull distribution, 117–18, 398
Weighting factors, 309–10, 352, 357
Whitehouse, G., 348
Wiener, N., 438
Wiener–Hopf equation, 438
Windows, 140
Word processors, 62–63
Work authorization, 31
Workbenches, 161–63
Work breakdown structure, 27–28
Workstations, 161–63, 322, 373, 378–79,
  473–75
WSEIAC, 294

Yourdon, E., 75, 368

Zwicky, F., 21–22, 43